M. Melkonian, R. A. Andersen,
and E. Schnepf

The Cytoskeleton of Flagellate and Ciliate Protists

Springer-Verlag Wien GmbH

Dr. Michael Melkonian
Botanisches Institut
Universität zu Köln
Köln, Federal Republic of Germany

Dr. Robert A. Andersen
Provasoli-Guillard Center for Culture of Marine Phytoplankton
Bigelow Laboratory for Ocean Sciences
West Boothbay Harbor, Maine, U.S.A.

Dr. Eberhard Schnepf
Lehrstuhl für Zellenlehre
Ruprecht-Karls-Universität Heidelberg
Heidelberg, Federal Republic of Germany

Originally published by Springer-Verlag Wien New York in 1991

Printed on acid-free paper

With 154 Figures

Special Edition of "Protoplasma", Vol. 164, 1991

ISBN 978-3-211-82294-4   ISBN 978-3-7091-6714-4 (eBook)
DOI 10.1007/978-3-7091-6714-4

Preface

Protists are fascinating organisms that have diversified evolutionarily to inhabit most of the biosphere. Their cytoskeleton is the primary reason that such a diversification is possible. Microtubules provided freedom from the constraint of prokaryotic cell walls and they were organized in various cytoskeletal configurations. The nine pairs plus two flagellum, a major component of many protistan cytoskeletons, gave a new meaning to the word motility beginning at least one thousand million years ago. Associated with the development of the flagellum were new contractile and noncontractile proteins that are necessary for control and manipulation of the flagella. As the new protists ventured out into new habitats and filled new niches, they evolved new ways of sensing their surrounding environment. Phototactic responses became tightly coupled with the flagellar apparatus. Chemotactic receptors formed on the cell surface, particularly on the flagellar membrane. The flagellum was also used for identifying mates for sexual reproduction and for gathering food for phagocytosis. In all of these changes, the cytoskeleton played a primary role.

Some ancient protistan lineages gave rise to the plants and animals. Over evolutionary time, the plants and animals ventured into new habitats and filled new niches, radiating to give rise to a diverse range of new organisms. But, unlike the protists, the plants and animals did not invent de novo new cellular features – they modified protistan characters, especially those of the cytoskeleton. It is this fact that makes study of the protistan cytoskeleton so relevant and important to plant and animal biology. Protists are excellent model systems for studying eukaryotic cytoskeletons, and considerable scientific effort has been spent investigating protistan cytoskeletons.

The study of protists is not without its troubles, however. Historically, the "protists" were known and studied as algae, fungi, and protozoa. Often, two or three sets of terms were invented for what is now known to be the same structure because there was little exchange of knowledge between the phycologists, mycologists, and protozoologists. For example, phycologists use the term "basal body" while protozoologists use "kinetosome"; mycologists were indifferent and used both terms!

As knowledge increased, it became more and more apparent that the organisms studied in the classical fields did not fit into monophyletic lineages. And when efforts were made to organize the protists into a phylogenetic classification, it became strikingly apparent that we needed a common vocabulary. Until such a common vocabulary exists, it will remain difficult to compare studies by scientists from different fields, it will be difficult to identify homologous structures among protists, and the historical barriers created by individual fields of mycology, phycology and protozoology will remain.

This issue specifically addresses the problem of a common vocabulary for the cytoskeleton. A Task Force of internationally known protistologists was created in 1987 during a special workshop organized by John O. Corliss at the 7th Meeting of the International Society of Evolutionary Protistology held in Egham, England. The first Task Force was given the charge of resolving problems dealing with the cytoskeleton, especially the flagellar/ciliar components. Michael Melkonian was elected chairman of this Task Force, and Robert A. Andersen, Donald J. S. Barr, Denis H. Lynn, Øjvind Moestrup and Michael A. Sleigh were chosen as committee members. Since then, the Task Force has worked towards a common vocabulary despite resistance both from within and outside the group. The National Science Foundation of the United States of America recognized the importance of this goal of a common vocabulary and the complementary goal of establishing a systematics that was not artifically induced by historical fields of science. They supported financially the original workshop and three subsequent scientific gatherings that allowed international debate in distant venues (International Union of Biological Sciences in Canberra, Australia, 1988; VIII International Congress of Protozoology in Tsukuba, Japan, 1989; 8th Meeting of the International Society of Evolutionary Protistology, College Park, Maryland, U.S.A., 1990). The exposure in Europe, Australia, Asia, and North America reached the majority of scientists working with protists.

The 1990 meeting of ISEP-8 brought to fruition the work by the cytoskeleton Task Force. Two scientific symposia were convened. The first brought forward the latest scientific discoveries in the field of cytoskeletal research, and the second applied these and past discoveries to protistan systematics. Following these two symposia, the Task Force reported its findings and opened debate on a common vocabulary for terminology and nomenclature of cytoskeletal structures. This issue is the written record of the Task Force meeting, the latest research, and the impact of cytoskeletal homologies on systematics.

We are deeply grateful to John O. Corliss whose brainchild led to this issue, and we are extremely grateful to the Systematic Biology Program of the National Science Foundation of the United States of America for supporting financially the work leading to this issue. Naturally, much remains to be done. The implementation and use of a common vocabulary requires sacrifices and extra effort by all workers in the field. But the fruits of harvest are worthy of the toil in the fields. The greater exchange of knowledge across mycology, phycology, and protozoology will benefit not only protistologists, but all eukaryotic systematists and cell biologists.

Finally, we would like to remind you that six other Task Forces were established at the 1987 meeting in Egham, all with the same goal as this Task Force but in areas as disparate as mitosis/cytokinesis to paleobiology. Each Task Force consists of internationally known scientists and each is working toward the common goal of advancing the science based on protists.

M. Melkonian, R. A. Andersen, and E. Schnepf

Contents

Protoplasma (1991) 164: 1–8

PROTOPLASMA

Terminology and nomenclature of the cytoskeletal elements associated with the flagellar/ciliary apparatus in protists

R. A. Andersen[1,*], **D. J. S. Barr**[2], **D. H. Lynn**[3], **M. Melkonian**[4], **Ø. Moestrup**[5], and **M. A. Sleigh**[6]

[1] Bigelow Laboratory for Ocean Sciences, West Boothbay Harbor, Maine, [2] Biosystematics Research Centre, Ottawa, Ontario, [3] Department of Zoology, University of Guelph, Guelph, Ontario, [4] Botanisches Institut, Universität zu Köln, Köln, [5] Institut for Sporeplanter, University of Copenhagen, Copenhagen, and [6] Department of Biology, University of Southampton, Southampton

Received April 24, 1991
Accepted May 10, 1991

Summary. The terminology used to describe the cytoskeleton of protists is sometimes redundant and confusing. The independent origins of protistology from studies on algae, fungi, and protozoans led to these problems. Additionally, recent studies indicate that seemingly unrelated taxa form new, monophyletic groups. However, homologous structures that support monophyly are difficult to identify when the literature is published in journals restricted to specific fields or when the terminology is in duplicate or triplicate. The purpose of this paper is to establish clear definitions, identify synonyms, and indicate homologies where possible.

Keywords: Flagellate/ciliate; Protists; Cytoskeleton; Flagellar apparatus; Kinetid; Terminology; Nomenclature.

Introduction

At the 7th Meeting of the International Society for Evolutionary Protistology, held at Royal Holloway College, England in 1987, Dr. John Corliss (University of Maryland, U.S.A.) led a discussion of terminological and nomenclaturial problems. The problems stem from a historical division of protistology into mycology, phycology and protozoology, causing a poor exchange of information and creating synonymous terms and specific names. To address the problems of terminology and nomenclature, a strategy involving Task Forces of leading scientists was suggested. Seven areas were identified and a "Task Force" was established for each.

The first Task Force is to examine "Flagellar Systems in Protists". The Task Force members are R. A. Andersen (Bigelow Laboratory for Ocean Sciences, U.S.A.), D. J. S. Barr (Biosystematics Research Centre, Canada), D. H. Lynn (University of Guelph, Canada), M. Melkonian (chair) (Universität zu Köln, Germany), Ø. Moestrup (University of Copenhagen, Denmark), and M. A. Sleigh (University of Southampton, England). The charge of the Task Force is to (1) bring together all the pertinent literature on the subject; (2) identify serious gaps in our present knowledge; (3) discuss possible homologies hidden in algal/fungal/protozoan segregation of data sets and determine what research is needed to clarify these points; (4) organize one or more symposia/workshops on selected specific topics (including areas of active research); (5) suggest invited contributors for these sessions; (6) be prepared to "run" such meetings at sites chosen; (7) be amenable to summarize and/or edit a sponsored report (chapter, book) on the collected knowledge within a reasonable amount of time following the meeting.

At the 8th Meeting of the International Society for Evolutionary Protistology, held at the University of Maryland, U.S.A. in 1990, the Task Force organized two symposia, "Biology of Protistan Flagellar/Cytoskeletal Systems" (M. Melkonian and R. A. Andersen, conveners) and "Flagellar/Cytoskeletal Systems in Evolutionary Studies of Protists" (R. A. Andersen and M. Melkonian, conveners), and a workshop entitled "Terminology and Nomenclature of the Cytoskeleton in Flagellate/Ciliate Protists". This chapter presents a summary of the workshop and the evaluations made

* Correspondence and reprints: Bigelow Laboratory for Ocean Sciences, Provasoli-Guillard Center for Culture of Marine Phytoplankton, McKown Point, West Boothbay Harbor, ME 04575, U.S.A.

by members of the Task Force. We address the terminology (definitions, synonyms, comments) and the nomenclature (labeling of systems that occur in multiple numbers) of the cytoskeleton in flagellate/ciliate protists. We did not include citations to the literature because of the enormous number of possible ones. A wide selection of relevant citations can be found in accompanying papers in this volume of *Protoplasma*. We use vernacular names for taxa. In most instances the taxonomic composition of the group will be clear to readers; there is debate concerning the most correct scientific name for some taxa and the hierarchial level of classification (hence the suffix of the name) for some taxa. One group, here identified as the chromophyte algae, may not be clear to readers. It includes groups such as the brown algae, diatoms, golden algae, haptophytes, and xanthophytes but it excludes specifically the cryptomonads, dinoflagellates, and euglenoids.

It is hoped that this account serves to identify specific terminological problems and to suggest specific solutions that will be acceptable to the scientific community. The following is subdivided into two parts, Terminology and Nomenclature. The paper is restricted to cytoskeletal elements associated with the flagellar/ciliary apparatus; the spindle cytoskeleton, though partly related, is excluded along with nonflagellar/ciliary cytoskeletal elements. In the descriptions we generally proceed from the flagellum/cilium tip to the cell interior (flagellar/ciliary roots).

Results

Terminology

Cytoskeleton

Definition: Intracellular network of protein filaments that is insoluble in non-ionic detergents.
Comments: The cytoskeleton gives the cell strength, rigidity and shape, and it is responsible for cell motility and intracellular movements. The cytoskeleton of flagellate/ciliate protists consists of (1) the flagellar apparatus/kinetid, and (2) cytoskeletal protein filaments not associated with the flagellar apparatus.

Flagellar apparatus/kinetid

Definition: An organellar complex consisting of one or more basal bodies/kinetosomes that may bear flagella/cilia, may have microtubular and fibrous roots associated with their bases, and may function in locomotion, feeding, sensation, and reproduction.
Comments: There is tremendous diversity in the elements that make up the flagellar apparatus/kinetid; some elements are homologs and some are analogs. Comparisons of the flagellar apparatus/kinetid of seemingly very different taxa have suggested the existence of new monophyletic assemblages.

Basal apparatus

Definition: Flagellar or ciliary apparatus exclusive of flagella/cilia.

Flagellum/Cilium

Definition: A long, cylindrical extension of a eukaryotic cell, bounded by the plasma membrane and containing an axoneme. A flagellum/cilium is a motility organelle that is mainly involved in cell movement by means of water propulsion, but can perform additional functions such as feeding, mating, and sensory perception.
Synonyms: eukaryotic flagellum, undulipodium.
Comments: The flagellum/cilium can be subdivided into different sub-compartments: (1) axoneme, (2) paraxonemal structures, (3) flagellar/ciliary matrix, (4) flagellar/ciliary membrane, (5) flagellar/ciliary surface. In the axial direction each sub-compartment of the flagellum/cilium can often be subdivided into 3 parts: tip, shaft, and transitional region/zone.

Axoneme

Definition: An axial array of (usually) 9 outer doublet and 2 central microtubules.
Comments: The basic architecture of the axoneme is presumably universal in eukaryotic cells and includes: outer doublet microtubules, central pair of microtubules, outer and inner dynein arms, nexin links, radial spokes, central sheath, etc. This does not mean that axonemes in different protists are identical in molecular or fine structural terms. Typically, the doublet microtubules consist of one complete microtubule with 13 subunits (A-tubule) and one partial microtubule with 10–11 subunits (B-tubule). Individual doublet microtubules can often be distinguished because variability exists in paraxonemal structures or cell surface components that enable specific doublet labeling. Some axonemal components may be modified or missing in the different protist groups. For example, the central pair of microtubules may be missing (especially in flagella/cilia that are not involved in motility), special dynein arms may be present, and septations may occur within the lumen of the B-tubules.

Flagellar/ciliary matrix

Definition: The cytosol of the flagellum/cilium, often lacking structural detail.

Paraxonemal structures (PAS)

Definition: Structures embedded in the flagellar matrix that are not part of the axoneme but often connected to specific axonemal doublets. We distinguish at least two major types, paraxonemal body and paraxonemal rod.
Synonyms: paraxial structures, paraflagellar structures.

Paraxonemal body (PAB)

Definition: Proteinaceous structure restricted to a certain area along the flagellum/cilium.
Synonyms: paraxial body, paraflagellar body (PFB), flagellar swelling.
Comments: A paraxonemal body occurs in chromophytes, euglenoids and kinetoplastids; no homology is implied. Some PABs are autofluorescent, emitting green light when excited with blue light. PABs may play a role in phototactic photoreception. Some PABs are paracrystalline, linked to both the axoneme and paraxonemal fibers. "Flagellar spines" are a special type found in male gametes of oogamous brown algae.

Paraxonemal rod (PAR)

Definition: Long, cylindrical structure (solid or hollow) that extends nearly the entire length of a flagellum/cilium, located between the axoneme and flagellar membrane and usually connected to the axoneme and flagellar membrane by specific links.
Synonyms: flagellar rod, paraxial rod, paraflagellar rod.
Comments: The paraxonemal rod occurs in dinoflagellates, euglenoids, kinetoplastids, pedinellids and silicoflagellates. It may be very different in ultrastructure and biochemical composition in the different groups of protists and no homology is implied. Examples are the PAR of *Euglena gracilis* that is a paracrystalline, hollow structure that is non-contractile, the PAR in the transverse flagellum of dinoflagellates (= paraxonemal fiber) that is solid, cross-striated, contractile and contains the protein centrin, and the PAR of the longitudinal flagellum of dinoflagellates such as *Gyrodinium* that is a hollow cylinder composed of helically arranged fibrils.

Flagellar/ciliary membrane

Definition: The extension of the plasma membrane that encloses the axoneme and flagellar/ciliary matrix.

Flagellar/ciliary surface

Definition: Structures associated with the outer surface of the flagellar/ciliary membrane, including fibrillar surface coat, hairs and scales.

Fibrillar surface coat

Definition: Fibrous component covering the entire surface of the flagellum/cilium.
Synonyms: tomentum.
Comments: Chlamydomonas and *Chlamydobotrys* have flagella/cilia bearing a fibrillar surface coat.

Flagellar scales

Definition: Organic structures of discrete size and shape, often covering the whole surface of the flagellum/cilium and generally assembled in the Golgi apparatus.
Comments: Flagellar scales usually differ in structure from scales present on the cell body proper. Different scale types are usually distinguished by descriptive terms like basket scales, canistrate scales, dendritic scales, flowerpot scales, knotted scales, limulus scales, man scales, pentagonal scales, spider-web scales, square-shaped scales, tree scales.

Flagellar hairs

Definition: Filamentous appendages usually arranged in one or more rows but not covering the entire surface of a flagellum/cilium.
Synonyms: flimmer, mastigoneme (some authors), tinsel.
Comments: Flagellar hairs of protists were first described in the 19th century using Loeffler's staining method for bacterial flagella. These hairy flagella were termed "Flimmergeisseln". Later, the term "mastigoneme" was introduced. Today, several very distinct types of flagellar hairs have been described.

Tubular flagellar hairs

Definition: Filamentous appendages consisting of at least a hollow shaft (> 15 nm diam.) often with one or more terminal filaments.
Synonyms: mastigoneme (some authors).
Comments: Tubular hairs are of three major types: bipartite hairs, tripartite hairs, and hair-scales. Bipartite and tripartite hairs are first assembled in the (chloroplast) endoplasmic reticulum or nuclear envelope while hair-scales are assembled in the Golgi apparatus; the former are composed principally of glycoproteins and the latter are of complex carbohydrates.

Bipartite hairs

Definition: Filamentous appendages consisting of only a cylindrical shaft and one or more terminal filaments; characteristic of the cryptomonads.
Synonyms: mastigoneme (some authors), retroneme.
Comments: Most cryptophytes have two slighty different types of bipartite hairs: on one flagellum the hairs are relatively short and in a single row along the length of the flagellum; on the other flagellum, hairs are longer and occur in two opposite rows along the length of the flagellum. The two types of hairs may also differ in the number and length of their terminal filaments. Some cryptophytes have other arrangements of hairs. Also, the shaft bears long and short filaments in *Hemiselmis* but apparently lacks such filaments in the other genera.

Tripartite hairs

Definition: Filamentous appendages consisting of a tapered base, a cylindrical shaft and one or more terminal filaments; characteristic of many chromophyte algae, Oomycete fungi and some heterotrophic flagellates.
Comments: The shaft bears long and short filaments in several groups but lacks filaments in others; hairs are usually arranged in two rows along the length of the flagellum and attached through the flagellar membrane to specific outer doublets.

Hair scales

Definition: Tubular flagellar hairs assembled in the Golgi apparatus and consisting of carbohydrates primarily, forming two rows along the longitudinal axis of the flagellum and attaching to axonemal B-tubules Numbers 4 and 8.
Comments: Hair scales occur on scaly flagella of green algae.

Non-tubular flagellar hairs

Definition: Filamentous appendages less than 15 nm in diameter.
Comments: Non-tubular flagellar hairs occur in various protistan groups, such as the dinoflagellates, euglenophytes, green algae (e.g., *Chlamydomonas*). In *Chlamydomonas* they consist of a single high-molecular mass glycoprotein.

Transitional region/transitional zone

Definition: The most proximal (basal) part of a flagellum/cilium adjacent to the basal body/kinetosome; comprising matrix, axoneme, and flagellar/ciliary membrane.
Synonyms: transition region, transition zone, flagellar transition region.
Comments: The transitional region is sometimes defined as the region between the proximal ends of the central pair of microtubules of the axoneme and the distal ends of the C-tubules of the basal body/kinetosome. However, we prefer a more general definition because some transitional structures cross over these boundaries. For example, the transitional helix of chromophytes and the helix of the euglenoid *Entosiphon* extend into the region containing the central pair of microtubules of the axoneme.
The transitional region may include three major components: (1) concentric fibers, (2) transitional plate, and (3) transitional fibers. It must be emphasized here that we do *not* imply a homology among taxa; we are only identifying positional and structural similarities.

Concentric fibers

Definition: Intra-axonemal structures consisting of filaments that often connect to the outer doublets of the axoneme. The filaments may be arranged as rings, a helix or a stellate structure.
Comments: We distinguish at least three types of concentric fibers: (1) concentric ring fibers, (2) the transitional helix, and (3) the stellate structure. Concentric fibers are the most variable of the transitional features, and they may have evolved independently more than once.

Concentric ring fibers

Definition: A stacked series of fine concentric rings (or possibly a helix) connecting A-tubules of the axonemal doublets.
Comments: This type occurs in chytrids, ciliates, euglenoids, plasmodiophoromycetes, and the amoeboflagellate *Naegleria*.

Transitional helix

Definition: A thick helix (or possibly stacked rings) that lies inside the outer doublets of the axoneme but apparently does not connect with them.
Synonyms: coiled fiber, Spiralkörper.
Comments: This type occurs in many but not all chromophyte algae and Oomycete fungi. A "double helix" occurs in some Oomycetes, hyphochytriomycetes, bicosoecids, slopalinds, and the xanthophyte *Heterococcus*. The "coiled fiber" of the green alga *Pyramimonas* may belong in this category.

Stellate structure

Definition: A cylinder consisting of a number of filaments that connect A-tubules of every second outer axonemal doublet with each other and in transverse flagellar section is seen as a star-like pattern.
Synonyms: star-shaped body, stellate pattern.
Comments: The stellate structure may consist of one or two parts (termed distal and proximal). If a transitional plate is intercalated between two parts of the stellate structure, the complex appears H-shaped in longitudinal flagellar section.

Transitional plate

Definition: Thin, plate-like structure oriented perpendiculary to the axoneme, usually filling the lumen of the axoneme and sometimes extending to the flagellar membrane.
Synonyms: axosome, basal plate, terminal cap, terminal partition, terminal plate, transverse diaphragm, transverse septum.
Comments: The transitional plate can vary significantly in its location, number and detailed structure.

Transitional fibers

Definition: Thin strands extending from a point between the A- and B-tubules to the plasma and flagellar membranes, beginning where the axoneme assumes the 9+2 arrangement and ending where the distal ends of the C-tubules terminate.
Synonyms: accessory spokes, anchoring fibers, curving arms, kinetosome props, links, props, struts.
Comments: In the flagellum/cilium, the transitional fibers are short and extend straight to the membrane. Below the transitional region they are necessarily longer and sometimes are whorled like a catherine wheel. In chytrids the transitional fibers below the transitional region are especially large, often with secondary fibers.

Basal body/kinetosome

Definition: Cylindrical structure (ca. 0.2 µm diameter) found at the base of a flagellum/cilium (or by itself in the area of a flagellar apparatus/kinetid), consisting of a continuation of the nine outer axonemal doublets (A, B) but with the addition of a C-microtubule to form a triplet. Proximal end containing a cartwheel structure and sometimes B- and C-tubule septations.
Comments: Basal bodies/kinetosomes, although often linked to flagella/cilia, may occur unattached to a flagellum. When unattached, they are termed barren, dormant or nascent, depending upon structure and state of development. Pairs of basal bodies/kinetosomes may be parallel, angled or antiparallel. The substructure of basal bodies/kinetosomes includes the cartwheel, lumen, and B-, C-tubule septations. Pairs of basal bodies/kinetosomes are interconnected by connecting fibers.

Nascent basal body/kinetosome

Definition: Nonflagellate/nonciliate, in the process of becoming full-length; sometimes lacking B- and C-tubules.
Synonyms: probasal body, prokinetosome.

Dormant basal body/kinetosome

Definition: Nonflagellate/nonciliate, fully formed with a full complement of triplet microtubules, eventually giving rise to a flagellum/cilium.
Synonyms: non-functional basal body, non-functional kinetosome.
Comments: These occur in many flagellate green algae (e.g., *Chlamydomonas, Pedinomonas*), in some euglenophytes, in some chromophytes. If the developmental status of a basal body/kinetosome that is not associated with a flagellum/cilium is unknown, it should simply be referred to as nonflagellate/nonciliate.

Barren basal body/kinetosome

Definition: Fully formed, having once given rise to a flagellum/cilium in an earlier developmental stage, but now nonflagellate/nonciliate; attached to the plasma membrane.
Synonyms: non-functional basal body, non-functional kinetosome, vestigial basal body, vestigial kinetosome.
Comments: These are found in the green alga *Monomastix* and the synurophyte *Mallomonas spendens*. If the developmental status of a basal body/kinetosome that is not associated with a flagellum/cilium is unknown, it should simply be referred to as nonflagellate/nonciliate.

Parallel basal body/kinetosome

Definition: A pair of basal bodies/kinetosomes that have the same orientation.

Angled basal body/kinetosome

Definition: A pair of basal bodies/kinetosomes in which the angle subtended by the long axes of the basal bodies is greater than 0° but less than 180°, most commonly approximately orthogonal.

Antiparallel basal body/kinetosome

Definition: A pair of basal bodies/kinetosomes that have opposite polarity and whose bases are adjacent or nearly so.

Cartwheel

Definition: A hub-like central axis with radiating arm-like structures extending to the A-tubule (rarely B-tubule) of each triplet, and appearing like spokes of a wheel when the basal body/kinetosome is viewed in cross-section, located at the proximal end of a basal body/kinetosome.
Synonyms: CW-pattern.
Comments: Sometimes cartwheels appear to be absent; this may be an evolutionarily derived state or the result of inadequate fixation.

Lumen

Definition: The central region of the basal body/kinetosome within the cylinder of nine triplet tubules. In the proximal part of the basal body/kinetosome the lumen is filled by the cartwheel.
Comments: The lumen may appear "empty" or it may contain various inclusions such as ribosome-like particles, dense spheres, dense bodies, fibrils and amorphous clumps.

B- and C-tubule septations

Definition: Strands bisecting the B- and C-tubules (viewed in transverse section) of some basal bodies/kinetosomes.
Comments: B-tubule septations sometimes occur along the full length of the axoneme.

Connecting fibers

Definition: Fibrillar or amorphous structure linking triplets of different basal bodies/kinetosomes with each other.
Synonyms: A variety of adjectives have been used – amorphous, banded, dense, fan-shaped, fibrillar, filamentous, plate-like, striated.
Comments: There is a bewildering array of these structures that defies classification, and at this time homologies between different protistan groups are uncertain. Centrin, a contractile protein, has been investigated at the molecular level and it occurs in the distal connecting fibers of many flagellate green algae.

Flagellar/ciliary roots

Definition: Fibrous, microtubular or amorphous structures originating at or near basal bodies/kinetosomes and terminating somewhere else in the cell but not at nearby basal bodies/kinetosomes.
Synonyms: flagellar rootlets.
Comments: The two major types are fibrous roots and microtubular roots.

Fibrous roots

Definition: Roots composed of a bundle of filaments, the thickness of which varies in roots of different origins. Frequently, but not always, the bundles appear cross-striated.
Synonyms: rhizoplast.
Comments: It is probable that the first types of flagellar or ciliary roots seen by light microscopists were fibrous roots. These roots have turned out to be structurally and chemically heterogenous in comparison to microtubular roots. In addition, cell and molecular biology studies may be necessary to determine their type by assessment of contractility, ion-binding characteristics, and ATPase activity as well as characterizing their proteins by antibody-binding tests.
In some cells the fibrous roots remain superficial (near the plasma membrane) throughout their length; the ciliate "kinetodesmal fibers", the superficial "transverse striated fibre" of some dinoflagellates or the "costa" of trichomonads are of this type. In other cells the striated roots extend deep into the cell cytoplasm, like the "rhizoplast" of many flagellates or the "parabasal fibers" of many metamonads, where they may make contact with such organelles as the nucleus, Golgi apparatus, plastid or mitochondrion.
Centrin, assemblin, and giardin may represent common types of proteins that are formed into fibrous roots with distinctive properties. It is clear that exciting discoveries are being made in studies on the fibrous roots of protists, but our knowledge is very elementary at present.
At this time there is not enough knowledge to classify all fibrous roots in the various protists into clear-cut categories, and this must await further biochemical characterization. At present, only two different types of fibrous roots, which occur in green algae, have been characterized biochemically in detail; these are System I and System II fibers.

System I fibers

Definition: Fibrous roots made of a bundle of 2 nm filaments, cross-striated with a periodicity of approximately 30 nm, largely composed of the noncontractile 34 kDa protein assemblin; usually associated with microtubular roots.

Synonyms: SMAC (= striated microtubule associated component). Possible synonyms: costa (some), kinetodesmal fiber, parabasal fiber, striated root (some), transverse striated fiber.

Comments: System I fibers are extremely stable mechanically and biochemically and are presumed to have a stress-absorbing function in the basal apparatus. In contrast to system II fibers, they do not connect with basal body triplets.

System II fibers

Definition: Fibrous roots made of a bundle of 4–8 nm filaments, often cross-striated with a periodicity of > 80 nm, composed at least in part of the Ca^{2+} modulated contractile protein centrin; not closely associated with root microtubules, but originating at basal body triplets.

Synonyms: R. fiber, rhizoplast (some), striated root (some).

Comments: System II fibers usually extend into the cell where they often link to the nucleus, or occasionally to other organelles (plastid, mitochondrion). Some system II fibers are superficial, especially in Ulvophyceae.

Microtubular roots

Definition: Flagellar roots consisting of one or more microtubules originating at or near a basal body/kinetosome; the microtubules having a greater stability when compared to cytoplasmic microtubules.

Synonyms: microtubular rootlets.

Comments: α-tubulin of microtubular roots is acetylated in at least some protists, and this posttranslational modification may characterize all microtubular roots. Acetylated α-tubulin stabilizes microtubules so that they fail to disassemble at low temperatures or high pressure.

Microtubular roots vary in number, with one or two roots per basal body/kinetosome being common. The paths of microtubular roots are usually superficial (near the plasma membrane), and their relative directions and curvatures are of primary importance in determining the morphology of a cell. The number of microtubules per root varies from one to over 80. When more than one microtubule is present in a root, the microtubules often run parallel for considerable distances and are associated laterally via intermicrotubular linkers.

Microtubular roots arise from microtubular organizing centers (MTOCs) along or near the basal bodies/kinetosomes. At this time little is known about these MTOCs. Microtubular roots change as flagella/cilia transform from immature to mature forms, but it is unknown if the MTOCs change during maturation or if there are separate MTOCs for immature and mature flagellar/ciliary roots.

Some microtubular roots contain MTOCs along their surfaces, and they in turn can give rise to cytoplasmic microtubules. These cytoplasmic microtubules are usually disassembled by cold or high pressure, suggesting they are not acetylated. However, some are stable in cold or high pressure. These cytoplasmic microtubules are usually involved in a vital cell function such as maintaining morphological contours for reproduction or feeding and forming or transporting scales or scale vesicles.

Haptonema

Definition: A filamentous appendage that is (usually) relatively long, consisting of an extension of the plasma membrane, a sheath of endoplasmic reticulum, and a core of several microtubules anchored near the basal bodies/kinetosomes.

Comments: Haptonemata occur on many motile prymnesiophytes (= haptophytes). Superficially, they resemble flagella/cilia when viewed with a light microscope, but their ultrastructure indicates they do not originate from axonemes. Haptonemata that are long may contract quickly to form a coil at the base of the flagella/cilia. The functions of haptonemata include attachment and food collection.

Nomenclature

Labeling of flagellum/cilium and basal body/kinetosome

Many flagellate/ciliate protists bear two flagella/cilia per cell (ciliates being an obvious exception); these two must be distinguished. This has been done with a variety of names in the individual groups: dorsal/ventral, anterior/posterior, hairy/smooth, left/right, short/long, cis/trans. The problem is to compare a specific flagellum/cilium between groups, or specifically, to establish homologies.

Recent studies on flagellar/ciliary development have added a new dimension to this problem. During cell division, the flagella have a semi-conservative replication such that each daughter cell receives an existing flagellum and each forms a new one. The newly formed flagella are termed immature, and during the following cell division they transform to a more mature state. The developmental process spans at least two cell cycles before a flagellum is fully mature; it remains mature for all subsequent cell divisions.

This flagellar/ciliary heterogeneity and the recognitions of flagellar transformation have led to a new means for identifying and labeling flagella/cilia and their basal bodies/kinetosomes. The oldest flagellum is Number 1, the next oldest is Number 2, etc. A uniflagellate cell may have either a Number 1 or a Number 2 flagellum, while a biflagellate cell will have both Number 1 and Number 2 flagella.

Labeling of doublets/triplets

The axonemal doublets can be labeled in some protists using either a bridge that may occur between doublets 5 and 6 or the missing dynein arm on doublet 1 (many green algae). Doublets are numbered sequentially in an anti-clockwise pattern when the flagellum is viewed from tip to base. Once the doublets are labeled, they can be followed into the cell and the triplets can be numbered accordingly.

Labeling of flagellar roots

To date, labeling of flagellar roots has been the most variable of all the cytoskeletal structures. Names have been descriptive (e.g., SRm, 4 r, Cr, 12 r in *Chilomonas*) or arbitrary (e.g., R 1, R 2, R 3, etc. in *Poterioochromonas*). Before the discovery of flagellar transformation, flagellar labeling was arbitrary in distantly related taxa. As a consequence, root homologies were difficult to establish. However, with consistent flagellar labeling (see above), it is possible to have consistent labeling of roots; it remains uncertain whether consistency in position and structure means homology.

Discussion

The proposals presented here represent an attempt to resolve terminological and nomenclaturial problems created in the past. It is our hope that workers will consider our remarks as they prepare future manuscripts. It is difficult to abandon terms that one has used or even created; this difficulty is shared by this Task Force as we are forced to compromise our own usage for the common good.

The problems are far from being resolved, however. We still cannot prove homologies for many structures. Furthermore, there remain thousands of protists that have not been examined in detail; undoubtedly, as these are investigated our beliefs will be adjusted. If this paper serves to guide others towards a protistological viewpoint that considers the findings of those outside one's field then we have achieved an important goal.

Acknowledgements

We appreciate the scholarly input of many colleagues who aided our efforts in preparing this paper, and in particular we thank Dr. Peter Beech, Dr. Keith Roberts and Prof. E. Schnepf for carefully reviewing the manuscript. We are also very grateful to Dr. David Schnidel and the National Science Foundation (U.S.A.) who have provided financial support for this and other Task Forces through the following grants: BSR 87-13924, BSR 88-15256, and BSR 89-13669 to J.O. Corliss and BSR 90-13232 to R.A. Andersen. Bigelow Laboratory for Ocean Sciences contribution number 91014.

Protoplasma (1991) 164: 9–11

PROTOPLASMA

Why do tubulin gene families lack diversity in flagellate/ciliate protists?

Carolyn D. Silflow*

Departments of Genetics and Cell Biology and of Plant Biology, University of Minnesota, St. Paul, Minnesota

Received October 12, 1990
Accepted December 11, 1990

Summary. Considerable amino acid sequence diversity is found among tubulin isotypes encoded by tubulin gene families in animal, higher plant, and fungal systems. In contrast, relatively little diversity is found among the isotypes produced by the gene families in a number of flagellate or ciliate protists. It is possible that proper assembly of the axoneme requires a homogeneous pool of tubulin subunits and that the axoneme thus provides a stringent selection against amino acid replacement substitutions among tubulin genes in these systems.

Keywords: Tubulin genes; Tubulin isotypes; Microtubules; Axonemes.

Tubulin multi-gene families

Genes encoding α- and β-tubulin are repeated in the genomes of many eukaryotic species, including protists (reviewed in Cleveland and Sullivan 1985, Silflow et al. 1987). Multiple expressed tubulin genes may provide a selective advantage by permitting the production of slightly different tubulin isoforms that function in specific microtubule arrays during the cell cycle or life cycle. Alternatively, multiple tubulin genes may facilitate the regulation of tubulin gene expression in response to various developmental signals or environmental stimuli (Raff 1984, Cleveland 1987).

DNA sequence data from the cloned tubulin genes of a large number of species are now available for comparison. Deduced amino acid sequences for both α-tubulin and β-tubulin are highly conserved in evolution, reflecting the ability of tubulins from any source to copolymerize in vitro. For example, the amino acid sequence of the major mammalian neuronal β-tubulin is more than 80% conserved with β-tubulins from a variety of organisms including *Drosophila melanogaster, Trypanosoma brucei, Chlamydomonas reinhardtii,* and *Neurospora crassa.* The most divergent tubulin genes described to date are those in fungal systems including *Saccharomyces cerevisiae* and *Schizosaccharomyces pombe*, in which the tubulins are 73–76% conserved with mammalian neuronal tubulins (reviewed in Burland et al. 1988).

Heterogeneity within tubulin gene families

An important conclusion drawn from many DNA sequencing is that within many species, one or more members of the tubulin gene families have diverged significantly. The most divergent genes within a family tend to be those whose expression is limited to specific tissues or specific life cycle stages (reviewed in Burland et al. 1989). The small gene families in fungal systems show considerable diversity. For example, in *S. cerevisiae*, the two α-tubulin genes encoded proteins that are 90% identical (Schatz et al. 1986); the two β-tubulin genes in *Aspergillus nidulans* are only 83% identical (May et al. 1987). Complete DNA sequence information is available for the β-tubulin gene family in the vertebrate chicken; most of the seven functional genes encode proteins that are greater than 90% conserved. A gene expressed only in hematopoietic tissues is more divergent, encoding a protein that is 83–84% conserved with the other isotypes (Monteiro and Cleveland 1988). In the insect *Drosophila melanogaster*, three of the four α-tubulin genes encode proteins that are 95–99% conserved; a fourth divergent gene encodes a maternally-expressed protein only 67% conserved with the other

* Correspondence and reprints: 250 BioScience Center, 1445 Gortner Ave., St. Paul, MN 55108, U.S.A.

isotypes (Theurkauf et al. 1986). Complete DNA sequence information on the six members of the α-tubulin gene family in *Arabidopsis thaliana* shows that the three α-tubulin isotypes produced range from 89 to 94% sequence identity, with the most divergent protein being expressed primarily in pollen (Ludwig et al. 1988, D.P. Snustad pers. comm.).

Homogeneity of tubulin gene families in flagellated/ciliated protists

In contrast to the heterogeneity observed in tubulin gene families in many species, a strikingly homogeneous pool of tubulin subunits is encoded by the tubulin gene families in a number of protistan species that have been examined. In *Chlamydomonas reinhardtii*, the two α-tubulin genes encode identical proteins, as do the two β-tubulin genes (Youngblom et al. 1984, Silflow et al. 1985, Silflow et al. unpubl. data). In the related unicellular alga *Polytomella agilis*, two of the β-tubulin genes encode identical proteins; a third gene encodes a protein differing by only one amino acid (Conner et al. 1989). The same observation of tubulin gene homogeneity has been made in the colonial alga *Volvox carteri* where the two β-tubulin genes encode identical proteins that differ in one amino acid from the *Chlamydomonas* β-tubulin. Likewise, the tubulins encoded by the two *Volvox* α-tubulin genes are identical, but differ in two amino acids from *Chlamydomonas* α-tubulin. The number of nucleotide differences between the two *Volvox* α-tubulin genes and between the two *Volvox* β-tubulin genes is approximately the same as the number of exchanges between the *Chlamydomonas* genes and the *Volvox* genes, but five to seven times the number of exchanges between the two *Chlamydomonas* genes (Harper and Mages 1988, Mages et al. 1988, Mages and Schmitt pers. comm.). These data suggest that some selective pressure may be preventing amino acid replacement substitutions in the *Volvox* genes.
A pattern of tubulin gene homogeneity has also been observed in protistan species distantly related to the Volvocalean species mentioned above. For example, the two β-tubulin genes encode identical proteins in *Stylonichia lemnae* (Conzelmann and Helftenbein 1987). In *Tetrahymena pyriformis*, the two predicted β-tubulin gene products differ by only two amino acids (Barahona et al. 1988). Although the complete gene family has not been analyzed, DNA sequencing of clones representing three α-tubulin genes in *Naegleria gruberi* indicated that all three genes encode identical proteins (Lai et al. 1988). Similary, restriction enzyme digestion of the tandemly repeated tubulin gene clusters in *Trypanosoma* and *Leishmania enrietti* have suggested little or no heterogeneity in these tubulin gene families (Landfear et al. 1983, Kimmel et al. 1985).

The axoneme as a constraint on tubulin gene evolution

A common feature of protistan systems with homogeneous tubulin gene families is the presence of flagellar or ciliary axonemes during a significant portion of the life cycle. These systems demonstrate that a variety of microtubule structures including axonemes, basal bodies and associated rootlet structures, mitotic and meiotic spindles are normally assembled from the same tubulin components. A similar demonstration was made in *Drosophila* for the testis-specific β2-tubulin which is the major B-tubulin component in sperm axonemes but also participates in meiotic spindles and cytoplasmic microtubules (Raff 1984). Except for post-translational acetylation of α-tubulin preferentially incorporated into axonemes (LeDizet and Piperno 1986), the composition of tubulin in axonemal microtubules must be similar to that in other microtubule structures in these cells.
It is possible that the presence of an axoneme in unicellular organisms provides selective pressure that prevents the amino acid sequence divergence of tubulins that assemble into the axoneme. A homogeneous backbone of tubulin dimers in the doublet microtubules may be essential for the precise placement and coordinated assembly of dynein arms, radial spokes, nexin links, and other proteins that interact with the microtubules. The large number of required interactions may decrease tolerance for amino acid variation in axonemal tubulin, as suggested by Little (1985) and Singhofer-Wowra et al. (1986). Support for this suggestion has recently been provided by a gene replacement experiment in *Drosophila,* where the divergent β3 gene was fused to the promoter of the testis-specific β2 gene and expressed in place of the normal β2 gene in testis tissue. Doublet axonemal microtubules were not assembled in the transformants where the divergent β3 protein was produced alone or where it represented more than 20% of the total β-tubulin (Hoyle and Raff 1990). These studies clearly indicated that the β3 protein lacks the ability to assemble into doublet microtubules and some types of cytoplasmic microtubules.
The conservation of amino acid sequences in tubulin isoforms known to assemble into axonemes is moderate. For example, the testis-specific β2 isoform in

Drosophila is only 87% conserved with *Chlamydomonas* β-tubulin (Rudolph et al. 1987). These differences may account in part for differences in axonemal microtubule structure and stability in different organisms. However, the available data suggest that within a single species, a homogeneous pool of tubulin subunits is required for assembly and function of axonemal microtubules and their associated structures. To examine this possibility, further analysis of tubulin isotypes found in axonemes from diverse species will be necessary. In addition, it will be possible to test utilization of divergent tubulin isotypes in axonemal microtubules using genetic engineering and transformation methods in a number of model systems.

Acknowledgements

This work was supported by a research grant from the U.S. Public Health Service.

References

Barahona I, Soares H, Cyrne L, Penque D, Denoulet P, Rodrigues-Pousada C (1988) Sequence of one α- and two β-tubulin genes of *Tetrahymena pyriformis*. J Mol Biol 202: 365–382

Burland TG, Paul ECA, Oetliker M, Dove WF (1988) A gene encoding the major beta tubulin of the mitotic spindle in *Physarum polycephalum* plasmodia. Mol Cell Biol 8: 1275–1281

Cleveland DW (1987) The multitubulin hypothesis revisited: what have we learned? J Cell Biol 104: 381–383

– Sullivan KF (1985) Molecular biology and genetics of tubulin. Annu Rev Biochem 54: 331–365

Conner TW, Thompson MD, Silflow CD (1989) Structure of the three β-tubulin-encoding genes of the unicellular alga, *Polytomella agilis*. Gene 84: 345–358

Conzelmann KK, Helftenbein E (1988) Nucleotide sequence and expression of two β-tubulin genes in *Stylonichia lemnae*. J Mol Biol 198: 643–653

Harper JF, Mages W (1988) Organization and structure of *Volvox* β-tubulin genes. Mol Gen Genet 213: 315–324

Hoyle HD, Raff EC (1990) Two *Drosophila* beta tubulin isoforms are not functionally equivalent. J Cell Biol 111: 1009–1026

Kimmel BE, Samson S, Wu J, Hirschberg R, Yarbrough LR (1985) Tubulin genes of the African trypanosome *Trypanosoma brucei rhodesience*: nucleotide sequences of a 3.7-kb fragment containing genes for α and β tubulins. Gene 35: 237–248

Lai EY, Remillard SP, Fulton C (1988) The α-tubulin gene family expressed during cell differentiation in *Naegleria gruberi*. J Cell Biol 106: 2035–2046

Landfear SM, Pratt DM, Wirth DF (1983) Tandem arrangement of tubulin genes in the protozoan parasite *Leishmania enrietti*. Mol Cell Biol 3: 1070–1076

Ledizet M, Piperno G (1986) Cytoplasmic microtubules containing acetylated α-tubulin in *Chlamydomonas reinhardtii*: spacial arrangement and properties. J Cell Biol 103: 13–22

Little M (1985) An evaluation of tubulin as a molecular clock. Bio Systems 18: 241–247

Ludwig SR, Oppenheimer DG, Silflow CD, Snustad DP (1987) The α 1-tubulin gene of *Arabidopsis thaliana*: primary structure and preferential expression in flowers. Plant Mol Biol 10: 311–321

Mages W, Salbaum JM, Harper JF, Schmitt R (1988) Organization and structure of *Volvox* α-tubulin genes. Mol Gen Genet 213: 449–458

May GS, Tsang ML-S, Smith H, Fidel S, Morris NR (1987) *Aspergillus nidulans* β-tubulin genes are unusually divergent. Gene 55: 231–243

Monteiro MJ, Cleveland DW (1988) Sequence of chicken cβ 7 tubulin: analysis of a complete set of vertebrate tubulin isotypes. J Mol Biol 199: 439–446

Raff EC (1984) The genetics of microtubule systems. J Cell Biol 99: 1–10

Rudolph JE, Kimble M, Hoyle HD, Subler MA, Raff EC (1987) Three *Drosophila* β-tubulin sequences: a developmentally regulated isoform (β 3), the testis-specific isoform (β 2), and an assembly-defective mutation of the testis-specific isoform (β 2t 8) reveal both an ancient divergence in metazoan isotypes and structural constraints for β-tubulin function. Mol Cell Biol 7: 2231–2242

Schatz PJ, Pillus L, Grisafi P, Solomon F, Botstein D (1986) Two functional α-tubulin genes of the yeast *Saccharomyces cerevisiae* encode divergent proteins. Mol Cell Biol 6: 3711–3721

Silflow CD, Chisholm RL, Conner TW, Ranum LPW (1985) The two α-tubulin genes of *Chlamydomonas reinhardtii* code for slightly different proteins. Mol Cell Biol 5: 2389–2398

Singhofer-Wowra M, Little M, Clayton L, Dawson P, Gull K (1986) Amino acid sequence data of α-tubulin from myxamoebae of *Physarum polycephalum*. J Mol Biol 192: 919–924

Theurkauf WE, Baum H, Bo J, Wensink PC (1986) Tissue-specific and constitutive α-tubulin genes of *Drosophila melanogaster* code for structurally distinct proteins. Proc Natl Acad Sci USA 83: 8477–8481

Youngblom J, Schloss JA, Silflow CD (1984) The two β-tubulin genes of *Chlamydomonas reinhardtii* code for identical proteins. Mol Cell Biol 4: 2686–2696

Protoplasma (1991) 164: 12–22

PROTOPLASMA

Transmembrane signaling in cilia and flagella

R. A. Bloodgood*

Department of Anatomy and Cell Biology, University of Virginia School of Medicine, Charlottesville, Virginia

Received September 20, 1990
Accepted December 11, 1990

Summary. Ciliary and flagellar membranes are dynamic. Ciliary and flagellar membranes have diverged widely during evolution and perform many specialized functions. Transmembrane signaling is an important component of the function of ciliary and flagellar surfaces in general. In this review, I discuss some of the functions performed by ciliary and flagellar surfaces and I present three different ciliary and flagellar signaling systems associated with rather different dynamic events performed by ciliary and flagellar surfaces. Two of these are associated with *Chlamydomonas* flagella and one is associated with vertebrate olfactory cilia. Calcium regulation of protein phosphorylation appears to be important in regulating glycoprotein movements in the *Chlamydomonas* flagellar membrane. Changes in levels of cAMP and cAMP-dependent protein phosphorylation are clearly central to the signaling associated with mating events in gametic flagella of *Chlamydomonas*, although calcium clearly has an important, if poorly understood, role to play. There is no known role for G proteins in flagellar membrane events in *Chlamydomonas*. In contrast, mammalian olfactory cilia possess an odorant activated, G protein regulated adenylate cyclase and conductance channels that are directly gated by cyclic nucleotides. A second class of odorants that do not affect adenylate cyclase activity appear to act through G protein activated phospholipase C and changes in IP_3 second messenger levels. These examples demonstrate the diversity in the signaling pathways associated with ciliary and flagellar membranes.

Keywords: Cilia; Flagella; Transmembrane signaling; Flagellar surface; Adhesion; Olfactory cilia; *Chlamydomonas*.

Abbreviations: CaPK-2 calcium-dependent protein kinase; db-cAMP dibutyryl cAMP; Fab fragment antigen binding; IgE immunoglobulin E; IP_3 myo-inositol trisphosphate; IP_4 myo-inositol tetrakisphosphate; OBP odorant binding protein; PIP_2 phosphoinositol bisphosphate; TFP trifluoperazine; WGA wheat germ agglutinin.

* Correspondence and reprints: Department of Anatomy and Cell Biology, Box 439, University of Virginia School of Medicine, Charlottesville, VA 22908, U.S.A.

Introduction

Cilia and flagella are eukaryotic cell organelles composed of a unique plasma membrane domain (termed the ciliary or flagellar membrane), a unique cytoskeletal

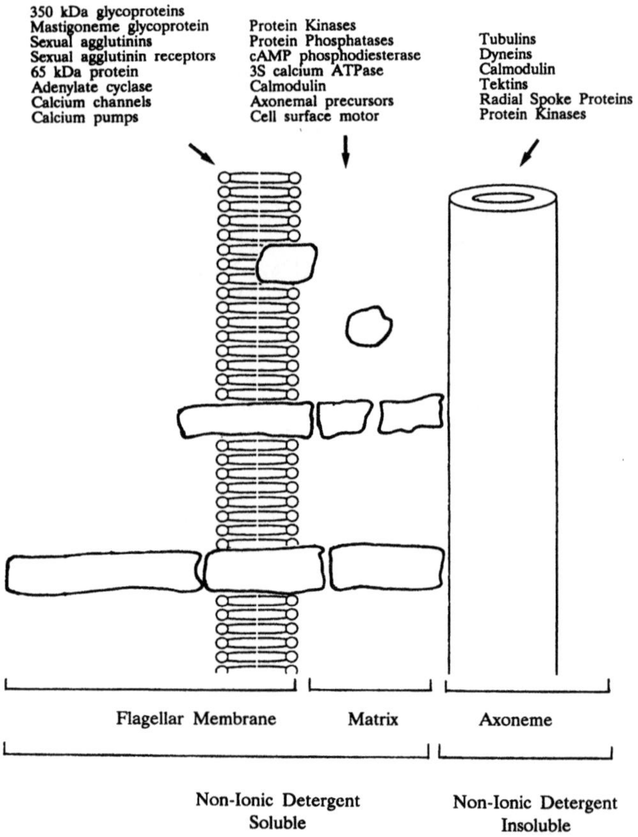

Fig. 1. Diagram illustrating the organization of the flagellum showing the three major compartments and examples of proteins found in each of these compartments

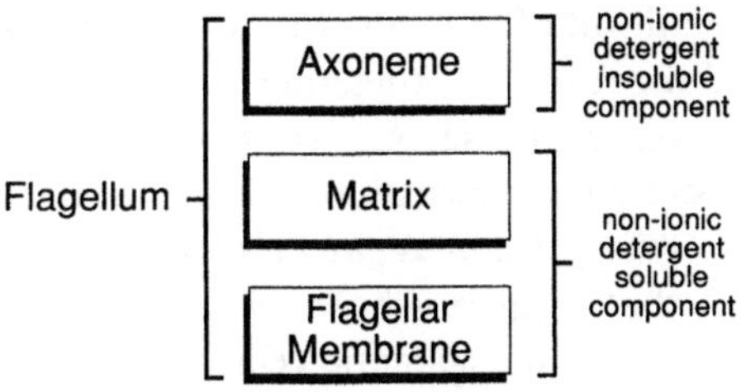

Fig. 2. Block diagram illustrating the 3 major compartments of the flagellum (axoneme, matrix, membrane) and how these are partitioned when flagella are treated with a non-ionic detergent such as Triton X-100 or Nonidet P-40

domain (termed the axoneme) and a unique soluble compartment (referred to as the matrix or flagelloplasm) (Figs. 1 and 2). Traditionally, cilia and flagella are isolated and then fractionated by extraction with nonionic detergents such as Triton X-100 or Nonidet P-40 yielding two fractions: an insoluble fraction which is primarily the axoneme and a soluble fraction which contains those components that were associated with the membrane and the matrix compartments prior to detergent treatment (Figs. 1 and 2). Most attention has been focused on functional, biochemical, and genetic studies of the axoneme (Luck 1984, Huang 1986) for two reasons: (1) initiation and propagation of bends was seen as the principal function of cilia and flagella and (2) isolated axonemes can be reactivated in the presence of ATP and magnesium to exhibit bending behavior strikingly similar to that seen in vivo. It is only recently that it has come to be appreciated that cilia and flagella perform a number of other functions, many of which are primarily associated with ciliary and flagellar membranes (Bloodgood 1990).

Indeed, it is of interest, from an evolutionary point of view, that there is much more diversity in the structure, function and biochemical composition of ciliary and flagellar surfaces than there is in ciliary and flagellar axonemes (Bloodgood 1990). This suggests that axonemal function is more critical for survival than the specialized functions of the ciliary/flagellar surface and also suggests that major modifications of the ciliary/flagellar surface need not impact on axonemal function; this latter observation is not surprising given the ability to reactivate isolated axonemes.

Functions of ciliary and flagellar surfaces

The known functions of ciliary and flagellar surfaces are listed in Table 1 and can be organized into four (albeit somewhat overlapping) general categories: (1) adhesion, (2) force transduction, (3) sensory functions,

Table 1. Functions of ciliary and flagellar surfaces

Adhesion
cell–substrate
cell–cell
cell–pathogen
Force transduction
whole cell gliding motility
flagellar realignment during mating
Sensory reception
mechanoreception
chemoreception
mating related signaling
Regulation of axonemal behavior
calcium concentration

and (4) regulation of axonemal behavior. It should not be surprising that these are also among the major properties of the general plasma membrane of eukaryotic cells. I will briefly describe these functional categories in somewhat more detail before focusing on transmembrane signaling pathways associated with cilia and flagella.

Adhesion

Ciliary and flagellar surfaces are involved in three kinds of adhesive events: (a) cell–substrate interactions, (b) cell–cell interactions, and (c) eukaryotic/prokaryotic interactions. These interactions run the whole gambit in terms of specificity of interaction. Examples of cell–substrate interactions include: (1) adhesion to solid substrates in soil that is necessary for whole cell gliding motility in *Chlamydomonas* (Lewin 1952, Bloodgood 1981, Reinhart and Bloodgood 1988), (2) anchorage of interstitial acoel turbellarian worms to substrates (Tyler 1973), and (3) transient adhesive interactions of oviduct ciliary surfaces with the gelatinous matrix of the cumulus mass during transport in the mammalian female reproductive tract (Norwood et al. 1978). Examples of cell–cell adhesion via ciliary and flagellar surfaces are found in the sexual interactions of algal flagellates such as *Chlamydomonas* (Snell 1985, van den Ende et al. 1990) and ciliate protozoa such as *Paramecium* (Watanabe 1990). In these cases, the adhesive interaction is mediated via ciliary and flagellar membrane glycoproteins and exhibits a high degree of specificity; sexual agglutination of flagellar surfaces in *Chlamydomonas* exhibits species specificity, gamete specificity and mating type specificity (Snell 1985). The third class of adhesive interactions mediated by ciliary/flagellar surfaces involves interactions between prokaryotic pathogens (such as *Bordetella pertussis*) and the surface of respiratory cilia (Tuomanen 1990).

Force transduction

The surface of the *Chlamydomonas* flagellum exhibits force transduction. This force can be applied to other flagellar surfaces (during flagellar realignment that occurs between two mating gametes; Mesland 1976), to inert objects such as polystyrene microspheres (that are moved along the flagellar surface; Bloodgood 1977, Bloodgood et al. 1979, Hoffman and Goodenough 1980) and to a solid substrate (resulting in whole cell gliding motility; Levin 1952, Bloodgood 1981). This force transduction phenomenon is not restricted to *Chlamydomonas* flagellar surfaces; it has also been observed (in the form of microsphere movements) for sea urchin embryo ciliary surfaces (Bloodgood 1980) and other cellular membrane extensions that contain a microtubule-based cytoskeleton, such as axopodia and reticulipodia (Bloodgood 1978, Kano and Ishii 1979, Suzaki and Shigenaka 1982, Bowser et al. 1984, Bowser and Bloodgood 1984).

Sensory functions

Cilia and flagella can act as sensory receptors and the ciliary/flagellar membranes play a central role in this function. Cilia can function as mechanoreceptors; the classical example of this is the "avoidance response" in *Paramecium* described by Jennings (1906). When the leading end of a swimming *Paramecium* strikes an object, mechanical deformation of the cilia occurs, there is a calcium action potential generated in the ciliary membrane, the intraciliary calcium rises, resulting in an alteration in axonemal bending properties (ciliary reversal) and the cell transiently swims backwards. Thurm (1968) described another form of mechanoreception in motile cilia from *Mytilus*; he demonstrated that an active stroke could be evoked in a resting cilium by bending the cilium passively through a small angle using a micromanipulator. Chemoreception involves specialized neuronal cells located in the olfactory epithelium; the primary sensory events occur in association with the apical cilia (and in particular the membrane-matrix compartments) on these cells (Lancet 1988, Anholt 1989). The signaling pathway associated with sensory transduction in these cells will be discussed in more detail below. Another, more diffuse form of sensory function of the surfaces of cilia and flagella involves the recognition of contact with inert surfaces or other cells and the resultant triggering of transmembrane signaling events. These pathways will be discussed in detail below for the cases of cell–cell contact during mating in *Chlamydomonas* gametes and for cell–substrate contact that triggers whole cell gliding motility in both vegetative and gametic cells of *Chlamydomonas*.

Regulation of axonemal behavior

While the axoneme is a machine that can function in vitro in the absence of any ciliary or flagellar membrane components or soluble proteins, events occurring in the flagellar membrane and the matrix compartment regulate the waveform produced by the axoneme in vivo. Calcium and cAMP are the major second messengers that regulate the activity of the axoneme. Intraciliary and intraflagellar free calcium concentrations are regulated by calcium pumps and voltage-dependent calcium channels located within the ciliary/flagellar membrane. cAMP levels are regulated by adenylate cyclase and cyclic nucleotide phosphodiesterase enzymes located within the membrane-matrix compartments. cAMP and calcium effects on axonemal behavior are often antagonistic and are thought to act by altering the levels of phosphorylation of axonemal and non-axonemal proteins (Brokaw 1987, Tash and Means 1987, Tash 1989, Stephens and Stommel 1989). While cAMP is most likely acting through stimulation of a cAMP-dependent protein kinase, calcium may be regulating levels of protein phosphorylation both through stimulation of calcium-dependent protein kinases and calcium-dependent protein phosphatases (Klee and Cohen 1988, Tash 1989). In most ciliary and flagellar systems, cAMP results in an increase in the beat frequency causing the cell to move forward faster while calcium results in a transient reversal of the waveform resulting in temporary backwards movement.

This brief review suggests that ciliary and flagellar surfaces perform many diverse functions but common to most of these is the involvement (or presumed involvement) of a transmembrane signaling pathway involving integral membrane proteins, small molecules acting as second messengers that either act directly to gate membrane channels or affect axonemal behavior or else operate indirectly by regulating protein phosphorylation and dephosphorylation. I will spend the remainder of this review highlighting three specific examples of signaling pathways associated with cilia and flagella. Two of these are associated with the flagella of the protistan organism *Chlamydomonas* while the third system is associated with the cilia of the vertebrate olfactory epithelium.

Chlamydomonas flagella

Chlamydomonas is a biflagellate green algal cell; the cell body plasma membrane is covered by a multilayered glycoprotein cell wall (Roberts et al. 1985) containing tunnels through which the flagella protrude. Hence the flagellar membrane domain is the only portion of the cell surface accessible for cell–cell contact and cell–substrate interactions, in contrast to the situation in most mammalian cells where the entire cell surface has the potential to participate in such events. Thus the flagellar membrane of *Chlamydomonas* has become highly specialized for three major functions: (1) mediating the initial interactions between gametic cells, (2) mediating whole cell locomotion in association with a solid surface (gliding motility), and (3) for regulating the axonemal behavior responsible for swimming through a liquid medium. Only in recent years has considerable attention been paid to understanding the signaling events associated with mating interactions and whole cell gliding motility. These two signaling systems differ in that one is specific to gametes and one is common to vegetative and gametic cells. The latter system regulates protein movements in the flagellar membranes, movements that appear to subserve both mating events and whole cell gliding motility (Bloodgood 1987, Bloodgood and Salomonsky 1989).

Signaling events associated with gametic interactions

Mating in *Chlamydomonas* involves one or more flagellar signaling events in which contact of the surfaces of plus and minus gametic flagella results in flagellar tip activation, flagellar alignment, cell wall release, and mating structure activation (Snell 1985, van den Ende 1985, van den Ende et al. 1988). In addition, normal mating in *C. eugametos* is known to be associated with redistribution of the sexual agglutinin molecules (glycoproteins known to be responsible for the initial specific flagellar interactions) to the flagellar tips (Homan et al. 1987, Musgrave and van den Ende 1987). Antibodies to flagellar membrane proteins (from either vegetative or gametic flagella) and the lectin concanavalin A in the case of *C. reinhardtii* and antibodies to the sexual agglutinin and the lectin wheat germ agglutinin in the case of *C. eugametos* can induce the full range of mating events in a population of unmated gametes (Claes 1975, 1977; Goodenough and Jurivich 1978; Homan et al. 1988; Kooijman et al. 1989) (Table 3). In *C. eugametos*, a WGA-binding glycoprotein is the integral membrane receptor for the sexual agglutinin; hence, it makes sense that crosslinking the agglutinin receptor with WGA will elicit the same responses as crosslinking the agglutinin itself (with antibodies to the sexual agglutinin). This is equivalent to the fact that crosslinking the IgE molecules on the plasma membrane of mast cells with antigen or crosslinking the IgE receptor on mast cells with specific antibodies both elicit the same responses, influx of calcium and exocytosis of secretory granule contents (Beaven and Ludowyke 1989). Crosslinking of flagellar membrane proteins appears to be an essential step in sexual signaling because monovalent Fab fragments are not effective (Goodenough and Jurivich 1978, Homan et al. 1988). This suggests that in normal mating interactions, cross-linking of the flagellar sexual agglutinins induces transmembrane signaling and an alteration in the level of a second messenger within the flagellum or cell body. What is the second messenger(s) utilized for signal transduction in the gametic *Chlamydomonas* flagellum? The published data speaking to this issue are summarized in Table 2. Calcium was the first canditate to be considered as a second messenger in gametic flagellar signaling. A number of indirect observations have suggested some involvement of calcium: (1) trifluoperazine (TFP) and W-7 (both calmodulin antagonists) do not affect gametic flagellar agglutination but inhibit all subsequent steps in mating (Detmers and Condeelis 1986), (2) li-

Table 2. Two types of second messengers implicated in *Chlamydomonas* gametic flagellar signaling

Calcium
A 23187 + Ca^{2+} induces wall release (Claes 1980)[a]
A 23187 + Ca^{2+} + ethanol induces mating structure activation (Schuring et al. 1990)[b]
Calcium in cell body redistributes during mating (Kaska et al. 1985)[a]
Lidocaine inhibits signaling (Snell et al. 1982)[a]
Calmodulin antagonists inhibit signaling (Detmers and Condeelis 1986)[a]
Mating associated with transient calcium efflux (Bloodgood and Levin 1983)[a]
cAMP
cAMP rises transiently during mating (Pijst et al. 1984)[b]
Exogenous cAMP + IBMX induces signaling (Pasquale and Goodenough 1987)[a]
Exogenous cAMP induces signaling (Kooijman et al 1990)[b]

[a] Observation made using *C. reinhardtii*
[b] Observation made using *C. eugametos*

docaine (xylocaine) does not affect gametic flagellar agglutination but prevents two of the major consequences of signaling (cell wall release and mating structure activation; Snell et al. 1982), (3) a transient increase in efflux of calcium occurs upon initiation of mating (Bloodgood and Levin 1983), (4) mating is accompanied by the redistribution of calcium in the cell body (Kaska et al. 1985). More recently, Schuring et al. (1990) have observed that a number of treatments (involving combinations of ethanol, A 23187, TFP and IP_3) that are expected to increase intracellular calcium trigger mating structure activation in *C. eugametos*.

Mating in both *C. eugametos* (Pijst et al. 1984) and *C. reinhardtii* (Pasquale and Goodenough 1987) is accompanied by a transient rise in intracellular cAMP. Treatment of unmated gametes of *C. eugametos* with wheat germ agglutinin (WGA; known to stimulate all mating responses) results in a rise in the intracellular cAMP level (Kooijman et al. 1990) presumably through an activation of adenylate cyclase and/or an inhibition of cAMP phosphodiesterase. Exposure of unmated gametes of *C. reinhardtii* and *C. eugametos* to exogenous db-cAMP (in combination with a phosphodiesterase inhibitor) results in an increase in flagellar agglutinability, flagellar tip activation, cell wall lysis and mating structure activation (Pasquale and Goodenough 1987, Goodenough 1989, Kooijman et al. 1990). It is significant that exogenous db-cAMP does *not* result in flagellar sexual agglutinin redistribution (Kooijman et al. 1990), something that occurs during normal mating (Homan et al. 1978) and in response to crosslinking of flagellar glycoproteins with WGA or an antibody to the sexual agglutinin (Kooijman et al. 1989, Homan et al. 1988). This suggests that crosslinking of the sexual agglutinin during normal mating may trigger another signaling pathway (responsible for flagellar membrane glycoprotein redistribution; see next section) in addition to elevating intracellular cAMP. Alternatively, it may be that crosslinking of flagellar membrane glycoproteins may be necessary for the intraflagellar motor to interact with the glycoproteins and to move them in the plane of the flagellar membrane. This requirement for plasma membrane protein crosslinking is also seen in the case of receptor "capping" in lymphocytes (de Petris 1977).

What is the mechanism by which a rise in intracellular cAMP activates later events in mating in *Chlamydomonas*? One obvious possibility is that the rise in intracellular cAMP activates a cyclic nucleotide-dependent protein kinase. Pijst et al. (1984) demonstrated that *C. eugametos* gametes contain a cAMP-dependent protein kinase and Hasegawa et al. (1987) have shown that *C. reinhardtii* vegetative flagellar axonemes possess a cAMP-dependent protein kinase activity. Pasquale and Goodenough (1987) reported that post-agglutination mating events in *C. reinhardtii* are inhibited by H-8, a protein kinase inhibitor that inhibits, among other things, cAMP-dependent protein kinase (Hidaka et al. 1984).

The presence of calcium-regulated adenylate cyclase activity in *Paramecium* cilia (Gustin and Nelson 1987; Schultz et al. 1987) and mammalian sperm (Gross et al. 1987) suggests one possible mechanism for integrating calcium and cAMP signaling systems in cilia and flagella. Pasquale and Goodenough (1987) reported that adenylate cyclase activity was highly enriched in the *C. reinhardtii* gametic flagellar membrane and that this enzymatic activity was inhibited by TFP, a calmodulin antagonist; they did not determine if calcium affected the enzyme activity. In contrast, Kooijman et al. (1990) did not find an enrichment in adenylate cyclase activity in the *C. eugametos* gametic flagellum. Additional roles for calcium could be in the activation of a calcium-dependent protein kinase or protein phosphatase. Antibodies to calcium-calmodulin-dependent protein phosphatase (calcineurin or protein phophatase 2 B) cross-react with a protein in the *Chlamydomonas* flagellum (Tash 1989).

Since there is some evidence for the involvement of calcium in mating related signaling and the source of this calcium appears to be intracellular (Bloodgood and Levin 1983, Kaska et al. 1985), it raises the possibility that inositol phospholipids could be regulating the release of calcium from intracellular stores in *Chlamydomonas*. *Chlamydomonas eugametos* gametes contain PIP_2 and IP_3 (Irvine et al. 1989). The presence of an active IP_3-3-kinase activity in *C. eugametos* (Irvine 1990) also suggests a possible role for IP_4 in calcium influx into the cells. Schuring et al. (1990) report that IP_3 activates *C. eugametos* gametes, but only in the presence of 3% ethanol, another agent thought to mobilize calcium.

Signaling related to glycoprotein redistribution

In both vegetative and gametic cells of *Chlamydomonas*, crosslinking of some, but not all, populations of flagellar membrane glycoproteins with antibodies or lectins results in activation of machinery for moving the crosslinked population of glycoproteins within the plane of the flagellar membrane (Goodenough and Jurivich 1978, Bloodgood et al. 1986, Bloodgood 1987,

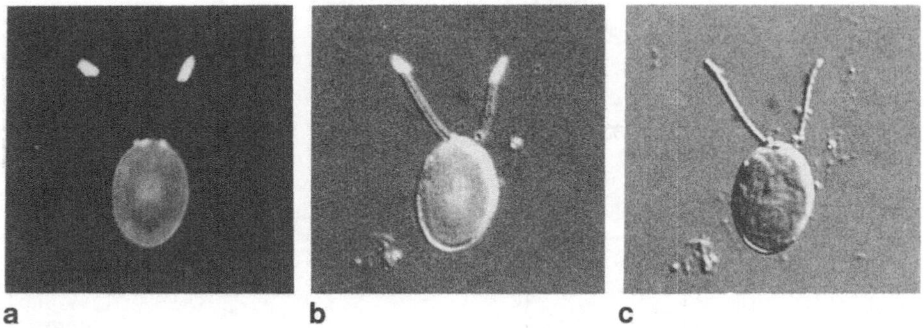

Fig. 3. Redistribution of glycoproteins along the flagellar membrane of a vegetative cell of *Chlamydomonas reinhardtii* induced by crosslinking with the FMG-1 monoclonal antibody. These cells were initially uniformly labeled with the antibody; the photographs were taken 10 min after warming the labeled cells to 25 °C. **a** Immunofluorescence image; **c** differential interference contrast image; **b** taken using both optical systems simultaneously. × 1500

Table 3. Multivalent ligand induced redistribution of *Chlamydomonas* flagellar glycoproteins

Species	Cell type	Ligand	Epitope	Glycoprotein
C. eugametos	gametes	WGA	CHO	agglutinin binding protein[a]
C. eugametos	gametes	mab 66.3	CHO	minus agglutinin[b]
C. reinhardtii	vegetative	Con A	CHO	con A binding glycoproteins[c]
C. reinhardtii	gametes	Con A	CHO	con A binding glycoproteins[c]
C. reinhardtii	vegetative	FMG-1 ab	CHO	350 kDa glycoproteins[c]
C. reinhardtii	gametes	FMG-1 ab	CHO	350 kDa glycoproteins[c]
C. reinhardtii	vegetative	FMG-3 ab	CHO	3–4 HMW glycoproteins[d]

In addition, several polyclonal antibodies to flagellar components also induce redistribution. Only multivalent ligands induce redistribution but not all multivalent ligands for flagellar membrane glycoproteins induce redistribution in *C. eugametos* gametes

[a] Kooijman et al. 1989
[b] Homan et al. 1988
[c] Bloodgood et al. 1986
[d] Bloodgood, unpubl. obs.

Homan et al. 1988, Kooijman et al. 1989) (Fig. 3 and Table 3). This process of active membrane protein movements, referred to as redistribution or tipping, is thought to underlie whole cell gliding motility (Bloodgood and Salomonsky 1989) and some of the dynamic events (such as the migration of flagellar contact sites) occurring in the flagellum during the course of mating (Homan et al. 1987, Musgrave and van den Ende 1987). In the case of these physiologically relevant processes, it is thought that flagella–flagella contact during mating and flagella–substrate interaction during gliding motility locally crosslink the membrane glycoproteins in the plane of the flagellar membrane. It is presumed that crosslinking of flagellar membrane glycoproteins activates a transmembrane signaling pathway that results in activation of intraflagellar machinery responsible for generating force to move the flagellar glycoproteins.

Recent results from the author's laboratory (Bloodgood and Salomonsky 1990, 1991) suggest that the signaling pathway involved in the redistribution of the major flagellar membrane glycoprotein common to vegetative and gametic cells involves calcium regulated changes in levels of flagellar protein phosphorylation. Antibody-induced glycoprotein redistribution is maximal at 20 μM free calcium and above but is inhibited if the free calcium concentration in the medium is lowered to 10^{-7} M or less (Bloodgood and Salomonsky 1990) (Fig. 4). This inhibition is reversible upon raising the extracellular calcium concentration. In addition, calcium channel blockers (diltiazem, D-600 and barium chloride) and calmodulin antagonists (TFP, W-7) reversibly inhibit the glycoprotein redistribution at permissive levels of calcium in the medium (Bloodgood and Salomonsky 1990).

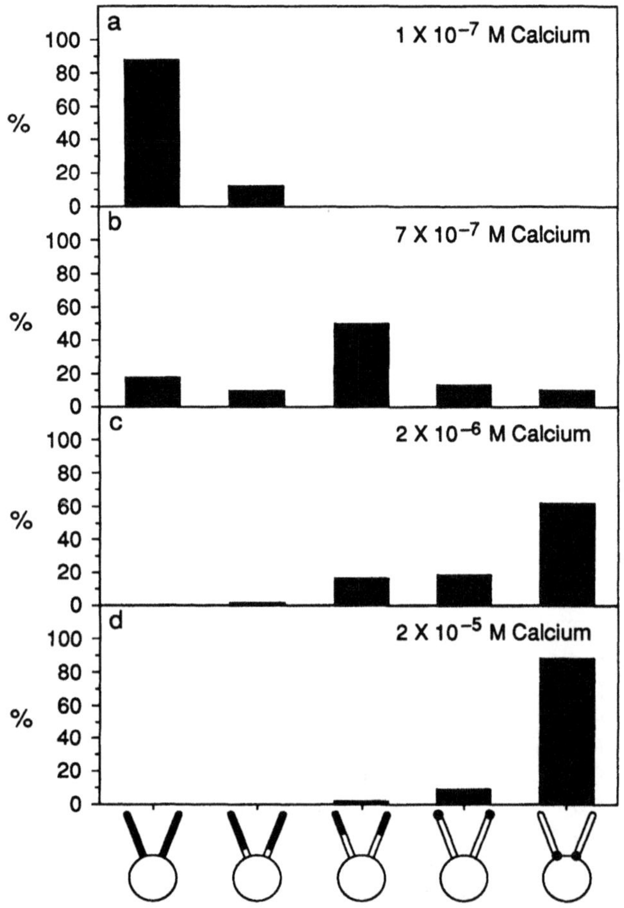

Fig. 4. Histogram showing the effect of free calcium concentration in the medium on antibody-induced glycoprotein redistribution in vegetative *Chlamydomonas reinhardtii*. 10^{-7} M free calcium does not permit any redistribution to occur while 20 μM free calcium allows maximal redistribution. 0.7 μM and 2 μM free calcium allow intermediate amounts of glycoprotein redistribution. Reproduced with permission of the Company of Biologists Ltd from Bloodgood and Salomonsky (1990)

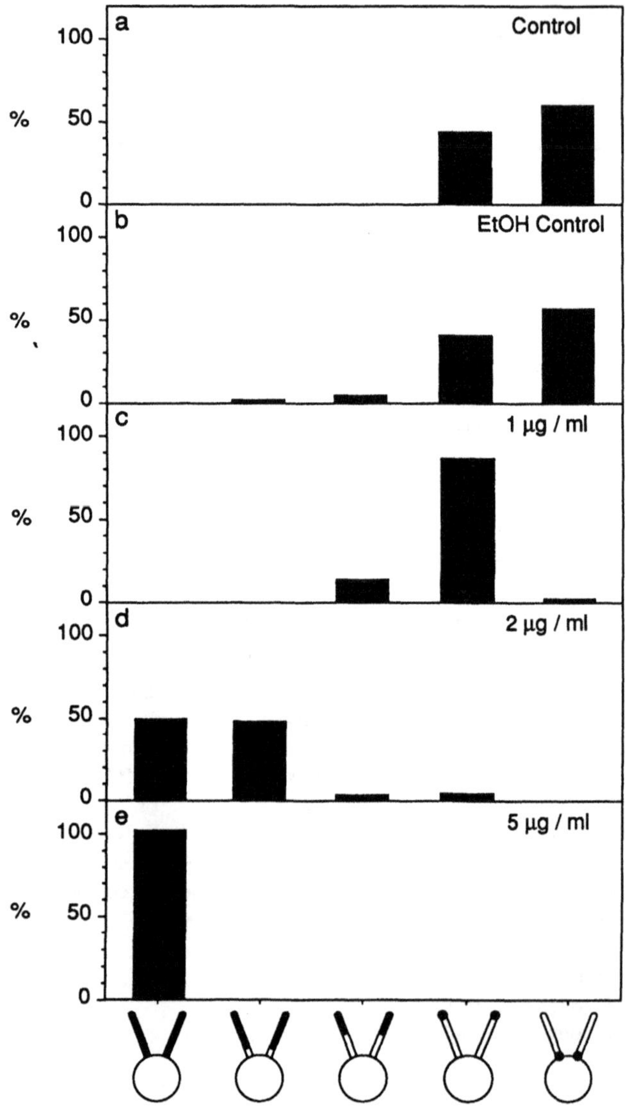

Fig. 5. Histogram showing the dose-dependent inhibition of antibody-induced flagellar glycoprotein redistribution in vivo by the protein kinase inhibitor staurosporine in the presence of 20 μM free calcium in the medium. The ethanol (EtOH) control shown represents the concentration of ethanol (2.5%) present in the highest drug concentration shown (5 μg/ml)

Antibody-induced flagellar glycoprotein redistribution is also inhibited by three protein kinase inhibitors: H-7, H-8, and staurosporine (Bloodgood and Salomonsky 1991) (Fig. 5). These observations strongly suggest that protein phosphorylation is involved in the signaling pathway whereby glycoprotein crosslinking activates the flagellar machinery responsible for glycoprotein movements. However, all of these drugs can inhibit multiple kinds of protein kinase so these observations do not allow us to unambiguously identify the specific protein kinase involved. The observation that glycoprotein redistribution is dependent on both extracellular calcium and protein phosphorylation led us to look at the calcium regulation of protein phosphorylation in vitro using non-ionic detergent extracts of flagella (containing the membrane and soluble compartments of the flagella). These in vitro studies revealed that the flagellar membrane-matrix fraction contains calcium-independent protein kinase activity, calcium-dependent protein kinase activity and protein phosphatase activity (Fig. 6). In the presence of 20 μM free calcium, there is a rapid phosphorylation followed by a slower dephosphorylation of a small group of phosphoproteins (Fig. 7). One or more of these phosphoproteins whose phosphorylation is regulated by calcium may regulate the activity of the flagellar machinery

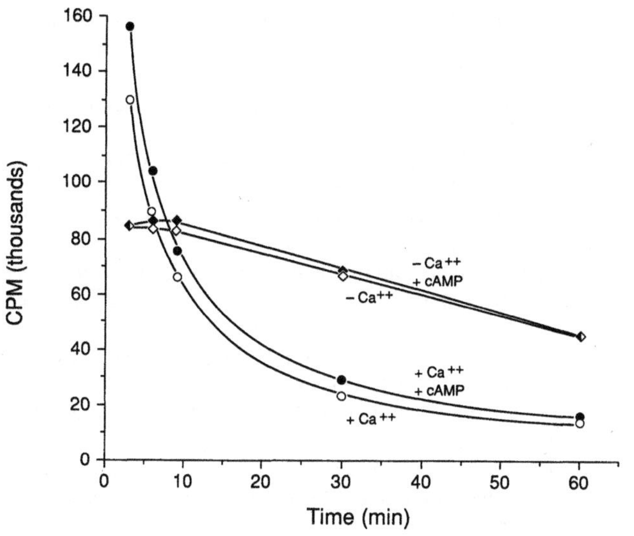

Fig. 6. Effect of calcium and cAMP on protein phosphorylation and dephosphorylation in a *Chlamydomonas reinhardtii* flagellar membrane-matrix extract. 20 µM free calcium (○,●) stimulated both the level of protein phosphorylation at 3 min and the subsequent rate of protein dephosphorylation. The inclusion of 150 µM cAMP (●,◆) in the reaction mixture had relatively little effect on phosphorylation and dephosphorylation at high or low calcium concentrations

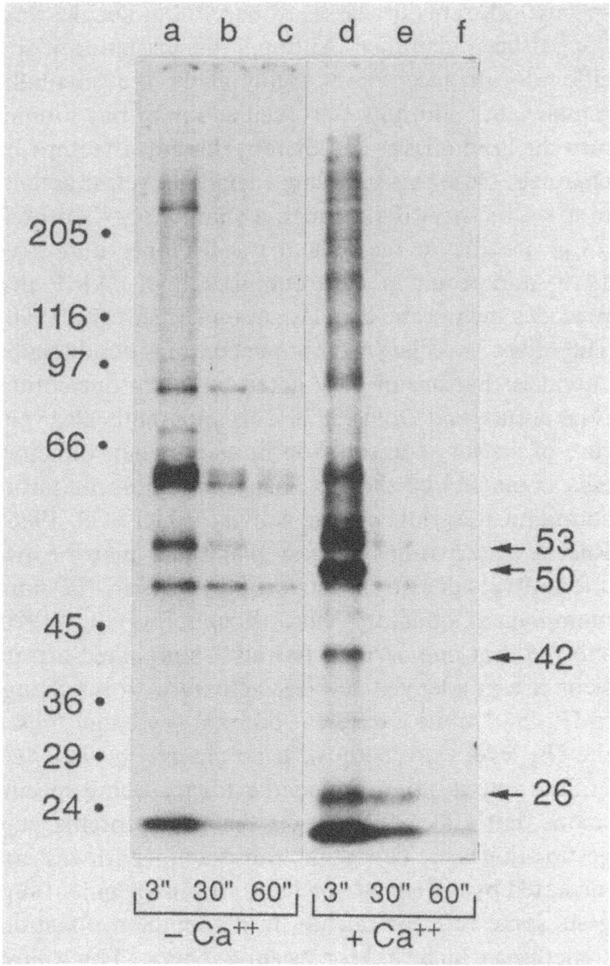

Fig. 7. Autoradiographic pattern from an 8% SDS-polyacrylamide gel of protein dephosphorylation in vitro in flagellar membrane-matrix extracts at low (< 10^{-8} M; *a–c*) and high (20 µM; *d–f*) free calcium concentrations. 20 µM free calcium stimulated the phosphorylation of a small group of polypeptides with apparent molecular weights of 26, 42, 50, and 53 kDa. Phosphoproteins with apparent molecular weights of 42 and 50 kDa are dephosphorylated more rapidly than the other phosphoproteins. Samples were taken at 3, 30, and 60 min after initiation of protein phosphorylation. The arrows on the right indicate proteins whose phosphorylation is stimulated by 20 µM free calcium. On the left, positions of the molecular weight standards

responsible for driving glycoprotein movements within the flagellar membrane. As yet, no clear function has been assigned for cAMP in the signaling pathway associated with antibody-induced glycoprotein movements. In vivo, exogenous db-cAMP alone has no effect on the antibody-induced redistribution and cannot replace the need for crosslinking of the glycoproteins with antibody or concanavalin A. The in vitro protein phosphorylation/dephosphorylation is not affected by addition of exogenous cAMP or cGMP or the presence of a peptide inhibitor of cAMP-dependent protein kinase (Kemp et al. 1976). The characteristics of the calcium-stimulated protein kinase activity observed in the detergent extracts of *Chlamydomonas* flagella most closely resembles those of a novel calcium-dependent protein kinase (CaPK-2) found in *Paramecium tetraurelia* cilia and cell bodies (Gunderson and Nelson 1987). Both the *Chlamydomonas* and the *Paramecium* protein kinase activities are calcium-dependent but lack the characteristics of calcium-calmodulin dependent protein kinase (unaffected by exogenous calmodulin and trifluoperazine) and protein kinase C (unaffected by phosphatidylserine and diacylglycerol).

Olfactory cilia

Vertebrate olfactory cilia are a remarkable example of the sensory roles that can be played by cilia and flagella (Lancet 1988, Anholt 1989, Reed 1990). The olfactory ciliary membrane is the site at which odorant molecules inititate the depolarization of the olfactory neurons. Although olfactory ciliary membranes are known to possess unique membrane proteins absent from the membranes of non-sensory respiratory cilia (Chen and Lancet 1984, Chen et al. 1986, Anholt et al. 1990) and a generalized odorant-binding protein (OBP) has been described (Snyder et al. 1989), membrane receptors for specific odorants or classes of odorant molecules have not yet been identified. Although the existence of specific odorant receptors is highly likely, the possibility remains that odorants exert their action by partitioning into the lipid bilayer and thereby directly affecting ion channels. Odorants stimulate adenylate cyclase activity in the olfactory cilia through a stimulatory G protein (G_{olf}) specific to the olfactory cell (Jones and Reed 1989) and result in an accumulation of cAMP that precedes membrane depolarization (Breer et al. 1990). The olfactory ciliary membrane contains conductance channels that are directly gated by cyclic nucleotides (Nakamura and Gold 1987). This apparently clear picture of sensory transduction in mammalian olfactory cilia is marred by the fact that some odorants fail to stimulate adenylate cyclase activity (Sklar et al. 1986). Recent observations suggest that there may be two alternative signal transduction pathways in the same mammalian olfactory cilia. Boekhoff et al. (1990) showed that one odorant (citralva) stimulated a transient rise in adenylate cyclase activity with no change in IP_3 level while a different odorant (pyrazine) raised the IP_3 level transiently with no change in adenylate cyclase activity; they provided evidence, using specific toxins that affect different classes of G proteins, suggesting that these two signal transduction pathways are mediated by different G proteins. IP_3 involvement suggests some role for calcium in mammalian olfaction, something which has been recently suggested by Anholt (1989).

Acknowledgements

This review is based on a talk presented in a symposium on the Biology of Protistan Flagellar/Cytoskeletal Systems organized by Dr. Michael Melkonian at the 8th meeting of the International Society for Evolutionary Protistology held in June, 1990, at the University of Maryland. The author's research discussed in this paper was supported by research grants from the National Science Foundation (DCB-8905530) and the Jeffress Trust (J-175). The able technical assistance of Ms. Nancy Salomonsky is acknowledged.

References

Anholt RRH (1989) Molecular physiology of olfaction. Am J Physiol 257: C 1043–1054

– Petro AE, Rivers AM (1990) Identification of a group of novel membrane proteins unique to chemosensory cilia of olfactory receptor cells. Biochemistry 29: 3366–3373

Beaven MA, Ludowyke R (1989) Stimulatory signals for secretion in mast cells and basophils. Adv Regul Cell Growth 1: 245–285

Bloodgood RA (1977) Motility occurring in association with the surface of the *Chlamydomonas* flagellum. J Cell Biol 75: 983–989

– (1978) Unidirectional motility occurring in association with the axopodial membrane of *Echinosphaerium nucleofilum*. Cell Biol Int Rep 2: 171–176

– (1980) Direct visualization of dynamic membrane events in cilia. J Exp Zool 213: 293–295

– (1981) Flagella-dependent gliding motility in *Chlamydomonas*. Protoplasma 106: 183–192

– (1987) Glycoprotein dynamics in the *Chlamydomonas* flagellar membrane. Adv Cell Biol 1: 97–130

– (ed) (1990) Ciliary and flagellar membranes. Plenum, New York

– Levin EN (1983) Transient increase in calcium efflux accompanies fertilization in *Chlamydomonas*. J Cell Biol 97: 397–404

– Salomonsky NL (1989) Use of a novel *Chlamydomonas* mutant to demonstrate that flagellar glycoprotein movements are necessary for the expression of gliding motility. Cell Motil Cytoskeleton 13: 1–8

– – (1990) Calcium influx regulates antibody-induced glycoprotein movements within the *Chlamydomonas* flagellar membrane. J Cell Sci 96: 27–33

– – (1991) Regulation of flagellar glycoprotein movements by protein phosphorylation. Eur J Cell Biol 54: 85–89

– Leffler EM, Bojczuk AT (1979) Reversible inhibition of *Chlamydomonas* flagellar surface motility. J Cell Biol 82: 664–674

– Woodward MP, Salomonsky NL (1986) Redistribution and shedding of flagellar membrane glycoproteins visualized using an anti-carbohydrate monoclonal antibody and concanavalin A. J Cell Biol 102: 1797–1812

Boekhoff I, Tareilus E, Strotmann J, Breer H (1990) Rapid activation of alternative second messenger pathways in olfactory cilia from rats by different odorants. EMBO J 9: 2453–2458

Bowser SS, Bloodgood RA (1984) Evidence against surf-riding as a general mechanism for surface motility. Cell Motil 4: 305–314

– Israel HA, McGee-Russell SM, Rieder CL (1984) Surface transport properties of reticulipodia: do intracellular and extracellular motility share a common mechanism? Cell Biol Int Rep 8: 1051–1063

Breer H, Boekhoff I, Tareilus E (1990) Rapid kinetics of second messenger formation in olfactory transduction. Nature 345: 65–68

Brokaw CJ (1987) Regulation of sperm flagellar motility by calcium and cAMP-dependent phosphorylation. J Cell Biochem 35: 175–184

Chen Z, Lancet D (1984) Membrane proteins unique to vertebrate olfactory cilia: candidates for sensory receptor molecules. Proc Natl Acad Sci USA 81: 1859–1863

– Pace U, Ronen D, Lancet D (1986) Polypeptide gb 95. A unique glycoprotein of olfactory cilia with transmembrane receptor properties. J Biol Chem 261: 1299–1305

Claes H (1975) Influence of concanavalin A on autolysis of gametes from *Clamydomonas reinhardtii*. Arch Microbiol 103: 225–230

– (1977) Non-specific stimulation of the autolytic system in gametes from *Chlamydomonas reinhardtii*. Exp Cell Res 108: 221–229

– (1980) Calcium ionophore-induced stimulation of secretory activity in *Clamydomonas renhardtii*. Arch Microbiol 124: 81–86

de Petris S (1977) Distribution and mobility of plasma membrane components on lymphocytes. In: Poste G, Nicolson GL (eds) Dynamic aspects of surface organization. North-Holland, Amsterdam, pp 643–728 [Cell surface reviews, vol 3]

Detmers PA, Condeelis JS (1986) Trifluoperazine and W-7 inhibit mating in *Chlamydomonas* at an early stage of gametic interaction. Exp Cell Res 163: 317–326

Goodenough UW (1989) Cyclic AMP enhances the sexual agglutinability of *Chlamydomonas* flagella. J Cell Biol 109: 247–252

– Jurivich D (1978) Tipping and mating-structure activation induced in *Chlamydomonas* by flagellar membrane antisera. J Cell Biol 79: 680–693

Gross MK, Toscano DG, Toscano WA Jr (1987) Calmodulin-mediated adenylate cyclase from mammalian sperm. J Biol Chem 262: 8672–8676

Gunderson RE, Nelson DL (1987) A novel Ca^{2+}-dependent protein kinase from *Paramecium tetraurelia*. J Biol Chem 262: 4602–4609

Gustin MC, Nelson DL (1987) Regulation of ciliary adenylate cyclase by Ca^{2+} in *Paramecium*. Biochem J 246: 337–345

Hasegawa E, Hayashi H, Asakura S, Kamiya R (1987) Stimulation of in vitro motility of *Chlamydomonas* axonemes by inhibition of cAMP-dependent phosphorylation. Cell Motil Cytoskeleton 8: 302–311

Hidaka H, Inagaki M, Kawamoto S, Sasaki Y (1984) Isoquinolinesulfonamides, novel and potent inhibitors of cyclic nucleotide dependent protein kinase and protein kinase C. Biochemistry 23: 5036–5041

Hoffmann JL, Goodenough UW (1980) Experimental dissection of flagellar surface motility in *Chlamydomonas*. J Cell Biol 86: 656–665

Homan W, Sigon C, van den Briel W, Wagter R, de Nobel H, Mesland D, Musgrave A, van den Ende H (1987) Transport of membrane receptors and the mechanics of sexual cell fusion in *Chlamydomonas eugametos*. FEBS Lett 215: 323–326

– Musgrave A, de Nobel H, Wagter R, de Wit D, Kolk A, van den Ende H (1988) Monoclonal antibodies directed against the sexual binding site of *Chlamydomonas eugametos* gametes. J Cell Biol 107: 177–189

Huang BP-H (1986) *Chlamydomonas reinhardtii*: a model system for the genetic analysis of flagellar structure and motility. Int Rev Cytol 99: 181–215

Irvine R (1990) Messenger gets the green light. Nature 346: 700–701

– Letcher AJ, Lander DJ Drobak BK, Dawson AP, Musgrave A (1989) Phosphoinositol (4,5) bisphosphate and phosphoinositol (4) phosphate in plant tissues. Plant Physiol 89: 888–892

Jennings HS (1906) Behavior of lower animals. Indiana University Press, Bloomington

Jones DT, Reed RR (1989) G_{olf}: an olfactory neuron specific-G protein involved in odorant signal transduction. Science 244: 790–795

Kano F, Ishii K (1979) Movement of the surface layer on the axopodium and its protoplasm in *Actinosphaerium*. Bull Fac Lib Arts Hosei Univ 31: 1–8

Kaska DD, Piscopo IC, Gibor A (1985) Intracellular calcium redistribution during mating in *Chlamydomonas reinhardtii*. Exp Cell Res 160: 371–379

Kemp BE, Benjamani E, Krebs EG (1976) Synthetic hexapeptide substrates and inhibitors of 3′:5′-cyclic AMP-dependent protein kinase. Proc Natl Acad Sci USA 73: 1038–1042

Klee CB, Cohen P (1988) The calmodulin-regulated protein phosphatase. In: Cohen P, Klee CB (eds) Calmodulin. Elsevier, Amsterdam, pp 225–248 [Molecular aspects of cell regulation, vol 5]

Kooijman R, de Wildt P, Beumer S, van den Vliet G, Homan W, Kalshoven H, Musgrave A, van den Ende H (1989) Wheat germ agglutinin induces mating reactions in *Chlamydomonas eugametos* by cross-linking agglutinin-associated glycoproteins in the flagellar membrane. J Cell Biol 109: 1677–1687

– – van den Briel W, Tan S-H, Musgrave A, van den Ende H (1990) Cyclic AMP is one of the intracellular signals during the mating of *Chlamydomonas eugametos*. Planta 181: 529–537

Lancet D (1988) Molecular components of olfactory reception and transduction. In: Margolis FL, Getchell TV (eds) Molecular neurobiology of the olfactory system. Plenum, New York, pp 25–50

Lewin RA (1952) Studies on the flagella of algae I. General observations on *Chlamydomonas moewusii* Gerloff. Biol Bull (Woods Hole, Mass) 103: 74–79

Luck JDL (1984) Genetic and biochemical dissection on the eukaryotic flagellum. J Cell Biol 98: 789–794

Mesland DAM (1976) Mating in *Chlamydomonas eugametos*. A scanning electron microscopical study. Arch Microbiol 109: 31–35

Musgrave A, van den Ende H (1987) How *Chlamydomonas* court their partners. Trends Biochem Sci 12: 470–473

Nakamura T, Gold GH (1987) A cyclic nucleotide-gated conductance on olfactory receptor cilia. Nature 325: 442–444

Norwood JT, Hein CE, Halbert SA, Anderson RGW (1978) Polycationic macromolecules inhibit cilia-mediated ovum transport in the rabbit oviduct. Proc Natl Acad Sci USA 75: 4413–4416

Pasquale SM, Goodenough UW (1987) Cyclic AMP functions as a primary sexual signal in gametes of *Chlamydomonas reinhardtii*. J Cell Biol 105: 2279–2292

Pijst HLA, van Driel R, Janssens PMW, Musgrave A, van den Ende H (1984) Cyclic AMP is involved in sexual reproduction of *Chlamydomonas eugametos*. FEBS Lett 174: 132–136

Reed RR (1990) How does the nose know? Cell 60: 1–2

Reinhart FD, Bloodgood RA (1988) Gliding defective mutant cell lines of *Chlamydomonas moewusii* exhibit alterations in a 240 kDa surface-exposed flagellar glycoprotein. Protoplasma 144: 110–118

Roberts K, Grief C, Hills GJ, Shaw PJ (1985) Cell wall glycoproteins: structure and function. J Cell Sci [Suppl] 2: 105–127

Schultz JE, Uhl DG, Klumpp S (1987) Ionic regulation of adenylate cyclase from the cilia of *Paramecium tetraurelia*. Biochem J 246: 187–192

Schuring F, Brederoo J, Musgrave A, van den Ende H (1990) Increase in calcium triggers mating structure activation in *Chlamydomonas eugametos*. FEMS Microbiol Lett 71: 237–240

Sklar PB, Anholt RRH, Snyder SH (1986) The odorant-sensitive adenylate cyclase of olfactory receptor cells. Differential stimulation by different classes of odorants. J Biol Chem 261: 15538–15543

Snell WJ (1985) Cell-cell interactions in *Chlamydomonas*. Annu Rev Plant Physiol 36: 287–315

– Buchanan M, Clausell A (1982) Lidocaine reversibly inhibits fertilization in *Chlamydomonas*: a possible role for calcium in sexual signaling. J Cell Biol 94: 607–612

Snyder SH, Sklar PB, Hwang PM, Pevsner J (1989) Molecular mechanisms of olfaction. Trends Neurosci 12: 35–38

Stephens RE, Stommel EW (1989) Role of cyclic adenosine monophosphate in ciliary and flagellar motility. In: Warner FD, Satir P, Gibbons IR (eds) Cell movement, vol 1. AR Liss, New York, pp 299–316

Suzaki T, Shigenaka Y (1982) Intra-axopodial particle movement and axopodial surface motility. In: Sakai H, Mohri H, Borisy GG (eds) Biological functions of microtubules and related structures. Academic Press, New York, pp 91–103

Tash JS (1989) Protein phosphorylation: the second messenger signal transducer of flagellar motility. Cell Motil Cytoskeleton 14: 332–339

Tash JS, Means AR (1987) Ca^{2+} regulation of sperm axonemal motility. Methods Enzymol 139: 808–823

Thurm U (1968) Steps in the transducer process of mechanoreceptors. Symp Zool Soc Lond 23: 199–216

Tuomanen E (1990) The surface of mammalian respiratory cilia. Interactions between cilia and repiratory microbes. In: Bloodgood RA (ed) Ciliary and flagellar membranes. Plenum, New York, pp 363–388

Tyler S (1973) An adhesive function for modified cilia in an interstitial turbellarian. Acta Zool 54: 139–151

van den Ende H (1985) Sexual agglutination in *Chlamydomonas*. Adv Microbiol Physiol 26: 89–123

– Klis FM, Musgrave A (1988) The role of flagella in sexual reproduction of *Chlamydomonas eugamentos*. Acta Bot Neerl 37: 327–350

– Musgrave A, Klis FM (1990) The role of flagella in the sexual reproduction of *Chlamydomonas* gametes. In: Bloodgood RA (ed) Ciliary and flagellar membranes. Plenum, New York, pp 129–147

Watanabe T (1990) The role of ciliary surfaces in mating in *Paramecium*. In: Bloodgood RA (ed) Ciliary and flagellar membranes. Plenum, New York, pp 149–171

Protoplasma (1991) 164: 23–37

PROTOPLASMA

Development of the flagellar apparatus during the cell cycle in unicellular algae

P. L. Beech*, Kirsten Heimann, and **M. Melkonian**

Botanisches Institut, Universität zu Köln, Köln

Received December 1, 1990
Accepted December 14, 1990

Summary. Recent evidence has shown that algal cells acquire different flagella and a heterogeneous basal apparatus through the prolonged development of these structures over more than one cell cycle. A system for numbering algal flagella and basal bodies, which is based on developmental studies, is discussed along with the various means by which the flagellar/basal body developmental cycle can be determined. We review the information now available on development of the separate components of the flagellar apparatus – this comes particulary from the Chlorophyta and the Chromophyta – and attempt to elucidate any information which may help in phylogenetic comparisons. New data is provided on developmental changes in the cartwheel part of the basal body and basal body-associated connecting fibrils in green algae.

Keywords: Algal flagellates; Basal body; Development; Cell division; Flagella; Flagellar apparatus; Flagellar developmental cycle.

Abbreviations: Bb basal body; d right (dexter) root; df right fibrils connecting Bb triplets to microtubular and/or fibrous roots; EM electron microscopy; F flagellum; IMF immunofluorescence microscopy; LM light microscopy; NBBC nucleus–basal body connector; s left (sinister) root; sf left fibrils connecting Bb triplets to microtubular and/or fibrous roots. See Nomenclature section of Introduction for the numbering of basal bodies and their flagella; the same numbers apply to Bb-associated d and s roots, and df and sf fibrils.

Introduction

Most unicellular, flagellate algae bear flagella that differ in size, structure and/or function. These differences are accompanied by a heterogeneity in the microtubular and/or fibrous roots that are attached to the flagellar basal bodies (Bbs). Even in those green algal (Chlorophyta) cells whose flagella and microtubular roots appear identical, the flagella are functionally different (Kamiya and Witman 1984) and opposite roots can differ in length (Lechtreck et al. 1989) and their association with the eyespot (Melkonian and Robenek 1984).

The differences in flagellar and flagellar root structure and function within a single cell require that, during division, a cell must be able to produce the correct complement of structures for each daughter. It has been known for many years that in the algae parental flagella and/or their Bbs are segregated to the daughter cells (e.g., *Polytomella agilis*, Doflein 1918); this is true in non-algal systems as well (e.g., Cleveland 1963). This semiconservative distribution, where each daughter cell receives one half of the parental flagellar apparatus and produces the other half de novo, is apparently a universal phenomenon, and new Bbs generally arise adjacent to an existing (parental) Bb. The developmental means by which cells acquire a heterogeneous flagellar apparatus have, until recently however, been unclear. If, in a biflagellate, the two flagella (and their Bbs) each represent a distinct type that matures in one cell cycle, then the two newly formed flagella/Bbs would be different to one another. Alternatively, if the two newly formed flagella/Bbs are the same, then one of the parental flagella/Bbs must change or transform to the other type so that flagellar apparatus heterogeneity is maintained in the daughter cells. This implies that flagella/Bbs require more than one cell cycle to mature and that distinct flagellar/Bb types are developmentally

* Correspondence and reprints: Botanisches Institut, Lehrstuhl 1, Universität zu Köln, Gyrhofstrasse 15, D-W-5000 Köln 41, Federal Republic of Germany.

related. Melkonian et al. (1987 a) have obtained evidence in support of the latter alternative; they showed that the anisokont flagella of *Nephroselmis olivacea* (Prasinophyceae) require more than one cell generation to fully develop. Newly formed Bbs each produce a short flagellum which, in the following cell cycle elongates to become a long flagellum; the long flagellum remains as such through subsequent cell cycles. This transformation of one flagellar/Bb type to another during cell division indicates that the separate flagella/Bbs of a cell, though they may appear to be different, are, through their developmental relatedness, inherently identical structures that express different morphologies depending on their age.

The flagellar developmental cycle has since been determined for species from several of the major algal groups (Table 1). It now seems likely that all flagellate algae, if not all eukaryotes, acquire dissimilar flagella and/or Bbs through the extended development of these organelles over more than one cell cycle.

We shall summarize a nomenclature for the numbering of flagella and Bbs that is based on flagellar developmental studies, and describe the various means by which flagellar developmental cycles have been determined. We shall then review the known patterns of flagellar development that occur during cell division in various algal groups: those that are evident from light microscopical (LM) observations will be discussed before those which have been elucidated by electron microscopy (EM). Following this, consideration will be given to what little is known of the changes that occur in the basal apparatus (the Bbs and their associated microtubular and fibrous roots) during flagellar development.

Nomenclature of basal body and flagellar numbering

Two groups (Heimann et al. 1989 b, Moestrup and Hori 1989) have recently proposed a system for numbering the flagella and Bbs of algal cells according to their developmental status. It is likely that at least in closely related organisms mature flagella/Bbs will be homologous to one another (and this is borne out by comparisons with other characteristics). The system is therefore seen as a possible means of clarifying phylogenetic relationships which otherwise may be obscure.

The term 'mature' will be used in reference to a Bb that will not, in subsequent generations, change its interphase position in the cell (with respect to the other Bbs or cell organelles) or produce a flagellum and/or roots that differ to those already present on that Bb. A mature flagellum is one which is associated with a mature Bb; it will not change its interphase characteristics (i.e., type of beat, orientation, relative length, presence or absence of hairs, etc.) through subsequent generations. Note, however, that a mature flagellum may undergo transient changes during cell division. In the extreme this can involve total flagellar retraction; here it may be better to think of a flagellum as being in a mature *state*.

The nomenclature labels flagella and Bbs in their mature developmental state as no. 1 and, for the reasons discussed by Moestrup and Hori (1989), the system is thus applicable to cells with any number of flagella. The mature flagellum/Bb is no. 1; the next younger flagellum/Bb in a cell is no. 2, and so on. In a quadriflagellate cell there are three developmental stages of flagella: two flagella are from newly formed Bbs and belong to the youngest generation. According to the nomenclature of Moestrup and Hori (1989) these flagella/Bbs would be nos. 3 and 4. We propose that such flagella/Bbs be both termed no. 3, and that they be distinguished from each other by a superscript that denotes the supposed organizing parental Bb (following Heimann et al. 1989 b). For example 3^1 would be more closely associated with Bb 1 (the mature Bb) and 3^2 with Bb 2 (a Bb that will not mature until the following cell cycle). Similarly, the newly formed Bbs in a biflagellate are termed 2^1 and 2^2 until they are segregated to the daughter cells. If it is not clear from which particular Bb a no. 2 or no. 3 Bb has probably arisen, then the system can still be used – as long as it is stated that the superscripts do not imply descent. The important thing is that the basic number of the flagellum/Bb indicates its developmental status with respect to the others of the same cell.

The notation used for a flagellum/Bb that has transformed from a no. 2 to a no. 1, but has not yet segregated to a daughter cell, is 1_2 (Heimann et al. 1989 b). Similary, in cells that have four flagella/Bbs at interphase, the two youngest Bbs to be transformed will be called 2_3 (e.g., Gaffal and el-Gammal 1990, Sluiman and Bloomers 1990).

We also note that according to this nomenclature it is possible to have a uniflagellate in which the single flagellum is termed F 2. That is, the flagellum arises from a Bb 2, and Bb 1 produces no flagellum (e.g., *Monomastix* sp., *Mallomonas splendens*, see below).

Means of determining the flagellar developmental cycle

Various methods have been used to determine the flagellar developmental cycle in algal cells (Table 1). Per-

Table 1. Species in which the flagellar developmental cycle has been determined

Division/class/species	Method[a]	Reference
Chlorophyta		
Prasinophyceae		
Nephroselmis olivacea	3	Melkonian et al. 1987 a
Pyramimonas octopus	5	Moestrup and Hori 1989
Chlorophyceae		
Spermatozopsis similis	1, 2, 6	Lechtreck et al. unpubl.
Chlamydomonas reinhardtii	1*	Holmes and Dutcher 1989
Incertae Sedis		
Pedinomonas tuberculata	2	Heimann et al. 1989 a
Monomastix sp.	2	Heimann et al. 1989 a
Glaucocystophyta		
Cyanophora paradoxa	1	Heimann et al. 1989 b
Chromophyta		
Chrysophyceae		
Epipyxis pulchra	1	Wetherbee et al. 1988
Synurophyceae		
Mallomonas splendens	1, 4	Beech and Wetherbee 1990 a, b
M. papillosa	1	Beech and Wetherbee 1990 a
Pedinellophyceae		
Pseudopedinella elastica	1	Heimann et al. 1989 a
Apedinella radians	7	Wetherbee et al. 1988
Prymnesiophyceae		
Pleurochrysis carterae	4	Beech et al. 1988
Euglenophyta		
Euglena mutabilis	7	Melkonian et al. 1987 a
Ploeotia costata	4	Farmer and Triemer 1988

[a] Method of determination:
1 Direct LM observations on the flagella of live, dividing cells;
*1** LM observations on Bb segregation at mitosis;
2 LM on fixed cells at various stages of the cell cycle;
3 statistical analysis of data gained by (2);
4 EM (thin sections) of Bb-associated roots in dividing cells;
5 EM (thin sections) of Bb-segregation at mitosis;
6 EM (wholemount preparations) of flagellar apparatuses from dividing cells;
7 cited as unpublished observations

haps the most satisfying and, to the uninitiated at least, most convincing method is that of direct observations on live cells (Wetherbee et al. 1988; Heimann et al. 1989 a,b; Beech and Wetherbee 1990a). The objective is to record from dividing cells the fate of the parental flagella and the emergence of the newly formed flagella. This requires high resolution light microscopy (Nomarski or phase contrast optics coupled with a flash mechanism; or computer enhanced image analysis and video recording) and immobilization of motile cells in a medium that will not interfere with the optics, cell viability, or flagellar motility (embedding in a weak solution of Agarose works well). It is necessary to distinguish between the individual flagella at all times. This is not an easy task, but variant flagellar orientations, beat patterns or apical swellings can act as handy markers. These latter features will, of course, be the changing characteristics of a transforming flagellum, and caution must be exercised. It can be virtually impossible to keep certain species alive (and dividing!) under a microscope for periods of more than a few minutes. In such cases the story can often be deduced from numerous images of fixed cells at various stages of division and flagellar development (e.g., *Pedinomonas tuberculata, Monomastix* sp.; Heimann et al. 1989 a).

A statistical approach has also been used to determine the developmental relationship between flagella. Melkonian et al. (1987 a), by comparing the lengths of individual flagella at different stages in the cell cycle of the anisokont prasinophyte *Nephroselmis olivacea*, obtained the first evidence that algal flagella require more than one cell generation to mature. Although fixed cells from an asynchronously dividing culture were used for this study, relative cell age could be reliably determined from morphological criteria.

The flagella of certain species do not persist during cell division and it is therefore impossible to directly observe the flagellar developmental cycle or elucidate it by statistical means. If, however, the parental Bbs in mitotic cells can be distinguished from the newly formed ones using EM, then comparison of the roots can reveal the Bb developmental cycle. For example, metaphase Bbs usually bear the roots that they will have in the daughter cells and thereby the flagellar developmental cycle can be inferred. Such an approach was used in analysis of the prymnesiophyte *Pleurochrysis carterae* (Beech et al. 1988).

This analysis of roots in dividing cells is only effective when dealing with organisms that have a markedly heterogenous basal apparatus; that is, where each of the Bbs in an interphase cell can be readily distinguished by the distinctly different roots that it bears. These conditions are not met for all protists. For example, in some green algae, where flagella can undergo no apparent change or else retract completely prior to division, the roots attached to opposing Bbs are often seemingly identical. Nevertheless, the flagellar/Bb developmental cycle in such species has been deduced by careful observation of the pattern in which the Bbs replicate and segregate to daughter cells. Moestrup and Hori (1989), by analysis of serial sections of dividing cells, elucidated the developmental relationship between the Bbs of an octoflagellate prasinophyte. Holmes and Dutcher (1989) used a combination of their LM and the EM of others to infer the Bb developmental cycle in *Chlamydomonas reinhardtii*.

Light microscopical observations on flagellar development

The transformation of one type of flagellum to another at cell division often requires gross morphological changes to the flagellum, such as: elongation, shortening or total retraction, loss of flagellar hairs, change in the pattern of beat, or change in orientation. One or a combination of these changes may occur in any one species, and two examples (Fig. 1 a and b) have been chosen to illustrate the variety in flagellar developmental patterns. Schemes A and B are not intended to define the flagellar developmental cycle for any par-

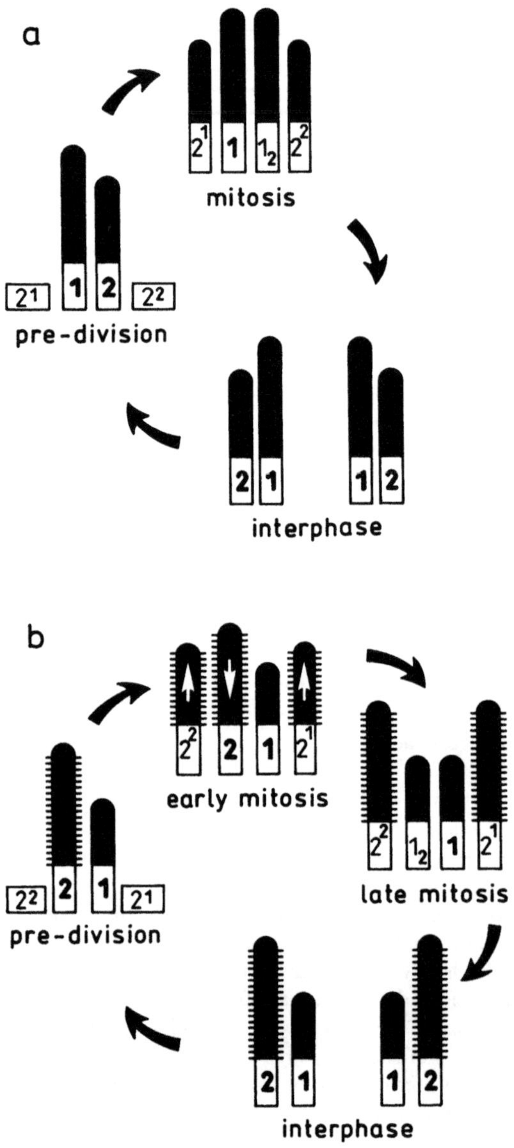

Fig. 1 a, b. Schematic representations of the flagellar developmental cycle. **a** Scheme A: *Nephroselmis olivacea* (based on Melkonian et al. 1987 a). Two new basal bodies (2^1, 2^2) are produced before cell division and, during mitosis, these Bbs give rise to flagella (F 2^1, F 2^2) which become the short flagella (F 2) of the daughter cells. At mitosis the parental short flagellum (F 2 at pre-division) elongates and is thereby transformed into F 1_2, a long flagellum (F 1) for a daughter cell; F 1 remains as such for subsequent cell generations. The different modes of flagellar beat and relative differences in the length of the long flagella are not shown (see text). **b** Scheme B: *Epipyxis pulchra* (based on Wetherbee et al. 1988). Prior to division, two new basal bodies (2^1, 2^2) are produced which, during mitosis, give rise to flagella (F 2^1, F 2^2) that will be the long, hairy flagella (F 2) of the daughter cells. At mitosis the parental, long, hairy flagellum (F 2) shortens and loses its hairs to become transformed into F 1_2, a short, smooth flagellum (F 1) for a daughter cell; F 1 remains as such for subsequent cell generations. A transient retraction and extension of F 1 during mitosis, and a difference in length between F 2^1 and F 2^2 are not shown (see text)

ticular group of organisms, though closely related species often fall under the same scheme.

Chlorophyta

Scheme A (Fig. 1 a) illustrates the basics of the flagellar/Bb developmental cycle of the prasinophyte *Nephroselmis olivacea* as described by Melkonian et al. (1987 a). Interphase cells of *N. olivacea* bear two flagella that are unequal in length. During swimming the shorter flagellum is anteriorly directed and displays the ciliary-type beat; the longer flagellum is posteriorly directed and beats in an undulatory fashion. Mitotic cells bear four flagella. The newly produced flagella ($F2^1$, $F2^2$) are of equal length and are the shortest – these flagella become the short flagella of the daughter cells. Meanwhile the parental short flagellum (F 2) elongates and changes its mode of beating to become a long flagellum ($F\,1_2$) with an undulatory-type beat. The parental long flagellum (F 1) remains as such through subsequent cell cycles. It should be noted that in *N. olivacea* flagella require three cell generations to fully mature; the newly elongated long flagellum elongates further in the third generation to attain its ultimate length (Melkonian et al. 1987 a). At interphase in *N. olivacea* Bb 2 bears a microtubular root that is associated with the eyespot (Moestrup and Ettl 1979).

Other prasinophytes have four or more flagella which are morphologically and dynamically indistinguishable from one another. An example is *Pyramimonas*: the flagella in this genus persist unchanged during mitosis as new flagella for the daughter cells emerge (e.g., Woods and Triemer 1981). Here then, analysis of the Bb developmental cycle is necessary to determine the relationship between the parental flagella of each daughter cell. This has been done by Moestrup and Hori (1989) for *Pyramimonas octopus*.

Modifications on the *Nephroselmis* type of flagellar development are evident in other green algae. *Pedinomonas tuberculata* bears a single flagellum (F 1) which is directed backwards during swimming and has an undulatory-type beat; a non-flagellate Bb is present in interphase cells and is Bb 2. At mitosis, two new Bbs (2^1, 2^2) are formed and a new flagellum emerges from Bb 2. This flagellum elongates to acquire the length of the existing flagellum (thereby Bb 2 becomes $Bb\,1_2$) and at cytokinesis one flagellum and a new non-flagellate Bb are partitioned to each of the daughter cells (Heimann et al. 1989 a). This developmental pattern follows Scheme A (Fig. 1 a), except that Bb 2 bears no flagellum at interphase.

Spermatozopsis similis is a *Clamydomonas*-like green unicell that, at interphase, bears two homodynamic flagella (ciliary-type beat) of slightly unequal length (Preisig and Melkonian 1984). Prior to cell division, two new Bbs are formed from which flagella ($F\,2^1$, $F\,2^2$) of equal length emerge. At the same time, the parental short flagellum (F 2) elongates to attain the length of the parental long flagellum (F 1) and is thereby transformed into $F\,1_2$. Following the semi-conservative distribution of flagella at cytokinesis, $F\,2^1$ and $F\,2^2$ elongate to become the short flagella of the daughter cells (Lechtreck et al. unpubl.). The flagellar developmental cycle of *S. similis* can also be illustrated using Scheme A (Fig. 1 a): Bb 1 is that which bears the longer flagellum, and Bb 2 the shorter flagellum. At interphase, Bb 2 has an eyespot-associated microtubular root (Reize and Melkonian 1988).

During vegetative reproduction in *Chlamydomonas reinhardtii*, two or more rounds of mitosis and cytokinesis are carried out within the confines of the parental cell wall. The two parental flagella, which are of equal length and show a ciliary-type beat, though their responses to calcium differ (Kamiya and Witman 1984), are retracted prior to the first mitosis and new flagella do not emerge until just before release of the sibling cells from the parental cell wall (Randall et al. 1967, Cavalier-Smith 1974). Analysis of the pattern of Bb segregation at mitosis in *C. reinhardtii* led Holmes and Dutcher (1989) to deduce that the Bb 2 is that which bears the eyespot-associated microtubular root.

The flagellar developmental cycle of the green uniflagellate *Monomastix* sp. has been determined by Heimann et al. (1989 a). The single flagellum (F 2) has an asymmetric ciliary-type beat and at interphase a non-flagellate Bb (Bb 1) is also present. The F 2 begins to shorten prior to mitosis; two new flagella ($F\,2^1$, $F\,2^2$) then emerge from newly formed Bbs and elongate to become the flagella of the daughter cells. The shortening F 2 is sloughed before it can completely retract; thereby Bb 2 is transformed into a $Bb\,1_2$. Bb 1 remains non-flagellate through subsequent generations and, because it previously had a flagellum, it is termed a barren Bb (see also, Andersen et al. 1991). This pattern of flagellar development, where flagella are suppressed from forming on mature Bbs, is clearly different to that described in any other green alga. It resembles the pattern of flagellar development in some chromophyte algae (see below), and compounds the fact that *Monomastix* occupies a most uncertain systematic position (Moestrup 1982, Heimann et al. 1989 a).

Glaucocystophyta

Cells of *Cyanophora paradoxa* have two heterodynamic flagella: the anteriorly directed flagellum has a ciliary-type beat and the posteriorly directed flagellum shows an undulatory-type beat. Both flagella persist during cell division as two new flagella ($F\,2^1$, $F\,2^2$) arise from newly formed Bbs (Heimann et al. 1989 b). The parental anterior flagellum (F 2), through a change in its mode of beat and a reorientation of its Bb, is transformed into a new posterior flagellum ($F\,1_2$) once the $F\,2^1$ and $F\,2^2$ have elongated to become the anterior flagella for each of the daughter cells. Unlike the previously discussed species, flagellar development in *Cyanophora* requires no change in length to the transforming F 2, only a change in orientation and the mode of beat.

Chromophyta

The unicells of many chromophyte algae have two flagella: one bears tripartite tubular hairs (cf. mastigonemes; see Andersen et al. 1991), is directed anteriorly, and has a sinusoidal pattern of beat; the other is smooth, often shorter and directed posteriorly, and beats in a less regular manner. The flagella of *Epipyxis pulchra* (Chrysophyceae) have such an appearance (though the short, smooth flagellum also points forward) and their developmental cycle has been elucidated by Wetherbee et al. (1988). Prior to mitosis in *Epipyxis*, both parental flagella decrease in length while two new flagella elongate from newly formed Bbs. The parental flagella shorten at different rates so that they simultaneously attain a minimum equivalent length (c. 25% the original length of the parental short flagellum). These become the short flagella of the daughter cells: i.e., the parental long flagellum (F 2) – through reduction in length, change in pattern of beat and loss of hairs – is transformed into a new short flagellum ($F\,1_2$); the parental short flagellum (F 1) remains as such for subsequent generations. Following cytokinesis these flagella presumably elongate to the normal interphase length of short, smooth flagella. One of the new flagella ($F\,2^1$) appears just before the other ($F\,2^2$) and retains this slight advantage in length through cytokinesis; $F\,2^1$ and $F\,2^2$ become the long flagella of the daughter cells. It is not known if this difference in length between the long flagella of daughter cells is maintained throughout interphase. Wetherbee et al. (1988) also noted that in *Epipyxis*, as in *Nephroselmis*, although two generations are required for a flagellum to attain its final type, differences exist between daughter cells and the complete maturation of a flagellum/Bb requires at least three cell cycles. Scheme B (Fig. 1 b) shows the flagellar developmental cycle in *Epipyxis*. For simplicity, the retraction and re-extension of the F 1 and the difference in length between $F\,2^1$ and $F\,2^2$ are not illustrated.

Direct observations on dividing cells have shown that *Mallomonas splendens* (Synurophyceae) has a flagellar developmental cycle similar to that in *Epipyxis* (Beech and Wetherbee 1990 a). Interphase cells are uniflagellate (the flagellum is anteriorly directed and bears tripartite hairs), but a barren Bb is also present (Beech and Wetherbee 1990 b). At preprophase two small flagella ($F\,2^1$, $F\,2^2$) arise from newly formed Bbs as the parental flagellum (F 2) shortens. Elongation of the new flagella continues throughout mitosis, with $F\,2^1$ being slightly longer that $F\,2^2$. The F 2 further continues to shorten so that by mid-cytokinesis it is totally retracted and Bb 2 has become transformed into $Bb\,1_2$. The parental barren Bb (Bb 1) remains barren throughout this and subsequent generations. The flagellar developmental cycle in *M. splendens* can be illustrated with Scheme B (Fig. 1 b); the difference being that in *M. splendens* no flagellum is attached to Bb 1. The difference in length between the elongating $F\,2^1$ and $F\,2^2$ is similar to what is seen in *Epipyxis* (above) and, likewise, it is not known if this difference persists throughout interphase.

The flagellar developmental cycles of the pedinellids *Pseudopedinella elastica* and *Apedinella radians* (Table 1) are virtually identical to that in *M. splendens* and therefore conform to Scheme B. In addition to an anteriorly directed, hairy flagellum (F 2) these species each have a barren Bb (Bb 1). The synurophyte *Mallomonas papillosa* has a non-emergent smooth flagellum associated with Bb 1. Therefore, though the flagellar developmental cycle of *M. papillosa* looks like that of *M. splendens* in the LM (Beech and Wetherbee 1990 a), it is probably more aligned with that of *Epipyxis*.

The flagellar developmental cycle has been determined in only one prymnesiophyte: *Pleurochrysis carterae* (Beech et al. 1988). Interphase cells bear two hairless flagella of slightly unequal length. Both flagella begin to shorten at preprophase so that by metaphase they are completely retracted; a pair of subequal flagella emerge from each nascent daughter cell at mid-cytokinesis. Thin-section analysis of the basal apparatuses of mitotic cells has shown that the Bb which produces the longer flagellum (i.e., the "left" flagellum in prymnesiophyte terminology; Beech and Wetherbee 1988) is Bb 1 (Beech et al. 1988; see below).

Synopsis

The flagellar developmental cycle is known for species from several of the major taxonomic groups and some recurrent patterns are now beginning to emerge.

Most green algae have flagella which appear to be homodynamic and morphologically similar, and therefore little obvious difference is apparent between flagella of differing age. For this reason uniflagellate green algae or those with unequal length flagella (*Pedinomonas, Nephroselmis, Spermatozopsis similis*) have been chosen for developmental studies. Flagellar transformation in all these organisms requires flagellar elongation. The retraction of flagella prior to division in *Chlamydomonas* is probably due to the walled condition of this alga. In the biflagellate green algae that have an eyespot-associated microtubular root (*Nephroselmis, Spermatozopsis, Chlamydomonas*) this root is always attached to the Bb 2 at interphase. Since the eyespot–root association is common in the green algae (see Melkonian and Robenek 1984) this feature may be generally useful for identifying the Bb 2.

In the chromophyte groups studied to date whose cells have a flagellum which bears tripartite hairs (Chrysophyceae, Synurophyceae, Pedinellophyceae; i.e., heterokont chromophytes) the F 2 is the hairy flagellum which, through retraction, becomes transformed into either a smooth (usually shorter) flagellum or a barren Bb. When a second flagellum is present, transformation results in a change in the flagellar beat, which may or may not be due simply to a loss of the tripartite hairs. In non-chromophyte groups where no flagellar hairs are present, comparisons can also be made between the pattern of beat and the developmental status of a flagellum. If a cell bears two heterodynamic flagella, the F 2 has the ciliary-type beat and the F 1 has the undulatory-type beat (e.g., *Nephroselmis* and *Cyanophora*). This pattern is also evident in uniflagellates: the cilia-like flagellum of *Monomastix* is an F 2, and the undulating flagellum in *Pedinomonas* is an F 1.

Developmental changes in the basal apparatus

It is not known what triggers a flagellum to transform from one morphological type to another during mitosis but, whatever the signal, it is likely to be transmitted through the Bb – the organizer of the flagellum. This is especially evident in cells which completely retract their flagella during mitosis (e.g., *Chlamydomonas, Pleurochrysis*), where only the Bbs (cf. Bbs and their flagella) persist between cell generations. Basal body transformation can be seen not only in the production of a different type of flagellum, but also in modifications to the Bb itself and to the roots and connecting fibres which are attached to the Bb.

Basal body development

The exact timing of formation of new Bbs for daughter cells is not well known in the algae. Studies on algal mitosis often simply state that new Bbs are formed alongside the parental Bbs during some period just before mitosis. In many species, however, Bbs are formed during early interphase (e.g., some euglenoids, see Farmer and Triemer 1988) or even the previous mitosis. The Bbs of *Chlamydomonas reinhardtii* are formed between mid-metaphase and telophase; at the end of mitosis they have a full complement of microtubular triplets but persist in a dormant state throughout cytokinesis and the subsequent interphase. Flagella are not produced until the following cell cycle or when all daughter cells are released from the parental cell wall (Gaffal 1988). Such Bbs in *Chlamydomonas* spp. and other green algae are referred to as dormant Bbs (see Andersen et al. 1991). In the Chrysophyceae also, there are reports of "extra" Bbs being regularly apparent at early interphase (Slankis and Gibbs 1972, Schnepf et al. 1977, Owen et al. 1990). Such reports have indicated (e.g., Schnepf et al. 1977) that often only a single extra Bb is present – thereby implying that new Bbs may not arise simultaneously; in *Ochromonas* sp. this has been confirmed (R.A. Andersen pers. comm.) with serial sections.

The details of Bb/centriole ontogeny have been best studied in ciliate and animal cell systems (for review see Vorobjev and Nadezhdina 1987). In these cells a nascent Bb (or nascent centriole) usually forms near the proximal end of, and perpendicular to, its parental Bb/centriole; once formed, the new Bb/centriole reorients to lie parallel to its parent. The cartwheel (a structure of nine-fold symmetry that exists in the proximal lumen of Bbs and centrioles) is one of the first structures to be formed. Associated with it are the A-subfibres of the Bb triplets and the B- and C-subfibres are then added to the outside of the A-subfibres. In some vertebrate cells the cartwheel disappears once the Bb/centriole is fully constructed and it has therefore been considered to be a key structural element in Bb assembly (e.g., Kalnins and Porter 1969). What is known of Bb assembly in algal flagellates (e.g., *Chlamydomonas*; Johnson and Porter 1968, Cavalier-Smith 1974, Gaffal 1988) largely conforms to this ontogenetic pattern. In some special cases, however, algal Bbs can be formed without any apparent association with a previously existing Bb (e.g., following zygospore germination in *Chlamydomonas;* Cavalier-Smith 1974; for other cell types, see Pickett-Heaps 1971). Also, though

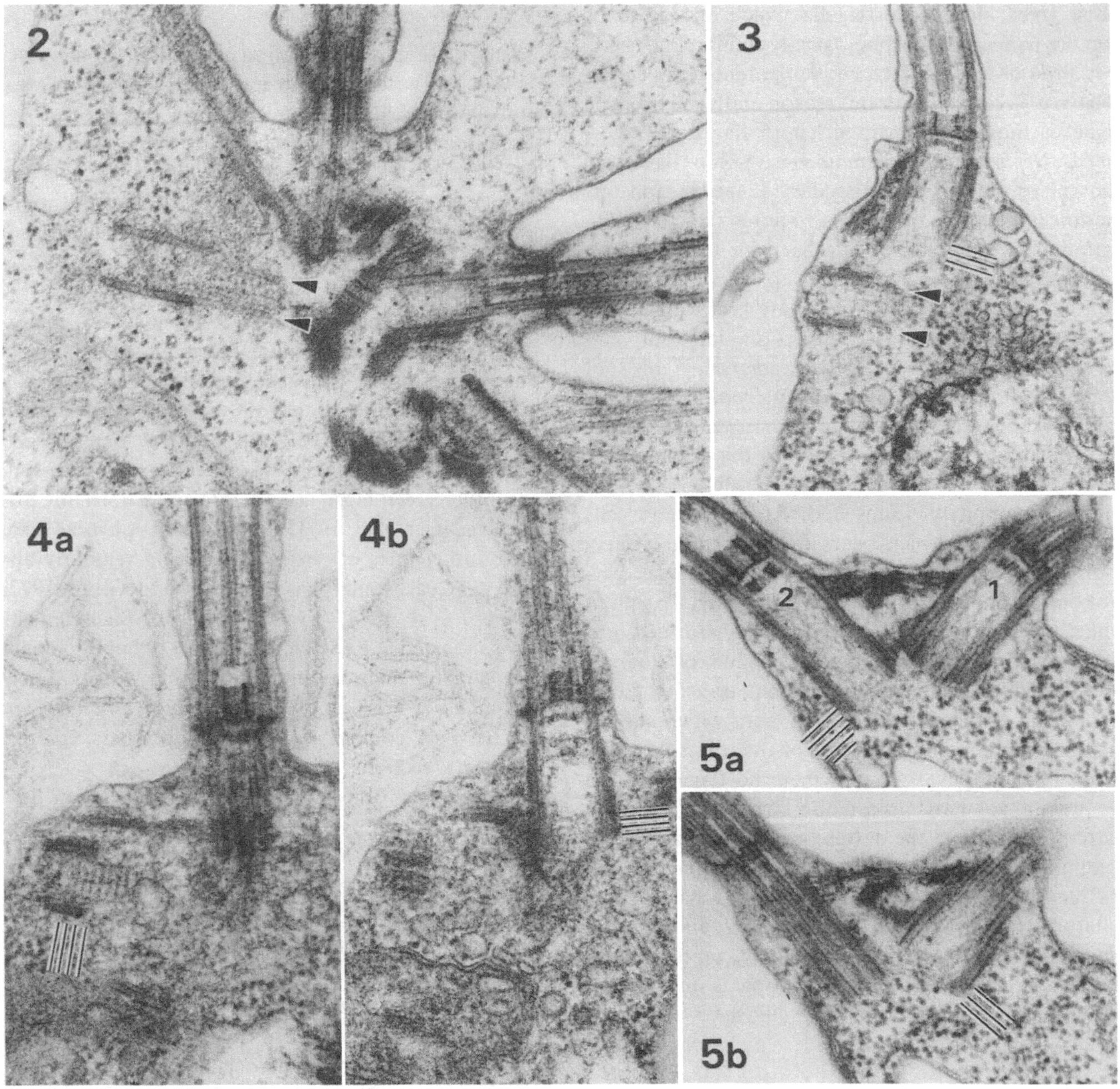

Fig. 2–5. Transmission electron micrographs of thin-sections through the basal apparatus of green algal flagellates. *P. parva* (SAG B 63–3; Schlösser 1982) and *C. geitleri* (SAG 6.73; Schlösser 1982) were cultured and fixed for EM as described in Preisig and Melkonian (1984) for *S. similis*. See Melkonian et al. (1987 b) for *S. exsultans*

Fig. 2. *Polytomella parva* Pringsheim: nascent Bb showing A-subfibres (◀) extending proximally from the other, more darkly staining, subfibres of the Bb triplets. Cartwheel material is prominent in the proximal half of this Bb. × 50000

Fig. 3. *Spermatozopis similis* Preisig and Melkonian: nascent Bb showing A-subfibres (◀) and cartwheel material as described for Fig. 2; note that the length of the cartwheel in the more mature, flagellar-bearing Bb (three distinct cartwheel tiers are marked) is approximately equal to the length of the cartwheel enclosed by complete triplets in the nascent Bb. × 55000

Fig. 4 a, b. Adjacent sections of *Chlamydomonas geitleri* Ettl: a dormant Bb showing all subfibres of the Bb triplets to be of equal length and subtending only four (marked in **a**) of the eight cartwheel tiers; the four cartwheel tiers (marked in **b**) of the more mature flagellar-bearing Bb are enclosed by the Bb triplets. × 60000

Fig. 5 a, b. Adjacent sections of *Spermatozopsis exsultans* Korshikov: Bb 2 (bearing the eyespot-associated root) shows 5 cartwheel tiers (marked in **a**), two of which are not enclosed by the Bb triplets; Bb 1 shows three cartwheel tiers (marked in **b**) which are all enclosed by the Bb triplets. × 60000

the cartwheel is one of the first discernable structures in algal Bb assembly (e.g., Johnson and Porter 1968), it persists in fully formed Bbs and displays variant sizes in different taxa (Moestrup 1982).

In several species of green algal flagellates we have observed pronounced differences in the length of the cartwheel between Bbs of differing age. The nascent Bbs (i.e., those under construction) in premitotic cells of *Polytomella parva* and *Spermatozopsis similis* have a cartwheel which is considerably longer than that in fully developed Bbs. Furthermore, the A-subfibres extend more proximally than the B- and C-subfibres and they subtend the cartwheel structure (Figs. 2 and 3). Though it is well known that the A-subfibres are the first of the Bb triplets to be formed (Johnson and Porter 1968, Cavalier-Smith 1974), such an extension of the A-subfibres proximal to the B- and C-subfibres has, to our knowledge, not previously been documented. Dormant Bbs in *Chlamydomonas geitleri* also show a cartwheel which protrudes proximally from the Bb (Fig. 4 a and b); however, the A-subfibres here are the same length as the B-and C-subfibres. It seems that only in nascent Bbs do the A-subfibres extend beyond the other subfibres.

As Bbs develop and produce flagella, differences in cartwheel length can also be correlated with different stages of Bb maturation. In *Spermatozopsis exsultans* a comparison of Bb 1 and Bb 2 reveals that the cartwheel in younger Bbs is longer than that in older, or more mature Bbs (Fig. 5 a and b). In this species, as well as in *S. similis* and *C. geitleri*, the only portion of the cartwheel that remains in more mature Bbs is that which is surrounded by the Bb triplets. Note also that the approximate length (or number of tiers) of this enclosed portion of the cartwheel appears to be the same for all the Bbs (be they nascent, dormant, or flagellar-bearing) of any one species (Figs. 3–5).

What is the reason for this decrease in cartwheel length as Bbs mature? It may be that once a Bb is constructed and the A-subfibres no longer cover that part of the cartwheel at the proximal end of the Bb, then this exposed (protruding) portion is subject to a gradual degradation which may not be completed until the Bb is fully mature.

Developmental changes in the Bb are also apparent in structures that are directly associated with the Bb triplets. Microtubular flagellar roots arise close to Bbs and, in the green algae, roots are known to originate alongside and attach via fibrils to specific Bb triplets (Melkonian 1984, Moestrup and Hori 1989). If, for a particular species, the developmental relationship between the Bbs is known and the numbering of the Bb triplets has been established (in the green algae this follows Hoops and Witman 1983, Melkonian 1984, Hori and Moestrup 1987), then the sequential, developmental modifications to fibrils connecting triplets to microtubular roots can be determined from interphase cells. *Spermatozopsis exsultans* is a quadriflagellate (for description, see Melkonian et al. 1987 b) with a microtubular root system very similar to that of the biflagellate *S. similis* (Melkonian and Preisig 1984, Lechtreck et al. 1989): two roots arise alongside and are attached to each of Bbs 1 and 2 (1 s, 1 d and 2 s, 2 d roots respectively; the eyespot-associated root is 2 s; Fig. 6). In

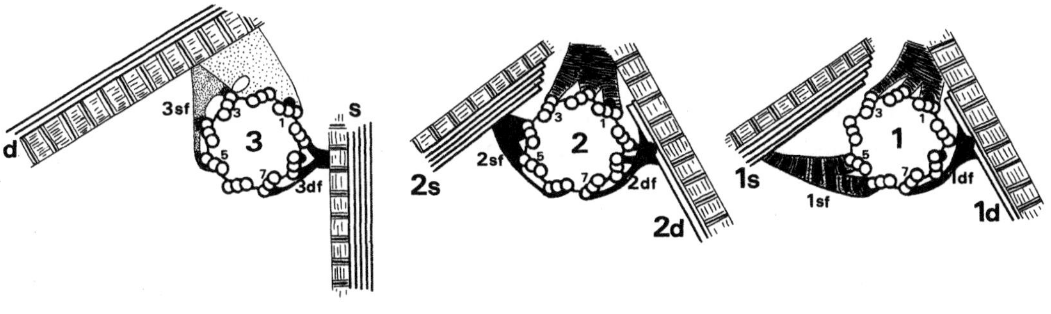

Fig. 6. Schematic representation of the structures associated with the proximal half of the Bbs (three maturation stages: *3, 2, 1*) in the quadriflagellate *Spermatozopsis exsultans*. Bbs are viewed from their distal ends and arranged so that particular Bb triplets (numbered according to Hoops and Witman 1983, Melkonian 1984, Hori and Moestrup 1987) all have the same relative orientation. Microtubular roots (2 s, 2 d, 1 s, 1 d) are labelled according to Moestrup and Hori (1989) and their system I fibres (cross-striated structures) are shown. Fibrils 3 sf and 3 df connect Bb 3 to system I fibres via triplets 3–5 and 7–9 (Bb 3^1 and Bb 3^2 have an identical arrangement of associated structures). Fibrils 2 sf and 2 df connect Bb 2 to the 2 s and 2 d microtubular roots via triplets 4–6 and 7–9. Fibrils 1 sf and 1 df connect Bb 1 to the 1 s and 1 d microtubular roots via triplets 5, 6, and 7–9. Note the change in appearance and triplet association of the sf fibrils in comparison to the relative constancy of the df fibrils. Proximal connecting fibres are attached to triplets 1–3 of Bb 2 and Bb 1; the stippled area in the same position on Bb 3 may represent a nascent, proximal connecting fibre. Original material upon which the diagrams are based was prepared as in Melkonian et al. (1987 b)

S. exsulants the two additional Bbs ($Bb\,3^1$ and $Bb\,3^2$) give rise to no microtubular roots but are nevertheless attached to system I fibres (see below). Each of the three Bb types in *S. exsultans* is connected to flagellar roots by two distinct fibrillar systems (Fig. 6). The left systems (3 sf, 2 sf, and 1 sf) are variable with respect to their appearance and the particular triplets from which they arise. On Bb 3 what we have labelled as 3 sf is not distinctly fibrillar but is rather an amorphous, darkly staining area which links triplets 3, 4, and 5 to the system I fibre of a d root. On Bb 2 the 2 sf fibrils have a true fibrillar appearance and arise from triplets 4, 5, and 6 to attach to the microtubules of the 2 s root. On Bb 1 the 1 sf fibrils consist of relatively large bundles of distinctly cross-striated fibres which arise from triplets 5 and 6 and follow a wide arc to attach to the microtubules of the 1 s root. The differences between the 2 sf and 1 sf fibrils may reflect developmental differences in the 2 s and 1 s roots: most obviously that the 2 s root is associated with the eyespot and the 1 s root is not. The amorphous appearance of the 3 sf may indicate that this fibrillar system is in a nascent state. It is possible that only in the next and following cell generations it is a completely functional Bb/root connector. The right systems (3 df, 2 df, and 1 df) always arise from triplets 7, 8, and 9. The overall appearance of these fibrils remains constant as the Bb matures through the three cell generations. In contrast to the 3 sf, the 3 df fibril is apparently fully formed and provides a good linkage to the system I fibre of the s root. A reason for the 3 sf and 3 df fibrils being connected to the system I fibres, rather than to the microtubules of the respective roots (as in Bbs 2 and 1), may lie in the fact that these fibrils will in the next and following cell generations attach to different types of roots. If microtubular roots have specific sites of recognition for connecting fibrils, then s microtubular roots might only be capable of attachment to sf fibrils (and d roots to df fibrils). Whatever the developmental changes in the roots (see below), it is apparent that the attachment sites for the fibrillar connectives, and the fibrils themselves, remain the same or undergo slight, progressive modifications as the Bbs mature. The data support the contention (e.g., Hoops and Witman 1983, Melkonian 1984) that Bb triplets are non-equivalent and that the inherent asymmetry of the Bb determines the patterning of the entire flagellar apparatus.

In many biflagellates, close analysis of the spatial relationship between the Bbs of a replicated set reveal that there are two distinct pairs. Each pair consists of an old and new Bb and probably represents a parental/daughter pair of Bbs. In many chlorophytes the orientation of the Bbs in parental/daughter pairs is such that the daughter Bb is situated to the right hand side of its parent (right and left positions as for the microtubular roots). This pattern is evident in, for example, *Chlamydomonas reinhardtii* (Gould 1975, Holmes and Dutcher 1989), *Brachiomonas submarina* (Segaar and Gerritsen 1989), and *Spermatozopsis similis* (Lechtreck et al. unpubl.). It is clear from these reports and others that the Bbs of a parental/daughter pair segregate together at mitosis. In certain green algae it has been proposed that in order for the correct orientation of the flagellar apparatuses in daughter cells to be established, there must be configurational rearrangements within each of the segregating Bb pairs (O'Kelly and Floyd 1984, Holmes and Dutcher 1989). Rotation of Bbs during basal apparatus segregation has also been postulated in synurophytes (Andersen 1985, Beech and Wetherbee 1990 b).

Development of microtubular roots

What little is known of the timing of formation of microtubular roots indicates that it generally commences (i.e., microtubules begin to elongate from root templates alongside a Bb) between preprophase and metaphase. One clear exception to this is in the euglenoid *Ploeotia costata*. Bb replication in this species occurs very soon after cytokinesis and the new Bbs, apparently at this early stage in the cell cycle, bear 'protoroots' (Farmer and Triemer 1988). A feature with perhaps more interesting variability, however, is the developmental status of a Bb when it first acquires roots. New roots in a variety of green algae first arise on Bbs that have been formed at least one cell generation previously. For example, in chlorophytes which have dormant Bbs, new roots are first detected on Bbs as they come out of dormancy at least one generation after formation (e.g., Gaffal 1988, Segaar and Gerritsen 1989). *Mantoniella squamata* is a small prasinophyte in which roots are associated only with the Bb of the long flagellum (this is likely to be the F 1; see Heimann et al. 1989 a). The other Bb (Bb 2) has a very short, ineffective flagellum. Only after the Bbs replicate at preprophase do the new roots arise from the transforming Bb 2 (Barlow and Cattolico 1981). In both these examples the production of roots is coincident with the production of a new flagellum from the same Bb. In *Pyramimonas* spp. (Prasinophyceae) on the other hand, though the Bbs with roots attached are also at least one generation old, all the Bbs bear flagella (e.g.,

Moestrup and Hori 1989). In many green algae, of course, all the Bbs that are present at interphase bear microtubular roots (e.g., *Nephroselmis olivacea*, Moestrup and Ettl 1979; *Spermatozopsis similis*, Melkonian and Preisig 1984). In these cells the roots presumably appear on the Bbs soon after the Bbs themselves are formed. An illustration of this is in *Pedinomonas*, where just before mitosis the newly formed Bbs exhibit traces of microtubular roots (Pickett-Heaps and Ott 1974); *Pedinomonas* also exemplifies that in a green flagellate roots can be present on a non-flagellate Bb (see also Heimann et al. 1989 a).

Microtubular roots are visible alongside new Bbs when the latter are first apparent at late interphase/preprophase in synurophytes (Andersen 1985, Beech and Wetherbee 1990 b). New roots are first detected at prometaphase in the prymnesiophyte *Pleurochrysis carterae* (here the new Bbs are formed at preprophase; Beech et al. 1988). Since in virtually all chromophytes only two Bbs are present at interphase (exceptions in the Chrysophyceae have been alluded to above) and both of these bear flagellar roots (see Andersen 1989 a, b; Preisig 1989), roots would presumably have been formed on Bbs soon after their formation or at least during the same cell cycle. The same would appear to be true for other groups such as the dinoflagellates and the cryptomonads (see Roberts 1991, Moestrup 1982).

What happens to the parental roots during cell division? In chlorophytes which have roots associated with at least two Bbs at interphase, various levels of root reduction and elongation are reported during mitosis and cytokinesis (Aitchison and Brown 1986, Gaffal 1988, Segaar and Gerritsen 1989, Gaffal and el-Gammal 1990, Sluiman and Blommers 1990). The general conclusion in these studies is that at least part of each parental root remains attached to its Bb throughout cell division, resulting in a semi-conservative distribution of roots (or root stubs) to daughter cells. It has recently been shown in *Brachiomonas submarina* (Segaar and Gerritsen 1989) and *Chlamydomonas reinhardtii* (Gaffal and el-Gammal 1990) that parental s roots, which elongate and associate with each other during basal apparatus segregation at cell division, represent the 'Metaphase Band' of Johnson and Porter (1968). This confirms a prediction of Holmes and Dutcher (1989). An association between opposite s roots was also noted during mitosis in *Mantoniella squamata* (Barlow and Cattolico 1981). However, since this species has only one s root at interphase (see above), the s root which is associated with the transforming parental Bb must be newly formed.

In biflagellate green algae which display a cruciate flagellar root system (Moestrup 1978) transformation of the 2 s and 2 d roots (i.e., those attached to the transforming Bb) at cell division requires little change in the final structure of the root. Examples of changes that nevertheless must occur are: loss of association with the eyespot for the 2 s root as it becomes a 1 s root (see Melkonian and Robenek 1984, Lechtreck et al. unpubl.); the 2 d root of many prasinophytes will develop an association with the scale duct as it becomes a 1 d root (Moestrup and Hori 1989, Inouye et al. 1990); in *Spermatozopsis similis* the 2 s and 2 d roots will acquire different final lengths in transforming to 1 s and 1 d roots (Lechtreck et al. 1989).

Not all green algae have a cruciate flagellar root system. The two Bbs of *Nephroselmis olivacea*, for example, are each associated with different types of microtubular roots (Moestrup and Ettl 1979). Melkonian et al. (1987 a) have discussed that, during flagellar/Bb transformation in this species, the Bb which produced the parental short flagellum (Bb 1_2) must acquire a left root while the number of microtubules in its right root is reduced.

Most chromophyte species have a heterogenous basal apparatus where markedly different roots are associated with each of the two Bbs (e.g., see Andersen 1989 b, Preisig 1989). Transformation of the roots associated with Bb 2 will therefore, as in *Nephroselmis*, entail structural changes to the roots themselves. In *Pleurochrysis carterae* (Prymnesiophyceae), the Bb developmental cycle has been determined through analysis of the mitotic changes in the microtubular roots (Beech et al. 1988). Once the two Bbs have replicated the parental roots in *P. carterae* undergo major reductions at early mitosis; by prometaphase each of the four Bbs bears rudiments of the roots that it will have in the daughter cells. The two parental Bbs can, at this stage, be distinguished because their flagella are still partially emergent. Both parental Bbs then have the same complement of roots, and the newly formed Bbs the other types of roots. It was thus demonstrated that in *P. carterae* newly formed Bbs bear simple roots (and produce a short flagellum), and in second and subsequent generations these Bbs bear compound roots (and produce a longer flagellum). It could not be determined in *Pleurochrysis* if the parental roots are partially disassembled but retained in a reduced form (i.e., semi-conservative replication requiring structural modifications to the roots on the transforming Bb), or if the parental roots are totally disassembled and all roots for the daughter cells are formed de novo (Beech et al.

1988). Whatever the case, we emphasize that transformation of the entire flagellar apparatus is probably effected by a transformation which first takes place in the Bb: in the case of flagellar roots, transformation of the site of root nucleation (adjacent the Bb triplets) will presumably bring about a change in the structure of the root.

Microtubular roots in interphase cells of the Synurophyceae are not attached to a Bb (Andersen 1985). The single microtubular root for each daughter cell in *Mallomonas splendens* is formed de novo at preprophase, alongside the newly formed Bbs. At the same time the parental root disassembles and does not reappear in the daughter cells (Beech and Wetherbee 1990 b). At cytokinesis the newly formed roots move away from their respective Bb and elongate toward the nuclear apex to attain their interphase appearance. For a root to be conserved during cell division it probably must remain attached to its Bb (Beech and Wetherbee 1990 b). The flagellar apparatus of the synurophytes may be unique among the algae in that it displays more than one absolute orientation (Andersen 1985). Though the reasons for this are likely to hinge on the sequence of development of the whole flagellar apparatus, it has been suggested that elucidation of the developmental behaviour of other synurophycean organelles (e.g., the chloroplasts) is required before the question can be fully answered (Beech and Wetherbee 1990 b).

Images of the replicated basal apparatus in the euglenoid, *Ploeotia costata*, show that the two parental Bbs bear the roots that the ventral Bb has in early interphase cells. Farmer and Triemer (1988) concluded from this that the dorsal Bb/flagellum transforms to the ventral Bb/flagellum during cell division. This concurs with a report by Melkonian et al. (1987 a) that the long (dorsal) flagellum of *Euglena mutabilis* shortens during division to transform into a short (ventral) flagellum for the next cell generation. The parental roots of *Ploeotia* are apparently conserved more or less intact during cell division and are distributed along with their Bbs to each of the daughter cells; little change is made to basic organization of the transforming (dorsal) root as it becomes an intermediate root since both have the same number of microtubules (Farmer and Triemer 1988, see also Surek and Melkonian 1986).

Development of fibrous roots

In addition to microtubular roots, the basal apparatus of most algal flagellates contains one or more fibrous roots. Green algal cells have two distinct types of fibrous roots which differ from one another on morphological, positional and biochemical grounds (Melkonian 1980, Melkonian et al. 1988, Lechtreck et al. 1989, Lechtreck and Melkonian 1991). The first type is commonly known as system I fibres (Micalef and Gayral 1972, Melkonian 1980; or striated microtubule-associated components, Floyd et al. 1980). They are closely associated with the microtubular roots and, as yet, little is known of their ontogeny or the changes they undergo during cell division. The second type is known as system II fibres (Micalef and Gayral 1972, Melkonian 1980), rhizoplasts, or nucleus-basal body connectors (NBBCs; Wright et al. 1985). They consist of bundles of filaments, often cross-striated, which extend from the Bbs to attach to an organelle (usually the nucleus) or the cell periphery. System II type fibrous roots were seen long before the advent of EM and have been found in a wide range of taxonomic groups. A calcium-modulated, contractile protein, called centrin, has been isolated from and shown to be the major component of the system II fibres of the green alga *Tetraselmis* (Salisbury et al. 1984). In recent years, antibodies to centrin have been used to document the distribution and dynamics of system II fibres in, particularly, the green algae (for reviews see Melkonian 1989, Salisbury 1989).

Centrin-containing system II fibres are probably ubiquitous in the green algae and the vast majority of species have more than one separate structure (Schulze et al. 1987). Some information on the ontogeny of green algal system II fibres has come from immunofluorescence (IMF) studies on the centrin cytoskeleton of dividing cells of *Chlamydomonas reinhardtii* (Salisbury et al. 1988, Wright et al. 1989). Interphase cells have a system II fibre (NBBC) descending from each of the two flagellate Bbs to the nuclear surface. The system II fibres contract at preprophase, pulling the nucleus towards the Bbs, resulting in two closely associated centrin foci at the cell apex. As mitosis progresses these foci separate so that one is localized to each of the spindle poles. At cytokinesis the centrin array for each of the daughter cells is established. That is, system II fibres extend from both of the flagellar bearing Bbs (Salisbury et al. 1988, Wright et al. 1989). This cycle, where the two parental system II fibres undergo dramatic morphological changes (and are perhaps entirely broken down) during mitosis to apparently contribute material to the spindle poles, is reminiscent of what has been described with EM in *Tetraselmis* by Stewart et al. (1974). These results indicate that the system II fibres in *Chlamydomonas* and *Tetraselmis* are formed after cell division and are largely consumed in the following mitosis. In *Pyramimonas* spp., however, the fate of the system II fibres may be somewhat different; Pearson and Norris (1975) state that system II fibres are formed on newly replicated Bbs prior to cell division; and Woods and Triemer (1981) noted no significant change in the appearance of the system II fibres (other than an association with the spindle poles) during mitosis. Clearly, further detailed work combining EM and IMF is needed before conclusions can be drawn on the ontogeny of green algal system II fibres. Although the above reports all show that system II fibres (and/or centrin) are somehow associated with spindle poles, Wright et al. (1989) have illustrated that a mutant of *Chlamydomonas* that lacks centrin-containing system II fibres (NBBCs) is capable of mitosis.

Fibrous roots connecting the Bbs to the nucleus, or some other organelle, are common in the chromophytes and are usually known as rhizoplasts. Since centrin has now been localized in these structures (Melkonian et al. 1991) we shall refer to them as system II fibres (see Andersen et al. 1991). The intimate association of the system II fibre and the spindle pole is well known for chromophytes such as *Ochromonas danica* (Slankis and Gibbs 1972, Bouck and Brown 1973) and *Hydrurus foetidus* (Vesk et al. 1984). System II fibres also constitute part of the spindle poles in *Mallomonas splendens* (Synurophyceae). In this species Beech and Wetherbee (1990 b) showed that system II fibres are formed de novo, at the proximal end of the newly formed Bbs, in each cell generation. The nascent system II fibres organize the spindle and, after mitosis, each changes its morphology to assume the interphase appearance. The parental rhizoplast is broken down at preprophase and apparently has little or no involvement with the spindle poles (Beech and Wetherbee 1990 b). It is not known if this rhizoplast cycle is common to heterokont chromophytes outside the Synurophyceae. In the Chrysophyceae, however, the single interphase system II fibre is generally reported to arise from the Bb of the long, hairy flagellum (Slankis and Gibbs 1972; Schnepf et al. 1977; Vesk et al. 1984; Andersen 1989 b, 1990; Owen et al. 1990). In all chrysophytes this flagellum is likely to be the F 2. If we assume that a mature system II fibre cannot transfer its association from one Bb to another, then in the Chrysophyceae, also, the parental system II fibre must undergo some disassembly as new fibres arise from the newly formed Bbs. The system II fibres of many heterokont chromophytes appear to be attached to both Bbs at interphase (e.g., Andersen 1985, 1989 a; Vesk and Moestrup 1987). This is also true for *M. splendens* (Beech and Wetherbee 1990 b). In these cases it may be that although the bulk of the parental rhizoplast disappears from the transforming Bb 2 (Bb 1_2), some fibres are still retained or reform on nucleating sites on the Bb 1.

We can postulate that in the heterokont chromophytes spindle organizers are coincident with the system II fibres and that they arise from newly formed Bbs. Since two spindle poles are required for effective mitosis, then in the heterokonts, two newly formed Bbs must also be present. This requires the presence of two parental Bbs and indicates, in our opinion, that the ancestor of the heterokont chromophytes had two Bbs at interphase. This is in agreement with the conclusions of Heimann et al. (1989 a).

In *Pavlova* (Prymnesiophyceae) a fibrous root (rhizoplast, system II fibre) arises from the Bb of the long, anteriorly directed flagellum and is associated with the spindle poles (Green and Hori 1988). The anterior Bb of the Pavlovales is considered, on the basis of its associated roots, to be the homologue of the right Bb of the non-pavlovalean prymnesiophytes (i.e., the Bb 2 in *Pleurochrysis carterae*; Green and Hori 1986) and the anterior Bb of the heterokont chromophytes (that bearing the long hairy flagellum – Bb 2; Preisig 1989). We suggest that the long, anteriorly directed flagellum in the Pavlovales arises from a newly formed Bb and is therefore F 2; the short, posteriorly directed flagellum will be the F 1. If this is true, then the system II fibre in *Pavlova* must develop in the same manner as that proposed above for the Synurophyceae and the Chrysophyceae.

Conclusions

The Bbs of algal cells that undergo vegetative division require more than one cell generation to fully mature. Basal body-associated structures (i.e., flagella, roots, fibrils) have particular forms depending on the level of maturity (developmental status) of the Bb. Heterogeneity in the flagellar apparatus exists because algal cells at interphase have Bbs at different stages of development. Thus a cell is capable of modifying a single flagellar apparatus component for different uses in different cell generations.

This developmentally regulated heterogeneity in the flagellar (centriolar) apparatus may be a common feature of all eukaryotic cells. Examples can be seen in centrioles of *Physarum* amoebae (Gely and Wright 1986), the kinetids of ciliates (Sleigh 1988) and the centrioles and their associated structures in mammalian cell lines (e.g., Rieder and Borisy 1982).

Acknowledgements

PLB thanks the Alexander von Humboldt Foundation for a supporting fellowship and funds were provided to MM by the Deutsche Forschungsgemeinschaft (Me 658/3-5). Thanks to Dr. R.A. Andersen for his helpful comments. PLB and MM were recipients of travel grants from NSF.

References

Aitchison WA, Brown DL (1986) Duplication of the flagellar apparatus and the cytoskeletal microtubule system in the alga *Polytomella*. Cell Motil Cytoskeleton 6: 122–127

Andersen RA (1985) The flagellar apparatus of the golden alga *Synura uvella*: four absolute orientations. Protoplasma 128: 94–106

– (1989 a) The Synurophyceae and their relationship to other golden algae. Nova Hedwigia 95: 1–26

– (1989 b) Absolute orientation of the flagellar apparatus of *Hibberdia magna* comb. nov. (Chrysophyceae). Nord J Bot 8: 653–669

– (1990) The three-dimensional structure of the flagellar apparatus of *Chrysophaerella brevispina* (Chrysophyceae) as viewed by high-voltage electron microscopy stereo pairs. Phycologia 29: 86–97

– Barr DJS, Lynn DH, Melkonian M, Moestrup Ø, Sleigh MA (1991) Terminology and nomenclature of the cytoskeleton of flagellate/ciliate protists. Protoplasma 164: 1–8

Barlow SB, Cattolico RA (1981) Mitosis and cytokinesis in the Prasinophyceae. I. *Mantoniella squamata* (Manton and Parke) Desikachary. Amer J Bot 68: 606–615

Beech PL, Wetherbee R (1988) Observations on the flagellar apparatus and peripheral endoplasmic reticulum of the coccolithophorid, *Pleurochrysis carterae* (Prymnesiophyceae). Phycologia 27: 142–158

– – (1990 a) Direct observations on flagellar transformation in *Mallomonas splendens* (Synurophyceae). J Phycol 26: 90–95

– – (1990 b) The flagellar apparatus of *Mallomonas splendens* (Synurophyceae) at interphase and its development during the cell cycle. J Phycol 26: 95–111

– – Pickett-Heaps JD (1988) Transformation of the flagella and associated flagellar components during cell division in the coccolithophorid *Pleurochrysis carterae*. Protoplasma 145: 37–46

Bouck GB, Brown DL (1973) Microtubule biogenesis and cell shape in *Ochromonas*. I. The distribution of cytoplasmic and mitotic microtubules. J Cell Biol 56: 340–359

Cavalier-Smith T (1974) Basal body and flagellar development during the vegetative cell cycle and the sexual cycle of *Chlamydomonas reinhardtii*. J Cell Sci 16: 529–556

Cleveland LR (1963) Functions of flagellate and other centrioles in cell reproduction. In: Levine L (ed) The cell in mitosis. Academic Press, New York, pp 3–53

Doflein F (1918) Beiträge zur Kenntnis von Bau und Teilung der Protozoenkerne. Zool Anz 46: 289–306

Farmer M, Triemer RE (1988) Flagellar systems in the euglenoid flagellates. BioSystems 21: 283–291

Floyd GL, Hoops HJ, Swanson JA (1980) Fine structure of the zoospore of *Ulothrix belkae* with emphasis on the flagellar apparatus. Protoplasma 104: 17–31

Gaffal FP (1988) The basal body-root complex of *Chlamydomonas reinhardtii* during mitosis. Protoplasma 143: 118–129

– el-Gammal S (1990) Elucidation of the enigma of the 'metaphase band' of *Chlamydomonas reinhardtii*. Protoplasma 156: 139–148

Gely C, Wright M (1986) The centriole cycle in the amoebae of the myxomycete *Physarum polycephalum*. Protoplasma 132: 23–31

Gould RR (1975) The basal bodies of *Chlamydomonas reinhardtii*: formation from probasal bodies, isolation, and partial characterization. J Cell Biol 65: 65–74

Green JC, Hori T (1986) The ultrastructure of the flagellar root system of *Imantonia rotunda* (Prymnesiophyceae). Br Phycol J 21: 5–18

– – (1988) The fine structure of mitosis in *Pavlova* (Prymnesiophyceae). Can J Bot 66: 1497–1509

Heimann K, Benting J, Timmermann S, Melkonian M (1989 a) The flagellar developmental cycle in algae: two types of flagellar development in uniflagellated algae. Protoplasma 153: 14–23

– Reize IB, Melkonian M (1989 b) The flagellar developmental cycle in algae: flagellar transformation in *Cyanophora paradoxa* (Glaucocystophyceae). Protoplasma 148: 106–110

Holmes JA, Dutcher SK (1989) Cellular asymmetry in *Chlamydomonas reinhardtii*. J Cell Sci 94: 273–285

Hoops HJ, Witman GB (1983) Outer doublet heterogeneity reveals structural polarity related to beat direction in *Chlamydomonas* flagella. J Cell Biol 97: 902–908

Hori T, Moestrup Ø (1987) Ultrastructure of the flagellar apparatus in *Pyramimonas octopus* (Prasinophyceae) I. Axoneme structure and numbering of peripheral doublets/triplets. Protoplasma 138: 137–148

Inouye I, Hori T Chihara M (1990) Absolute configuration analysis of the flagellar apparatus of *Pterosperma cristatum* (Prasinophyceae) and consideration of its phylogenetic position. J Phycol 26: 329–344

Johnson UG, Porter KR (1968) Fine structure of cell division in *Chlamydomonas reinhardtii*: basal bodies and microtubules. J Cell Biol 38: 403–425

Kalnins VI, Porter KR (1969) Centriole replication during ciliogenesis in the chick tracheal epithelum. Z Mikrosk Anat Forsch 100: 1–30

Kamiya R, Witman GB (1984) Submicromolar levels of calcium control the balance of beating between the two flagella in demembranated models of *Chlamydomonas*. J Cell Biol 98: 97–107

Lechtreck K-F, McFadden GI, Melkonian M (1989) The cytoskeleton of the naked green flagellate *Spermatozopsis similis*: isolation, whole mount electron microscopy, and preliminary biochemical and immunological characterization. Cell Motil Cytoskeleton 14: 552–561

– Melkonian M (1991) An update on fibrous flagellar roots in green algae. Protoplasma 164: 38–44

Melkonian M (1980) Ultrastructural aspects of basal body associated fibrous structures in green algae: a critical review. BioSystems 12: 85–104

– (1984) Flagellar apparatus ultrastructure in relation to green algal classification. In: Irvine DEG, John DM (eds) Systematics of the green algae. Academic Press, London, pp 73–120

– (1989) Centrin-mediated motility: a novel cell motility mechanism in eukaryotic cells. Bot Acta 102: 3–4

– Preisig HR (1984) Ultrastructure of the flagellar apparatus in the green flagellate *Spermatozopsis similis*. Plant Syst Evol 146: 145–162

– Robenek H (1984) The eyespot apparatus of flagellated green algae: a critical review. Prog Phycol Res 3: 193–268

– Beech PL, Katsaros C, Schulze D (1991) Centrin mediated cell motility in algae. In: Melkonian M (ed) Algal cell motility. Chapman and Hall, New York (in press)

– Schulze D, McFadden GI, Robenek H (1988) A polyclonal antibody (anti-centrin) distinguishes between two types of fibrous flagellar roots in green algae. Protoplasma 144: 56–61
– McFadden GI, Reize IB, Preisig HR (1987 b) A light and electron microscopic study of the quadriflagellate green alga *Spermatozopsis exsultans*. Plant Syst Evol 158: 47–61
– Reize IB, Preisig HR (1987 a) Maturation of a flagellum/basal body requires more than one cell cycle in algal flagellates: studies on *Nephroselmis olivacea* (Prasinophyceae). In: Wiessner W, Robinson DG, Starr RC (eds) Algal development, molecular and cellular aspects. Springer, Berlin Heidelberg New York Tokyo, pp 102–113
Micalef H, Gayral P (1972) Quelques aspects de l'infrastructure des cellules végétatives et des cellules reprodutrices d'*Ulva lactuca* L. (Chlorophycees). J Microsc 13: 417–428
Moestrup Ø (1978) On the phylogenetic validity of the flagellar apparatus in green algae and other chlorophyll a and b containing plants. BioSystems 10: 117–144
– (1982) Flagellar structure in algae: a critical review, with new observations particulary on the Chrysophyceae, Phaeophyceae (Fucophyceae), Euglenophyceae, and *Reckertia*. Phycologia 21: 427–528
– Ettl H (1979) A light and electron microscopical study of *Nephroselmis olivacea* (Prasinophyceae). Opera Bot 49: 1–39
– Hori T (1989) Ultrastructure of the flagellar apparatus in *Pyramimonas octopus* (Prasinophyceae) II. Flagellar roots, connecting fibres, and numbering of individual flagella in green algae. Protoplasma 148: 41–56
O'Kelly CJ, Floyd GL (1984) Flagellar apparatus absolute orientations and the phylogeny of the green algae. BioSystems 16: 227–251
Owen HA, Mattox KR, Stewart KD (1990) Fine structure of the flagellar apparatus of *Dinobryon cylindricum* (Chrysophyceae). J Phycol 26: 131–141
Pearson BR, Norris RE (1975) Fine structure of cell division in *Pyramimonas parkeae* Norris and Pearson (Chlorophyta, Prasinophyceae). J Phycol 11: 113–124
Pickett-Heaps JD (1971) The autonomy of the centriole: fact or fallacy? Cytobios 3: 205–214
– Ott DW (1974) Ultrastructural morphology and cell division in *Pedinomonas*. Cytobios 11: 41–58
Preisig HR (1989) The flagellar base ultrastructure and phylogeny of chromophytes. In: Green JC, Leadbeater BSC, Diver WL (eds) The chromophyte algae: problems and perspectives. Clarendon Press, Oxford, pp 167–187
– Melkonian M (1984) A light and electron microscopical study of the green flagellate *Spermatozopsis similis* spec. nova. Plant Syst Evol 146: 57–74
Randall JT, Cavalier-Smith T, McVittie A, Warr JR, Hopkins JM (1967) Developmental and control processes in the basal bodies and flagella of *Chlamydomonas reinhardtii*. Dev Biol [Suppl] 1: 43–83
Reize IB, Melkonian M (1988) Absolute orientations of basal bodies in green algae evaluated by light microscopy. Bot Acta 101: 192–195
Rieder CL, Borisy GG (1982) The centrosome cycle in PtK_2 cells: asymmetric distribution and structural changes in the pericentriolar material. Biol Cell 44: 117–132
Roberts KR, Roberts JE (1991) The flagellar apparatus and cytoskeleton of the dinoflagellates: a comparative overview. Protoplasma 164: 105–122
Salisbury JL (1989) Centrin and the algal flagellar apparatus. J Phycol 25: 201–206
– Baron AT, Sanders MA (1988) The centrin-based cytoskeleton of *Chlamydomonas reinhardtii*: distribution in interphase and mitotic cells. J Cell Biol 107: 635–641
– – Surek B, Melkonian M (1984) Striated flagellar roots: isolation and partial characterization of a calcium-modulated contractile organelle. J Cell Biol 99: 962–970
Schlösser UG (1982) Sammlung von Algenkulturen Göttingen. Ber Deutsch Bot Ges 95: 181–276
Schnepf E, Deichgräber G, Röderer G, Herth W (1977) The flagellar root apparatus, the microtubular system and associated organelles in the chrysophycean flagellate *Poterioochromonas malhamensis* Peterfi (syn. *Poterioochromonas stipitata* Scherffel and *Ochromonas malhamensis* Pringsheim). Protoplasma 92: 87–107
Schulze D, Robenek H, McFadden GI, Melkonian M (1987) Immunolocalization of a Ca^{2+}-modulated contractile protein in the flagellar apparatus of green algae: the nucleus-basal body connector. Eur J Cell Biol 45: 51–61
Segaar PJ, Gerritsen AF (1989) Flagellar roots as vital instruments in cellular morphogenesis during multiple fission (sporulation) in the unicellular green flagellate *Brachiomonas submarina* (Chloromonadales, Chlorophyta). Crypt Bot 1: 249–274
Slankis T, Gibbs SP (1972) The fine structure of mitosis and cell division in the chrysophycean alga *Ochromonas danica*. J Phycol 8: 243–256
Sleigh MA (1988) Flagellar root maps allow speculative comparisons of root patterns and their ontogeny. BioSystems 21: 277–282
Sluiman HJ, Blommers PCJ (1990) Basal apparatus behavior during cellular division (sporulation) in the coccoid green alga *Clorosarcina*. Protoplasma 155: 66–75
Stewart KD, Mattox KR, Chandler CD (1974) Mitosis and cytokinesis in *Platymonas subcordiformis*, a scaly green monad. J Phycol 10: 65–79
Surek B, Melkonian M (1986) A cryptic cytostome is present in *Euglena*. Protoplasma 133: 39–49
Vesk M, Moestrup Ø (1987) The flagellar root system in *Heterosigma akashiwo* (Raphidophyceae). Protoplasma 137: 15–28
– Hoffman LR, Pickett-Heaps JD (1984) Mitosis and cell divisions in *Hydrurus foetidus* (Chrysophyceae). J Phycol 20: 461–470
Vorobjev IA, Nadezhdina ES (1987) The centrosome and its role in the organization of microtubules. Int Rev Cytol 106: 227–293
Wetherbee R, Platt SJ, Beech PL, Pickett-Heaps JD (1988) Flagellar transformation in the heterokont *Epipyxis pulchra* (Chrysophyceae): direct observations using image-enhanced light microscopy. Protoplasma 145: 47–54
Woods JK, Triemer RE (1981) Mitosis in the octoflagellate *Pyramimonas amylifera* (Chlorophyta). J Phycol 17: 81–90
Wright RL, Adler SA, Spanier JG, Jarvik JW (1989) Nucleus-basal body connector in *Chlamydomonas*: evidence for a role in basal body segregation and against essential roles in mitosis or in determining cell polarity. Cell Motil Cytoskeleton 14: 516–526
– Salisbury JL, Jarvik JW (1985) A nucleus-basal body connector in *Chlamydomonas reinhardtii* that may function in basal body localization or segregation. J Cell Biol 101: 1903–1912

Protoplasma (1991) 164: 38–44

PROTOPLASMA

An update on fibrous flagellar roots in green algae

K.-F. Lechtreck and **M. Melkonian***

Botanisches Institut, Universität zu Köln, Köln

Received December 31, 1990
Accepted January 16, 1991

Summary. In flagellate green algae two types of fibrous flagellar roots can be distinguished: system I fibres, cross-striated bundles of 2 nm filaments (striation periodicity about 30 nm), which are associated with flagellar root microtubules, and system II fibres, contractile bundles of 4–8 nm filaments which are often cross-striated (striation periodicity variable but greater than 80 nm). The major protein of system II fibres is centrin, a Ca^{2+}-modulated phosphoprotein, which is a member of the EF-hand protein family. The major protein of system I fibres (of several *Chlamydomonas*-type green algae) is a 34 kDa phosphoprotein, named assemblin. Because of the solubility characteristics of system I fibres and the properties of their major protein (paracrystal-formation in vitro, several isoelectric variants, heptad motifs in parts of the amino acid sequence), assemblin is presumably related to the k-m-e-f class of α-helical fibrous proteins.

Keywords: Flagellate green algae; Fibrous flagellar roots; System II fibres; System I fibres; Centrin; Assemblin; Phosphorylation.

Abbreviations: NBBC nucleus-basal body connector; SMAC striated microtubule-associated component; k-m-e-f class keratin-myosin-elastin-fibrinogen class; EF-hand protein family Ca^{2+}-binding proteins containing one to several Ca^{2+}-binding motifs consisting of a peculiar helix-loop-helix configuration; PVDF polyvinylidene difluoride.

Introduction

The flagellar basal apparatus of flagellate green algae consists of 2^n (n is usually 1–3) basal bodies plus various basal body associated components. Among the latter the principal components are connecting fibres and flagellar roots. Additional minor components link special basal body triplets with flagellar roots (see Beech et al. 1991). Flagellar roots originate at or near basal bodies, usually run for a considerable distance into the cytosol and terminate somewhere in the cell. Basically two types of flagellar roots can be distinguished, consisting either of microtubules (microtubular roots) or of filaments (fibrous roots). The occurrence, structure, and disposition of flagellar roots in green algae has been the subject of several reviews over the last twelve years (Moestrup 1978, 1982; Melkonian 1980, 1984; O'Kelly and Floyd 1984). Since the last major review of fibrous flagellar roots in green algae was published (Melkonian 1980) one type of fibrous root was isolated, purified and its protein composition determined (Salisbury et al. 1984). Subsequently, centrin was identified as a Ca^{2+}-modulated contractile protein in fibrous roots of this type in many flagellate green algae (Schulze et al. 1987), and also as a universal component of the basal apparatus/centriole/centrosome-complex of many eukaryotic cells (Salisbury et al. 1986, Baron and Salisbury 1988). In his review on fibrous flagellar roots of green algae Melkonian (1980) distinguished two major types of fibrous roots: system I and system II fibres (initially termed as such by Micalef and Gayral 1972). While system II fibres are mainly composed of centrin, the chemical composition of system I fibres has remained unknown until very recently when pure preparations of system I fibres were obtained from the green flagellate *Spermatozopsis similis* (Lechtreck and Melkonian unpubl. results). This short review presents an update and a comparative analysis of the two principal types of fibrous flagellar roots in flagellate green algae including new data on the protein composition of both root types.

* Correspondence and reprints: Botanisches Institut, Lehrstuhl I, Universität zu Köln, Gyrhofstrasse 15, D-W-5000 Köln 41, Federal Republic of Germany.

System II fibres

System II fibres are contractile fibrous flagellar roots of flagellate green algae, which are not closely associated with flagellar root microtubules (review by Melkonian 1980). These structures extend from basal bodies into the cytoplasm where they often contact the nucleus. In the latter case they presumably represent the rhizoplast as visualized in stained cells by light microscopists (review by Salisbury 1988). In isolated cytoskeletons the nuclear remnants often remain associated with the basal apparatus through the system II fibres, which in this case have been termed nucleus-basal body connectors (NBBCs; Wright et al. 1985).
Contractility of these structures was first shown for the massive system II fibres of *Tetraselmis striata*, which exhibit a Ca^{2+}-dependent contraction when exposed to elevated concentrations of Ca^{2+} (Salisbury and Floyd 1978). The major protein of enriched fractions of isolated system II fibres of *T. striata* is a 20 kDa protein which undergoes a Ca^{2+}-dependent electrophoretic mobility shift in alkaline urea gels (Salisbury et al. 1984). The 20 kDa protein, which can be extracted from cytoskeletons in 500 mM sodium chloride (Huang et al. 1988 a), consists of two isoelectric variants (α, β), the more acidic one (β) is a phosphoprotein (Salisbury et al. 1984) (Fig. 9). The 20 kDa protein has been termed centrin (Salisbury et al. 1988; synonym: caltractin, Huang et al. 1988 a).
System II fibres, which consist of bundles of 4–8 nm filaments often exhibit cross-striations with a variable periodicity of greater than 80 nm (review by Melkonian 1980) (Figs. 1 and 2). Elevated intracellular Ca^{2+}-concentrations ($\geqslant 10^{-7}$ M Ca^{2+}) lead to a supercoiling of these filaments and contraction of the whole fibre within milliseconds. Contraction of system II fibres in vitro does not require addition of external ATP. Re-extension of these centrin-containing fibres is much slower (in the range of seconds to minutes and in the case of the *Tetraselmis* rhizoplast up to an hour) and involves a lowering of the Ca^{2+}-concentration in the vicinity of the fibre. In permeabilized cell models of *Chlamydomonas reinhardtii* continuous cycles of contraction/reextension of system II fibres require the presence of external ATP (Salisbury et al. 1987). Whether a phosphorylation/dephosphorylation cycle of centrin is necessary for contraction/reextension of system II fibres as originally suggested (Salisbury et al. 1984), is now doubtful as experimental conditions are known where the two processes are uncoupled (Martindale and Salisbury 1990).

A polyclonal antibody raised against centrin from *Tetraselmis* cross-reacts with system II fibres of many green algal taxa (Schulze et al. 1987). In *C. reinhardtii*, where the two system II fibres distally separate into several branches over the nuclear surface, contraction can be stimulated in vitro (Wright et al. 1985). In isolated cytoskeletons of the marine green flagellate *Dunaliella bioculata* we have determined that at 5×10^{-8} M Ca^{2+} the NBBCs begin to contract to about 50% of their original length (Merten and Melkonian unpubl. obs.). The functional significance of system II fibre contraction is still unresolved although many suggestions have been made (reviews by Salisbury 1989, Melkonian et al. 1991). Among the suggested functions are: initiator and/or coordinator of flagellar beat, shock absorber and force dissipator, signal transducer for flagellar shedding, signal transducer for gene induction, mediator for correct segregation of basal bodies, and essential component of the mitotic machinery. Recent studies on a number of *C. reinhardtii* mutants known to be defective for basal body assembly and/or localization (i.e., variable flagella-number mutants, particularly vfl-2 with no detectable centrin in the NBBC-region; Wright et al. 1989) support the view that the system II fibre (NBBC) plays a role in determining basal body localization and/or segregation. These observations argue against essential roles of system II fibres in mitosis or in regulating gene expression during flagellar regeneration as suggested by Salisbury et al. (1988). Similarly in *Polytomella parva* a NBBC is apparently absent (Schulze et al. 1987, Melkonian et al. 1988), yet mitosis and cytokinesis are not impaired.
In addition to its presence in system II fibres, centrin has also been found in two other structures of the flagellar basal apparatus of green algae, namely in the distal connecting fibres (McFadden et al. 1987, Melkonian et al. 1988) and in the flagellar transition region (Schulze et al. 1987, Melkonian et al. 1988, Sanders and Salisbury 1989). In *Spermatozopsis similis*, a naked green flagellate, the distal connecting fibre undergoes a reversible calcium-dependent contraction during the photophobic response of this organism (McFadden et al. 1987). In isolated cytoskeletons of *S. similis* (Lechtreck et al. 1989) contraction of the distal connecting fibre can be induced at the same elevated Ca^{2+}-concentration (i.e., $\geqslant 5 \times 10^{-8}$ M Ca^{2+}) as system II fibre contraction (McFadden et al. 1987).
Centrin from *C. reinhardtii* has been cloned and the complete amino acid sequence has been deduced from the cDNA sequence (Huang et al. 1988 b). The deduced

amino acid sequence of centrin shows a linear relatedness with calmodulin from the same species (45% sequence identity) and with the CDC 31 gene product from *Saccharomyces cerevisiae* (50% sequence identity). The predicted secondary structure of centrin indicates that the protein contains four homologous Ca^{2+}-binding domains that conform to the helix-loop-helix motif or EF-hand structure found in calmodulin and related Ca^{2+}-modulated proteins (Moncrief et al. 1990). In contrast to calmodulin and other related EF-hand proteins, centrin is phosphorylated and a cytoskeletal protein. The additional N- and C-terminal do-

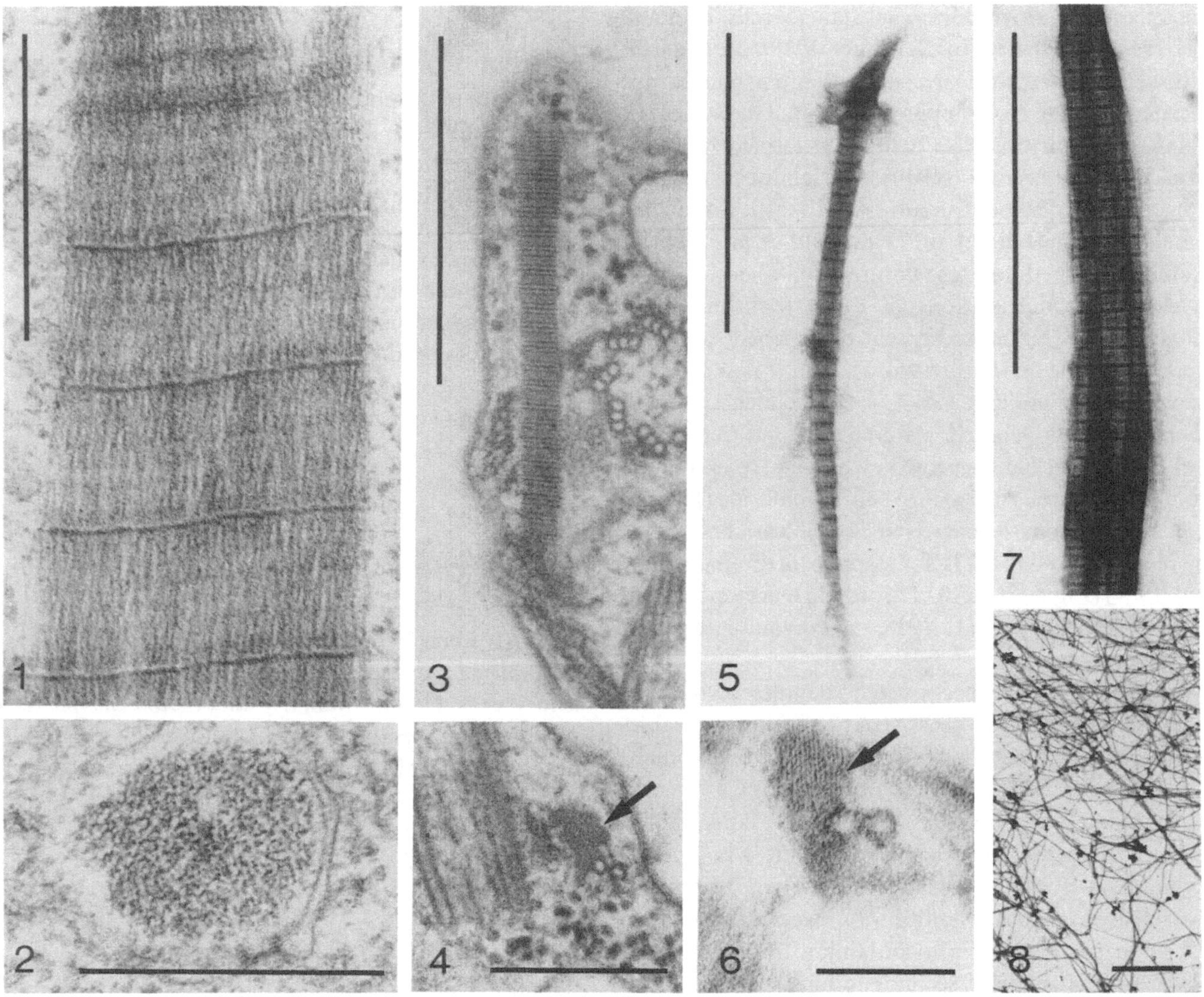

Figs. 1–8. Electron micrographs of system I and system II fibrous flagellar roots

Figs. 1 and **2.** Thin sections of system II fibres of *Tetraselmis cordiformis* in longitudinal section (Fig. 1) and cross-section (Fig. 2). Fixation of the cells was with 0.5% OsO_4 following the protocol of Melkonian (1982). Bars: 0.5 μm

Figs. 3 and **4.** Thin sections of system I fibre of *Spermatozopsis exsultans* in longitudinal (Fig. 3) and cross-sections (Fig. 4). Fixation of the cells was according to Melkonian et al. (1987). Arrow denotes the system I fibre. Bars: Fig. 3, 0.5 μm; Fig. 4, 0.25 μm

Fig. 5. Isolated system I fibre of *Spermatozopsis similis* following ultrasonication of NaCl (2 M)-extracted isolated basal apparatuses (Lechtreck et al. 1989). Whole mount stained with uranyl acetate. Bar: 0.5 μm

Fig. 6. Cross-section of a system I fibre from an isolated cytoskeleton of *Spermatozopsis similis*. Arrow indicates system I fibre (regularly spaced electron dense sheets can be seen). Isolated cytoskeletons were fixed according to McFadden et al. (1987). Bar: 0.1 μm

Figs. 7 and **8.** Assemblin paracrystals. Isolated basal apparatuses of *Spermatozopsis similis* (Lechtreck et al. 1989) were extracted with 2 M NaCl. The NaCl-insoluble residue was further extracted with 2 M urea. When the urea-soluble fraction is dialysed against 150 mM KCl paracrystals of assemblin are formed. Whole mount, negatively stained with uranyl acetate. Bars: Fig. 7, 0.5 μm; Fig. 8, 2 μm

mains of centrin, by which it is discriminated from calmodulin, could be responsible for its cytoskeletal properties.

System I fibres

System I fibres are microtubule associated, striated fibres with a striation periodicity of about 30 nm (Micalef and Gayral 1972, Melkonian 1980) (Figs. 3–6). These structures have also be termed "striated microtubule-associated components" or SMAC (Floyd et al. 1980). Since system I fibres consist of bundles of filaments which are stable in vitro without the accompanying root microtubules (Hoffmann 1970, Marano et al. 1985), the term system I fibre is preferable over SMAC. Although in situ system I fibres are apparently always closely associated with flagellar root microtubules, either over- or underlying the microtubules (review by Melkonian 1980) (Figs. 4 and 6), evidence has been presented that during flagellar reversal the system I fibres of *Spermatozopsis similis* detach from the root microtubules and remain associated with the contracted distal connecting fibre (Melkonian and Preisig 1984). Variability in cross-striation periodicity and hence contractility of system I fibres has not been reported.

A polyclonal antibody against centrin, the principal protein of system II fibres, does not label system I fibres in various flagellate green algae (Melkonian et al. 1988), indicating that system I and system II fibres are biochemically distinct.

In contrast to system II fibres, little is known about the chemical composition of system I fibres. Basal apparatus preparations of *Dunaliella bioculata* enriched in system I fibres revealed three major protein bands of 31 kDa, 50 kDa, and 76 kDa on SDS-gels (Marano et al. 1985). A detailed analysis of the system I fibres of *Spermatozopsis similis* (Lechtreck and Melkonian unpubl. obs.) led to the following results.

Each cell of *S. similis* contains two opposing system I fibres. Thin sections of *S. similis* previously revealed only a single continuous system I fibre (Melkonian and Preisig 1984). Similar observations had also been made in *Polytomella agilis* (Brown et al. 1976) and *Chlamydomonas reinhardtii* (Goodenough and Weiss 1978). We have found that the system I fibres have oblique proximal ends (Fig. 5) and thus opposing (rotated by 180°) system I fibres possibly interlock with an uninterrupted cross-striation pattern; this gives the image of a continuous fibre.

System I fibres are composed of several layers of laterally aligned 2 nm filaments (Figs. 6 and 7). The number of layers and their width (i.e., number of filaments within one layer) decrease from their proximal (near the basal bodies) to their distal regions. The fibres exhibit a complex striation pattern with a periodicity of 30 nm suggesting a gap-overlap-structure comparable to collagen filaments (Burgeson 1988).

System I fibres are insoluble in non-ionic detergents and in high-salt solutions (2 M NaCl), but can be dissolved with ionic detergents, potassium iodide (1 M) or urea (> 2 M), and with low salt buffers (< 10 mM). Dialysis of urea-solubilized system I fibres against a "reassembly buffer" yields reconstituted fibres which closely resemble the native system I fibres in ultrastructure and striation pattern (Figs. 7 and 8). After several cycles of disassembly and reassembly, paracrystals are formed which are almost pure polymers of a 34 kDa protein. We have named this 2 nm filament-forming protein, which is the main protein of the system I fibres of *S. similis*, assemblin. Assemblin consists of at least 4 acidic (pH 5–6) isoforms (α, β, γ, δ) of which the two most acidic ones (γ, δ) are phosphoproteins (Fig. 9).

A polyclonal antibody raised against assemblin labels the system I fibres of *S. similis* in indirect immunofluorescence and pre- and postembedding immunogold electron microscopy. A comparative immunofluorescence analysis in several flagellate green algae using the anti-assemblin antibody labelled epitopes along microtubular flagellar roots in *Chlamydomonas reinhardtii, Polytomella parva, Hafniomonas reticulata, Nephroselmis olivacea*, and in several species of *Dunaliella* and *Asteromonas*. Proteins of the same apparent size as assemblin from *S. similis* were detected by immunoblotting in *C. reinhardtii, P. parva*, and *N. olivacea*, but in *Dunaliella bioculata* a 31 kDa protein is immunoreactive. Parts of the amino acid sequence of assemblin were established by gas-phase microsequencing of the N-terminal region of the protein and of proteolytic fragments. The proline rich N-terminal domain and a 29 amino acid long internal domain show some sequence similarities (about 40% identity) to the N- and C-terminal domains of β-giardin (Holberton et al. 1988). The giardins are a group of 29–38 kDa proteins with several isoelectric variants that occur in the cytoskeleton of *Giardia lamblia* (Diplomonadida), an enteric parasite of the alimentary tracts of vertebrates (Peattie et al. 1989). β-Giardin (30 kDa) is the principal protein of the microribbons of *G. lamblia*. The microribbons are cross-striated, microtubule associated bundles of 2 nm filaments (Holberton 1981) which can be reconstituted in vitro and share similar solubility properties with the system I fibres of flagellate green algae

(Crossley and Holberton 1983, 1985). More than 95% of the β-giardin sequence is free of proline residues and shows heptad motifs, suggesting a continuous α-helical coiled-coil structure of the molecule (Holberton et al. 1988). Several heptad motifs were found in the sequenced fragments of assemblin, and, with the exception of the N-terminus and one other fragment, no prolines were detected. High portions of α-helical coiled-coils are characteristic of the k-m-e-f class of fibrous proteins (Cohen and Parry 1986). Most of these proteins share other characteristics, such as solubility properties, mechanical strength, and ability to form filaments/paracrystals in vitro independent of ATP hydrolysis.

System I fibres may therefore belong to a widespread group of cytoskeletal elements whose function could be related to adding mechanical support and elasticity to the cytoskeleton. In this context we note that system I fibres are well developed and very long in the naked cell wall-less taxa *Spermatozopsis, Dunaliella, Polytomella*, and *Hafniomonas* in comparison to the much less developed and relatively short system I fibres of *C. reinhardtii* and *N. olivacea*, taxa in which cell coverings (cell walls, scalelayers) possibly dissipate most of the forces caused by flagellar beating.

Comparison between system I fibres and system II fibres

System I and system II fibres are biochemically and ultrastructurally distinct types of flagellar roots in green algae. A comparison of their properties is presented in Table 1.

Table 1. Comparison between system I and system II fibrous flagellar roots of flagellate green algae

Characteristics	System I fibres	System II fibres
Striation periodicity	30 nm	80 nm
Filament diameter	2 nm	4–8 nm
Microtubule associated	+	–
Contractility	–	+
Solubility properties		
2 M urea	+ (soluble)	+
high salt	– (insoluble)	+
low salt	+	?
Major protein	assemblin	centrin
kDa	31–34	20
isoforms	4	2
pI	5.4–5.8	5.2–5.3
phosphorylated isoforms	2	1
Main characteristic	paracrystal-forming	Ca^{2+}-modulated
Related proteins	(β-giardin)	calmodulin
Protein family	(k-m-e-f class)	EF-hand

The possible role of protein phosphorylation in fibrous roots

Both centrin and assemblin are phosphoproteins. In mitosis the centrin-based cytoskeleton of *Chlamydomonas reinhardtii* is reorganized (Salisbury et al. 1988; for review, see Beech et al. 1991). Similarly system I fibres undergo reorganization during cell division since we have been unable to detect intact system I fibres during cell division in *Spermatozopsis similis* (unpubl. obs.). The reorganization of centrin- and assemblin-filaments during cell division may be controlled by protein hyperphosphorylation, a mechanism by which intermediate filaments and nuclear lamins are converted to a soluble form (Geisler and Weber 1988, Inagaki et al. 1987, Peter et al. 1990, Heald and McKeon 1990, Ward and Kirschner 1990).

Centrin is the cytoskeletal protein with the highest phosphate turnover in *S. similis*. Centrin in *Tetraselmis striata* is also phosphorylated with a high turnover (Salisbury et al. 1984). While previously the phosphorylation of centrin was discussed in relation to the contraction/re-extension cycle of the system II fibres (Salisbury et al. 1984, 1988), new data now indicate that this may not be true (Martindale and Salisbury 1990). Centrin phosphorylation is rapidly responsive to changes in the cell's environment (Ca^{2+}-influx, heat shock, pH-shock, ethanol treatment; Martindale and Salisbury 1990). Heat shock, for example, causes a reduction in centrin phosphorylation in the absence of system II fibre contraction in *T. striata*.

Centrin differs from most other known EF-hand proteins in that it is a phosphoprotein and a cytoskeletal structural protein. The assembly of several cytoskeletal proteins is regulated by protein phosphorylation (Ottaviano and Gerace 1985, Inagaki et al. 1988). It is thus possible that phosphorylation/dephosphorylation of centrin could be related to the filament-forming, cytoskeletal properties of centrin rather than to its contractile function which is modulated by Ca^{2+}.

In interphase cells of *S. similis* assemblin is the cytoskeletal protein with the second-highest phosphate turnover (after centrin). Phosphorylation of assemblin and centrin, the principal proteins of the two types of fibrous flagellar roots in flagellate green algae, is accordingly not restricted to dividing cells. We may therefore assume that phosphorylation/dephosphorylation of fibrous root proteins during interphase may serve other functions than filament assembly/disassembly. It is, for example, possible that phosphorylation regulates the elasticity of fibrous roots, which could in turn influence their function as stress absorbers. The elasticity of the microtubule associated protein tau, which forms paracrystals in vitro, increases when the protein is dephosphorylated and decreases by phosphorylation (Hagestedt et al. 1989). The preliminary analysis of phosphorylation of assemblin isoforms indicates that two or more phosphorylation sites with different turnover rates are present. Similarly, intermediate filament proteins have several phosphorylation sites with different turnover characteristics (Geisler and Weber 1988). For nuclear lamins it is known that phosphorylation sites are differentially regulated; cdc 2-dependent phosphorylation, which leads to disassembly of the nuclear lamina in vitro, occurs at specific phosphorylation sites (M-sites; Peter et al. 1990) whereas protein kinase A and C phosphorylate different sites. It is conceivable that in assemblin too, different enzymes operate at different stages in the cell cycle. Phosphorylation/dephosphorylation of assemblin during interphase could lead to changes in the conformation of the molecule, thereby interrupting the α-helical coiled-coil regions and thus affecting the elastic properties of the protein. During cell division phosphorylation/dephosphorylation of assemblin at sites involved in the interaction between different molecules could perhaps lead to assembly/disassembly of the filaments. These questions need to be addressed in the future for the principal proteins of both types of fibrous flagellar roots of flagellate green algae.

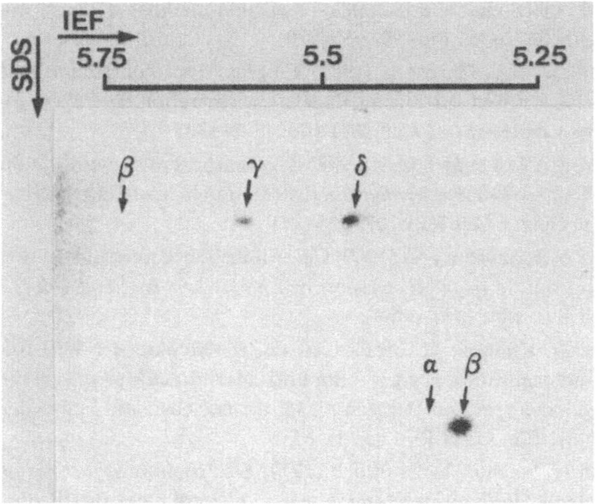

Fig. 9 Autoradiography of cytoskeletons of *Spermatozopsis similis* separated by two-dimensional gel electrophoresis and blotted on PVDF-membrane. Cells of *S. similis* were incubated with carrier-free $^{32}PO_4$ (0.1 mCi/ml) for 45 min prior to cell lysis/detergent extraction (2% Triton X-100). 2 D-gel electrophoresis and electroblotting of isolated cytoskeletons was essentially as described by O'Farrell (1975) and Towbin et al. (1979). The positions of β-assemblin and α-centrin were deduced from the corresponding Aurodye-stained (Janssen) protein blot (not shown). Only a portion of the autoradiogram is shown corresponding to an apparent molecular mass range of 15–40 kDa. Because of cathode drift α-assemblin is not focused within the pH-range of this particular 2 D-gel. The pH-range of the IEF gel has been calculated by direct pH-measurement of gel slices from a parallel control gel

Acknowledgements

This study was supported by the Deutsche Forschungsgemeinschaft (Me 658/3–5). KFL and MM were recipients of travel grants from NSF.

References

Baron AT, Salisbury JL (1988) Identification and localization of a novel, cytoskeletal, centrosome-associated protein in PtK 2 cells. J Cell Biol 107: 2669–2678

Beech PL, Heimann K, Melkonian M (1991) Development of the flagellar apparatus during the cell cycle in unicellular algae. Protoplasma 164: 23–37

Brown DL, Massalski A, Patenaude R (1976) Organization of the flagellar apparatus and associated cytoplasmic microtubules in the quadriflagellate alga *Polytomella agilis*. J Cell Biol 69: 106–125

Burgeson RE (1988) New collagens, new concepts. Annu Rev Cell Biol 4: 551–577

Cohen C, Parry DAD (1986) α-helical coiled-coils – a widespread motif in proteins. Trends Biochem Sci 11: 245–249

Crossley R, Holberton DV (1983) Selective extraction with sarkosyl and repolymerization in vitro of cytoskeleton proteins from *Giardia*. J Cell Sci 62: 419–438

– – (1985) Assembly of 2.5 nm filaments from giardin, a protein associated with cytoskeletal microtubules in *Giardia*. J Cell Sci 78: 205–231

Floyd GL, Hoops HJ, Swanson JA (1980) Fine structure of the zoospore of *Ulothrix belkae* with emphasis on the flagellar apparatus. Protoplasma 104: 17–31

Geisler N, Weber K (1988) Phosphorylation of desmin in vitro inhibits formation of intermediate filaments; identification of three kinase A sites in the aminoterminal head domain. EMBO J 7: 15–20

Goodenough UW, Weiss RL (1978) Interrelationships between microtubules, a striated fiber, and the gametic mating structure of *Chlamydomonas reinhardtii*. J Cell Biol 76: 430–438

Hagestedt T, Lichtenberg B, Wille H, Mandelkow EM, Mandelkow E (1989) Tau protein becomes long and stiff upon phosphorylation: correlation between paracrystalline structure and degree of phosphorylation. J Cell Biol 109: 1643–1651

Heald R, McKeon F (1990) Mutation of phosphorylation sites in lamin A that prevent nuclear lamina disassembly in mitosis. 61: 579–589

Hoffman LR (1970) Observations on the fine structure of *Oedogonium*. VI. The striated component of the compound flagellar "roots" of *O. cardiacum*. Can J Bot 48: 189–196

Holberton DV (1981) Arrangement of subunits in microribbons from *Giardia*. J Cell Sci 47: 167–185

– Baker DA, Marshall J (1988) Segmental α-helical coiled-coil structure of the protein giardin from the *Giardia* cytoskeleton. J Mol Biol 204: 789–795

Huang B, Watterson DW, Lee VD, Schibler MJ (1988 a) Purification and characterization of a basal body-associated Ca^{2+}-binding protein. J Cell Biol 107: 121–131

– Mengersen A, Lee VD (1988 b) Molecular cloning of cDNA for caltractin, a basal body-associated Ca^{2+}-binding protein: homology in its protein sequence with calmodulin and the yeast CDC 31 gene product. J Cell Biol 107: 133–140

Inagaki M, Gonda Y, Matsuyama M, Nishizawa K, Nishi Y, Sato C (1988) Intermediate filament reconstitution in vitro. The role of phosphorylation on the assembly-disassembly of desmin. J Biol Chem 263: 5970–5978

– Nishi Y, Nishizawa K, Matsuyama M, Sato C (1987) Site-specific phosphorylation induces disassembly of vimentin filaments in vitro. Nature 328: 649–652

Lechtreck K-F, McFadden GI, Melkonian M (1989) The cytoskeleton of the naked green flagellate *Spermatozopsis similis*: isolation, whole mount electron microscopy, and preliminary biochemical and immunological characterization. Cell Motil Cytoskeleton 14: 552–561

Marano F, Santa-Maria A, Krishnawamy S (1985) The flagellar apparatus of *Dunaliella*: isolation of basal body-flagellar root complex. Protoplasma 127: 82–92

Martindale VE, Salisbury JL (1990) Phosphorylation of algal centrin is rapidly responsive to changes in the external milieu. J Cell Sci 96: 395–402

McFadden GI, Schulze D, Surek B, Salisbury JL, Melkonian M (1987) Basal body reorientation mediated by a Ca^{2+}-modulated contractile protein. J Cell Biol 105: 903–912

Melkonian M (1980) Ultrastructural aspects of basal body associated fibrous structures in green algae: a critical review. BioSystems 12: 85–104

– (1982) Effect of divalent cations on flagellar scales in the green flagellate *Tetraselmis cordiforms*. Protoplasma 111: 221–233

– (1984) Flagellar apparatus ultrastructure in relation to the green algal classification. In: Irvine DEG, John DM (eds) Systematics of the green algae. Academic Press, London, pp 73–120

– Preisig HR (1984) Ultrastructure of the flagellar apparatus in the green flagellate *Spermatozopsis similis*. Plant Syst Evol 146: 145–162

– McFadden GI, Reize IB, Preisig HR (1987) A light and electron microscopic study of the quadriflagellate green alga *Spermatozopsis exsultans*. Plant Syst Evol 158: 47–61

– Schulze D, McFadden GI, Robenek H (1988) A polyclonal antibody (anti-centrin) distinguishes between two types of fibrous flagellar roots in green algae. Protoplasma 144: 56–61

– Beech PL, Katsaros C, Schulze D (1991) Centrin-mediated cell motility in algae. In: Melkonian M (ed) Algal cell motility. Chapman and Hall, New York (in press)

Micalef H, Gayral P (1972) Quelques aspects de l'infrastructure des cellules vegetatives et des cellules reprodutrices d'*Ulva lactuca* L. (Chlorophycees). J Microsc 13: 417–428

Moestrup Ø (1978) On the phylogenetic validity of the flagellar apparatus in green algae and other chlorophyll a and b containing plants. BioSystems 10: 117–144

– (1982) Flagellar structure in algae: a review, with new observations particularly on the Chrysophyceae, Phaeophyceae (Fucophyceae), Euglenophyceae, and *Reckertia*. Phycologia 21: 427–528

Moncrief ND, Kretsinger RH, Goodman M (1990) Evolution of EF-hand calcium-modulated proteins. I. Relationships based on amino acid sequences. J Mol Evol 30: 522–562

O'Farrell PH (1975) High resolution two-dimensional electrophoresis of proteins. J Biol Chem 250: 4007–4040

O'Kelly CJ, Floyd GL (1984) Flagellar apparatus absolute orientations and the phylogeny of the green algae. BioSystems 16: 227–252

Ottaviano Y, Gerace L (1985) Phosphorylation of the nuclear lamins during interphase and mitosis. J Biol Chem 260: 624–632

Peattie DA, Alonso RA, Hein A, Caulfield JP (1989) Ultrastructural localization of giardins to the edges of disk microribbons of *Giardia lamblia* and the nucleotide and deduced protein sequence of alpha giardin. J Cell Biol 109: 2323–2335

Peter M, Nakagawa J, Doree M, Labbe JC, Nigg EA (1990) In vitro disassembly of the nuclear lamina and M phase-specific phosphorylation of lamins by cdc 2 kinase. Cell 61: 591–602

Salisbury JL (1988) The lost neuromotor apparatus of *Chlamydomonas*: rediscovered. J Protozool 35: 574–577

– (1989) Algal centrin: calcium-sensitive contractile organelles. In: Coleman AW, Goff LJ, Stein-Taylor JR (eds) Algae as experimental systems. AR Liss, New York, pp 19–37

– Floyd GL (1978) Calcium-induced contraction of the rhizoplast of a quadriflagellate green alga. Science 202: 975–977

– Baron AT, Coling DE, Martindale VE, Sanders MA (1986) Calcium-modulated contractile proteins associated with the eucaryotic centrosome. Cell Motil 6: 193–197

– – Surek B, Melkonian M (1984) Striated flagellar roots: isolation and partial characterization of a calcium-modulated contractile organelle. J Cell Biol 99: 962–970

– Sanders MA, Harpst L (1987) Flagellar root contraction and nuclear movement during flagellar regeneration in *Chlamydomonas reinhardtii*. J Cell Biol 105: 1799–1805

– Baron AT, Sanders MA (1988) The centrin-based cytoskeleton of *Chlamydomonas reinhardtii*: distribution in interphase and mitotic cells. J Cell Biol 107: 635–641

Sanders MA, Salisbury JL (1989) Centrin-mediated microtubule severing during flagellar excision in *Chlamydomonas reinhardtii*. J Cell Biol 108: 1751–1760

Schulze D, Robenek H, McFadden GI, Melkonian M (1987) Immunolocalization of a Ca^{2+}-modulated contractile protein in the flagellar apparatus of green algae: the nucleus-basal body connector. Eur J Cell Biol 45: 51–61

Towbin H, Staehlin T, Gordon J (1979) Electrophoretic transfer of proteins from polyacrylamide gels to nitrocellulose sheets: procedure and some applications. Proc Natl Acad Sci USA 76: 4350–4354

Ward GE, Kirschner MW (1990) Identification of cell cycle-regulated phosphorylation sites on nuclear lamin C. Cell 61: 561–577

Wright RL, Adler SA, Spanier JG, Jarvik JW (1989) Nucleus-basal body connector in *Chlamydomonas*: evidence for a role in basal body segregation and against essential roles in mitosis or in determining cell polarity. Cell Motil Cytoskeleton 14: 516–526

– Salisbury J, Jarvik JW (1985) A nucleus-basal body connector in *Chlamydomonas reinhardtii* that may function in basal body localization or segregation. J Cell Biol 101: 1903–1912

Protoplasma (1991) 164: 45–53

PROTOPLASMA

Mechanisms of flagellar propulsion

A biologist's view of the relation between structure, motion, and fluid mechanics

M. A. Sleigh*

Department of Biology, University of Southampton, Southampton

Received September 27, 1990
Accepted December 14, 1990

Summary. Flagellar propulsion takes place in the viscosity-dominated realm of low Reynolds number fluid dynamics. Volumes of fluid are carried in a capture zone around the moving regions of the flagellum, and the flagellar motion achieves propulsion because some of that water is shed from the capture zone, either from the flagellar tip in typical flagellar motion or to the side reached at the end of the effective stroke in the case of ciliary motion. Helical flagellar motion is in principle more efficient than planar beating, and the rotation caused by the former introduces complications in propulsion that may be advantageous, e.g., in *Euglena*, or disadvantageous, e.g., in a fixed cell. The presence of a surface near to the moving organelle restricts the fluid motion, but this effect enhances ciliary propulsion. There is a great variety of beat patterns, functionally adapted hydrodynamically or in other ways for locomotion, feeding, and other more restricted roles.

Keywords: Motility; Swimming; Feeding; Cilia; Reynolds number; Hydrodynamics; Boundary layer.

Abbreviations: Re Reynolds number; C_N coefficient of resistance to normal motion; C_T coefficient of resistance to tangential motion; l length; v velocity; ρ fluid density; μ fluid viscosity; L an element of flagellar length moving at velocity V_L; V_W velocity of a wave; V_N velocity of element L in perpendicular (normal) direction; V_T velocity of element L in tangential direction; F_N force in normal direction; F_T force in tangential direction; F_P propulsive force; F_D drag force; E effective stroke; R recovery stroke; ω angular velocity of flagellum; Ω angular velocity of body.

Introduction

The essential components of the eukaryotic flagellum comprise the 9 + 2 microtubule array of the axoneme and its surrounding cytoplasm, enclosed in a specialised extension of the cell membrane. Organelles with this construction are capable of performing a great diversity of shape changes (e.g., Sleigh 1981), which can be used to propel the cells to which they are attached, or to propel water around the cells. Flagellar shape changes are believed to depend upon active sliding movements of the outer nine doublets of the axoneme relative to one another, mediated by activity of the mechanochemical ATPase dynein; the mechanism of axonemal bending will not be discussed here (for a recent review, see Satir and Sleigh 1990). Some flagella participate in a form of gliding locomotion in which large flagellar membrane glycoproteins, capable of binding to external structures, are moved in a directed manner in the plane of the flagellar membrane; this form of propulsion will not be considered here, but has been reviewed recently by Bloodgood (1990).

The diversity of flagellar beat patterns depends largely upon variations in the timing of active inter-doublet sliding within and along the axoneme (Sugino and Naitoh 1982). The three principal beat patterns are (1) propagated planar undulations of approximately sinusoidal form produced by regular alternations of more or less equal active sliding at opposite sides of the axoneme, (2) propagated helical undulations produced, it is assumed, by active sliding progressing sequentially around the 9 doublets, with one circuit during the formation of each wave, and (3) lateral (or ciliary) beating in which a more or less synchronous sliding of doublets along one side bends the axoneme stiffly one way and

* Correspondence and reprints: Department of Biology, School of Biological Sciences, University of Southampton, Bassett Crescent East, Southampton SO9 3TU, U.K.

a propagated wave of sliding of doublets at the other side returns the axoneme to its original shape by an unrolling motion (Sleigh 1974). It is the main function of this paper to review the characteristics of propulsion produced by the flagella of flagellates using these three patterns, and to examine how flagellar appendages and other variants of flagellar structure may affect flagellar function.

The fluid mechanics of flagellar propulsion has attracted attention from a number of applied mathematicians and physicists since G. I. Taylor (1952) pointed out a theoretical approach to its analysis. The researches of G. J. Hancock and of people associated with M. E. J. Holwill, T. Y. Wu, J. Lighthill, and more recently J. R. Blake have contributed greatly to our understanding in ways that will be referred to later. However, many of their papers are in mathematical journals not normally used by biologists, and an attempt has been made here to extract those data most relevant for an understanding of biological aspects of flagellar propulsion and adaptation. Earlier reviews in biological publications by Holwill (1966) and Roberts (1981) provide more details.

Principles of propulsion by flagella and cilia

The question of Reynolds number

Two types of forces are concerned in the propulsion of organisms by swimming or flying (Shapiro 1961): these are inertial forces, forces associated with the motion of a mass of fluid, and viscous forces, forces associated with the sliding of layers of fluid over each other. Familiar examples like the swimming of an adult fish or the flying of a bird depend almost entirely on inertial forces, but swimming by flagella or cilia is dominated by viscous forces, and as a result involves phenomena not in accord with our everyday experience of how swimming is achieved by the large scale acceleration of fluid. The ratio of inertial forces to viscous forces gives the Reynolds number (Re). This is a dimensionless quantity that can be calculated from

$$\mathrm{Re} = \rho v l/\mu,$$

where ρ and μ are respectively the density and viscosity of the fluid through which a structure of length l is moving at velocity v. For a reasonably large fish v and l are large and Re = 1000 or more, but for a flagellate or a spermatozoon v and l are small and Re = 10^{-2} or less. Clearly, for a fish where inertial fores are some 1000 times the viscous forces, the latter can be neglected, but for a small flagellate where inertial forces are one hundredth of viscous forces, the inertial forces can be neglected. In discussing the hydrodynamics of flagella and cilia we are therefore concerned only with viscous forces, and thus with the relationships of surfaces with the surrounding fluid. The very small inertial component means that momentum is negligible and a swimming cell stops almost instantly when the flagella or cilia cease to beat. Organisms of intermediate size and swimming speed may have an Re near 1; here both inertial and viscous forces must be taken into account, and the motion is hydrodynamically complex. Mathematicians and fluid dynamicists have had considerable success in understanding propulsion at both high and low Re, but few advances have been made in the intermediate range where Re = 1 to 10 or so.

Behaviour of the fluid when a flagellum or cilium moves in water

When a surface moves relative to the surrounding water, the fluid immediately in contact with the surface moves with it in a "no-slip" zone (Shapiro 1961, Blake and Sleigh 1974). As one moves further from the surface the drag exerted on the fluid by the moving surface decreases until a point is reached where the influence of the movement of the surface on the fluid is negligible. The extent of this "boundary layer" within which drag is felt by the overlying fluid is markedly influenced by the fluid viscosity, being more extensive in fluids of higher viscosity. A similar boundary layer exists above a stationary surface over which a fluid is moving. When a structure the size of a flagellum or cilium moves in water, a zone of water of appreciable thickness relative to the structure is carried around it, the immediately adjacent water moving at the speed of the structure, and that further away being progressively less tightly linked until a point is reached where the motion of the fluid becomes imperceptible.

Another way of looking at the interaction between the structure and the surrounding fluid is to consider the resistance offered by the fluid to the movement of the structure through it. At very low Re we need only consider viscous (surface) forces. A structure displays a certain (surface) coefficient of viscous resistance to movement through the fluid which is directly proportional to the viscosity and inversely proportional to $\log_n$ (radius/length) (Gray and Hancock 1955). As might be expected, the effect of the structure on the water depends on its orientation in relation to the movement, but because this is a surface feature the effect is only modest. For a smooth flagellum the normal surface coefficient (C_N) (responsible for resistance when the structure is moving perpendicular to its axis) is about twice the tangential coefficient (C_T) (related to movement along the axis). This difference in the resistance offered to a flagellum, and hence the force the flagellum exerts on the water, when moving with different orientations, is important in relation to the propulsive effects of different parts of flagellar and ciliary beat cycles. The fluid is moved all round the structure at these low Reynolds numbers, but the influence of the moving structure on the surrounding fluid is twice as great in front and behind as it is at an equal distance to the sides.

An alternative way of calculating forces and flows where Stokes' equations apply was introduced by Hancock (1953), using slender body theory to replace the flagellum by a series of "Stokeslets" and "potential dipoles" along the flagellar centre line; this approach was also followed by Lighthill (1976) and Higdon (1979 a, b, c) to give more exact solutions than resistance coefficients. This more complex approach will not be followed here, although conclusions derived by this method are mentioned later.

The motion of water relative to a flagellum along which helical waves of constant radius and constant pitch are being propagated at constant frequency (Fig. 1 a) is simplest to describe because its pattern is similar throughout the length of the flagellum (neglecting end effects, and the effects of a cell body, the latter causing profound complications to be described later). In this case all elements along the flagellum push against the water continuously at the same angle, i.e., in the direction of wave propagation and at an angle inclined in the direction of rotation of the flagellar helix (Fig. 1 c); this tends to continuously propel water spirally towards the end of the flagellum at a constant speed (Fig. 1 d). The reaction on the flagellum propels it in the opposite direction to the waves, but also induces a torque which will be considered later. Normal and tangential forces (dependent on C_N and C_T and the velocities in normal and tangential directions) both contribute to both the forward propulsion and torque (Fig. 1 c).

By contrast, when a planar wave is propagated along the flagellum, the motion of each element of the flagellum is constantly changing, and so, consequently, is the motion of the surrounding water which is being moved by that element of the flagellum (Fig. 2 a). This wave appears approximately sinusoidal, and this form tends to be assumed in calculations, but waves formed

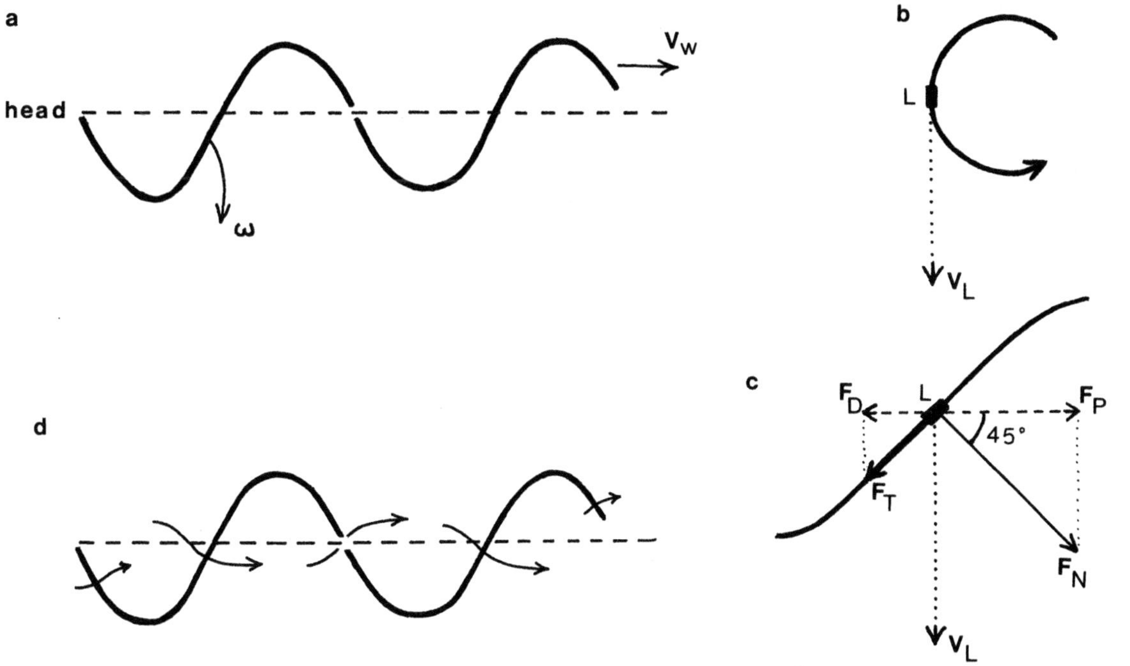

Fig. 1. Helical flagellar pattern. **a** A flagellum formed into a left-handed helix turns with constant angular velocity and propagates waves at velocity V_W. **b** and **c** An element L of the flagellum moves with a velocity V_L at a tangent to the helix and transverse to the axis of progression, as seen from the flagellar tip in **b** and from the side in **c**. The force on the water F_N is given by $C_N V_N L$ and the force F_T is given by $C_T V_T L$, depending on the velocities V_N and V_T at which the element L moves in the normal and tangential directions, respectively. F_N will have a component F_P propelling water in the direction of the wave and F_T has a component F_D, the drag force. **d** The motion of water produced by each element of the flagellum will be obliquely to the right and in the direction of rotation, resulting in the generation of a spiral flow

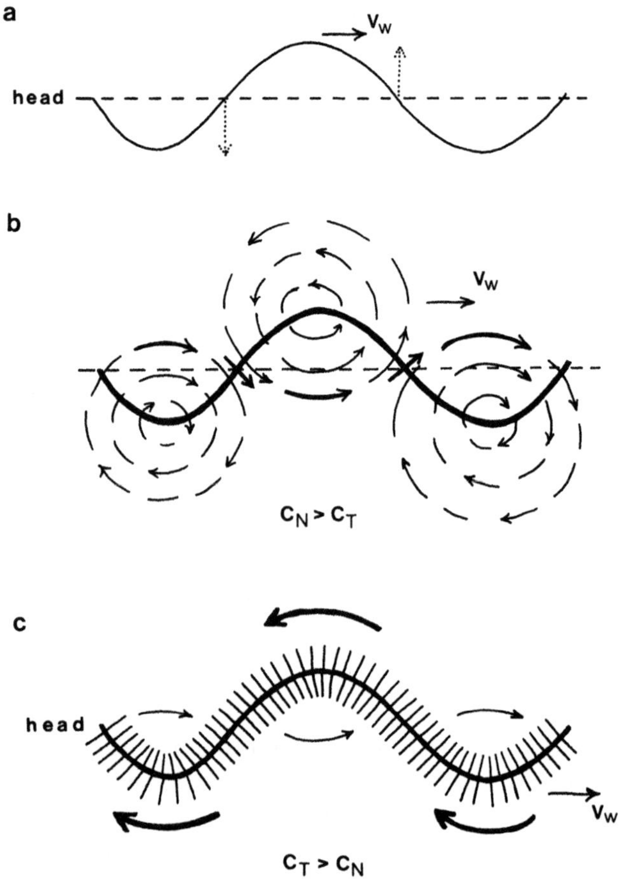

Fig. 2. Planar flagellar pattern. **a** The planar flagellum propagates approximately sinusoidal waves at velocity V_W; individual elements of the flagellum move first to one side and then to the other, the transverse velocity varying cyclically. **b** The water flow produced by an element of the flagellum is also directed alternately to one side and then the other, with periods of no lateral movement at the wave crests; the result is a series of local circulations that link to generate an axial flow in the direction of the wave along a smooth flagellum. **c** When mastigonemes are present on a flagellum, the form of the circulation created by the planar flagellar undulations is modified because $C_T > C_N$, and the dominant water flow around the flagellum is in the opposite direction to the propagation of waves along the flagellum

from arcs and straight lines (Brokaw 1965) or from meanders (Silvester and Holwill 1972) may more correctly represent their shape. The water movement around the flagellum can be portrayed in the plane of beat by normal and tangential forces with components in longitudinal and transverse directions (as in Fig. 1 c). Because C_N is about twice C_T, the lengths of flagellum between wave crests exert substantial forces in the direction of the waves, but near the wave crests, where there is no normal motion but only axial motion, drag forces opposing the flow will dominate. The resultant water propulsion is in the direction of the waves, and the local circulations of water induced as the flagellar waves are propagated towards the flagellar tip are as shown in Fig. 2 b (Higdon 1979 a). The longitudinal forces exerted by each wave produce a forward reaction on the flagellum, whereas the transverse forces of each half wave are in opposite directions and will tend to cancel each other out, although they generate lateral oscillations of the leading end (Holwill 1966).

When waves of either configuration are propagated along a flagellum from base to tip they will tend to shear water towards and then shed it from the flagellar tip. If the flagellated cell is free to move, the flagellum will undulate its way through the viscous water as well as propelling some water backwards in real space.

Flagella often carry external appendages that can affect water propulsion. The anterior flagellum of heterokont flagellates carries stiff mastigonemes, about 20 nm thick and about 1 µm long, in two rows. These flagella have a planar beat, and if the mastigonemes project rigidly from the flagellar surface they will increase C_T to greater than C_N, thereby causing the water movement to be reversed so that the resultant flow is in a direction opposite to the propagation of the flagellar waves (Fig. 2 c). If the mastigonemes project all round the flagellum $C_T/C_N = 1.2$, whilst if the mastigonemes are

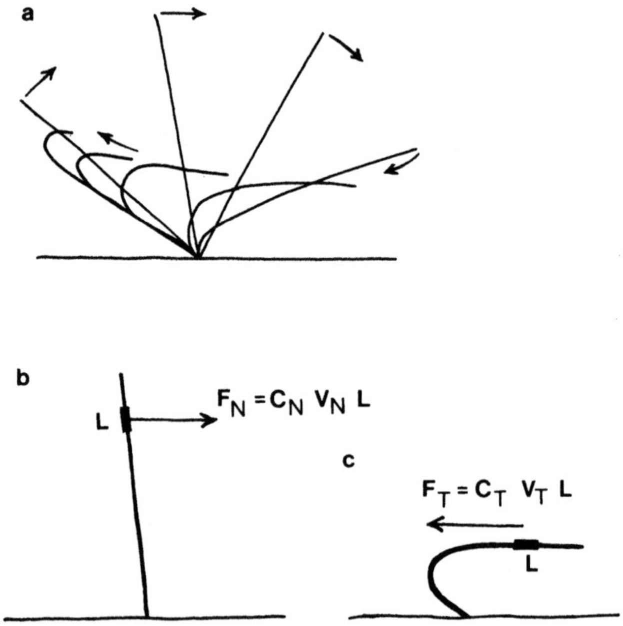

Fig. 3. Ciliary beat pattern. **a** The cycle of movement of a cilium composed of the effective stroke E and the recovery stroke R. **b** In the effective stroke an element L moves at velocity V_N and exerts a force F_N on the water. **c** In the recovery stroke an element L moves at a velocity V_T and exerts a force F_T on the water; this is less than F_N because $C_N = 2C_T$ and $V_N > V_T$

in two rows in the plane of beat $C_T/C_N = 1.8$ (Holwill and Sleigh 1967), so that reverse propulsion of water is much more effective with mastigonemes in the latter configuration. The presence of slender flexible hairs, or scales, on flagella is likely to produce small increases in both C_N and C_T and probably somewhat enhances propulsion by a helical flagellum (Lighthill 1976).

During a cilium-type beat (Fig. 3 a) the motion of the organelle is very different during the two active phases of the beat, and consequently the motion of the surrounding water is also different. During the effective stroke the cilium swings around its basal anchorage (Fig. 3 b) and, because it remains nearly straight, all elements of its length remain approximately perpendicular to the direction of movement. During the recovery stroke the cilium "unrolls" as a bend travels towards the tip, drawing the distal part longitudinally through the water (Fig. 3 c). In the effective stroke the dominant forces are normal to the axis of the smooth-surfaced organelle and only the larger C_N is important, whereas in the recovery stroke the dominant forces are tangential to the axis and only the smaller C_T is important. There is also a difference in velocity in the two strokes. Typically the velocity in the effective stroke is higher than that in the recovery stroke, but in the *Mytilus* abfrontal cilium (Baba and Hiramoto 1970) and in walking hypotrichs (Machemer and Sugino 1986) beat patterns have been described in which the "recovery" stroke is of much shorter duration than the "effective" stroke. The difference in the forces exerted in the two strokes normally produces a net propulsion of water in the direction of the effective stroke, although the presence of the cell surface (see below) probably has a greater influence in determining the different extent of water movement produced by effective and recovery strokes.

Motion produced by flagellar activity

The propulsion of a cell by a flagellum or cilia

It will be simplest to consider first the case of a cell propelled by a planar beat of a smooth flagellum. The waves propagated along the flagellum exert forces on the water which are resisted by the water, so that a forward thrust is imparted to the cell. At constant velocity this thrust is balanced by viscous drag on the cell body (the net thrust is zero). Various parameters of flagellar size and activity, as well as the cell size, affect swimming speed and propulsive efficiency (Holwill 1966, Higdon 1979 a, Roberts 1981). The speed of propulsion is directly proportional to the speed with which waves are propagated along the flagellum (= frequency × the length of flagellum in each wave). At a given beat frequency and with a flagellar length of at least 10 times the body radius, the ratio of flagellar length to its radius has little effect on either swimming speed or propulsive efficiency; with a ratio of flagellar length to its radius of 100 or so, and a ratio of flagellar length to body radius of 10 or so (optimum between 20 and 40), the propulsive efficiency is greatest when the wavelength is between 4 and 6 times the amplitude. Larger cell bodies produce greater detrimental effects at high beat amplitudes because of interference of flow around the cell body; short flagella lose efficiency because of this interference, whilst very long flagella lose efficiency because of their increased drag. Power consumption is least when there is only one wave on the flagellum, but increases only slowly as the number of waves on the flagellum increases. These relationships are in general more readily observed with spermatozoa because smooth protistan flagella seldom show a simple sinusoidal beat (for example, the amplitude and/or the wavelength may increase towards the tip) and their flagella are often shorter than 10 times the body radius. Although flagella bearing mastigonemes show a planar beat, they appear to follow slightly different rules, but have not been examined so fully; they usually have several waves within their length.

For small values of amplitude/wavelength, propulsion by helical waves is in theory about twice as fast as by planar waves at the same beat frequency. The optimum flagellar length is therefore shorter for helical waves than for planar waves; this is because all elements of the helical flagellum contribute to forward propulsion whereas some elements of the planar flagellum produce only drag (Higdon 1979 c). Helical waves are also almost twice as efficient as planar waves, especially with a larger cell body (Holwill 1966, 1974). However, the propulsion of a cell by a helically beating flagellum is more complex because the flagellar waves produce a turning moment about the axis of propulsion which causes the cell body itself to rotate with a couple such that its motion opposes the turning due to the flagellum (Holwill 1966, Chwang and Wu 1971) (Fig. 4 a). In equilibrium swimming the turning moments of the cell body and the flagellum must balance, so a large body will turn slowly and a small body must turn quickly to balance the flagellar torque. Clearly, when the cell body rotates the flagellar axoneme must rotate around its internal axis in the same direction as the cell body turns, since there is no rotary joint at the flagellar base comparable to that of the prokaryote flagellum. The

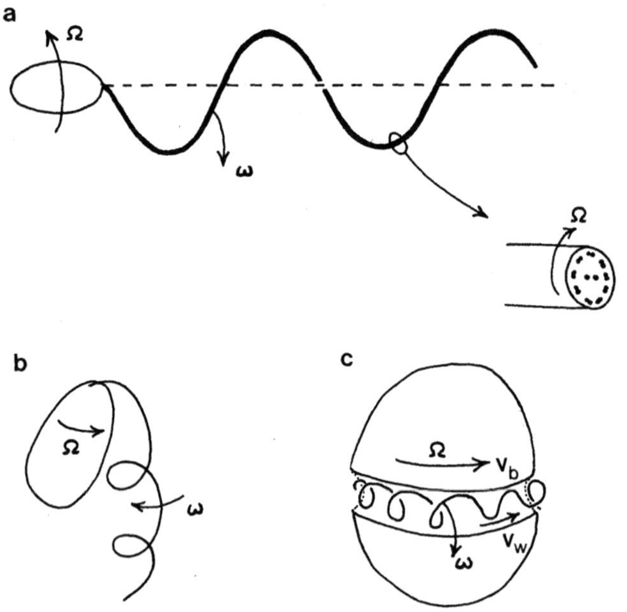

Fig. 4. Cells propelled by helical waves. **a** When the flagellum turns helically with angular velocity ω, the cell body turns in the opposite direction at an angular velocity Ω, so that the flagellum must also rotate around its own internal axis at an angular velocity Ω; the effective angular velocity of the flagellum is ω − Ω. **b** The flagellum of *Euglena* turns back through 180°; the effective angular velocity of the flagellum is here ω + Ω. **c** The body of a dinoflagellate turns at velocity V_b in the direction of propagation of the waves on the transverse flagellum at velocity V_W; this enhances the propulsive effect of the transverse flagellum

rotation of the cell body (at an angular velocity Ω) in the opposite direction to the helical motion of the flagellum (at an angular velocity ω) reduces the speed at which waves move relative to the water (to ω − Ω), and will reduce propulsive efficiency; the slower the body turns, the less the loss, but with a larger body more work must be done to propel the body forwards against the resistance of the water. On the other hand, if there is no cell body on a helical flagellum, one could theoretically approach the point where the flagellum will rotate around the axis of the flagellar helix at an angular velocity Ω, and no forward motion is seen − it will not even appear to propagate waves (Taylor 1952, Chwang and Wu 1971). The most efficient helical flagellar wave has a wavelength about 6 or 7 times the amplitude, and a pitch angle near 45°, although faster swimming speeds can be achieved (less economically) at lower wavelength/amplitude ratios. Optimal propulsion is achieved with a ratio of flagellar length to cell body radius of about 10 with a smaller cell body and about 12 when the cell body radius is 20 to 50 times the flagellar radius; the optimum number of waves increases as this ratio increases and optimum wavelength decreases as the ratio increases (Higdon 1979 c). The torque increases with flagellar length and with wave amplitude; increasing the number of waves on a given flagellum reduces amplitude, thereby reducing the torque and increasing the effective rotation of the flagellar helix. Thicker flagella tend to have an increased power consumption, but flagellar thickness has little influence on swimming speed.

Where larger cells are propelled, special mechanisms may be invoked, e.g., the flagellum of *Euglena* is turned back from the anterior end of the cell and propagates helical base-to-tip waves; because the flexible flagellum bends through 180° near its base, the rotation of the cell body is communicated to the flagellum so as to increase its helical motion and increase the speed with which waves move in relation to the water (Holwill 1966, Lighthill 1976) (Fig. 4 b). In dinoflagellates such as *Heterocapsa* the transverse flagellum clearly has a left-handed helix (Taylor 1987) (Fig. 4 c), and this helix must therefore turn anticlockwise (as seen from the flagellar tip) as waves are propagated towards the tip; the cell body rotates clockwise (seen from the posterior end) (Gaines and Taylor 1985), presumably because of flagellar torque, thereby increasing the effective rate of propagation of the flagellar helix. Since a helical wave can be more efficient in propulsion than a planar wave, it is not surprising to find that many flagellates and sperm whose flagellar undulations are basically planar show a tendency towards a three dimensional beat near the tip (Holwill 1966); this causes a slow rotation and aids directed locomotion by reducing effects of pitching and yawing.

The flagellar wave is less effective near the cell surface because water propulsion is restrained by the surrounding boundary layer within which the overlying water is progressively more difficult to move as one approaches the cell surface. While this effect is disadvantageous for propulsion by an undulating flagellum, it is exploited by organelles with a ciliary beat because it can accentuate the differences in water transport between the effective and recovery strokes. During the effective stroke the cilium is extended well away from the cell surface and can transport a large conical volume of water, but during the recovery stroke the cilium moves close to cell surface where the restraint of the boundary layer profoundly reduces the water transported to a narrow cylindrical zone (Blake and Sleigh 1974) (Fig. 5 a); this effect acts in addition to the dominance of the larger C_N in the effective stroke and the

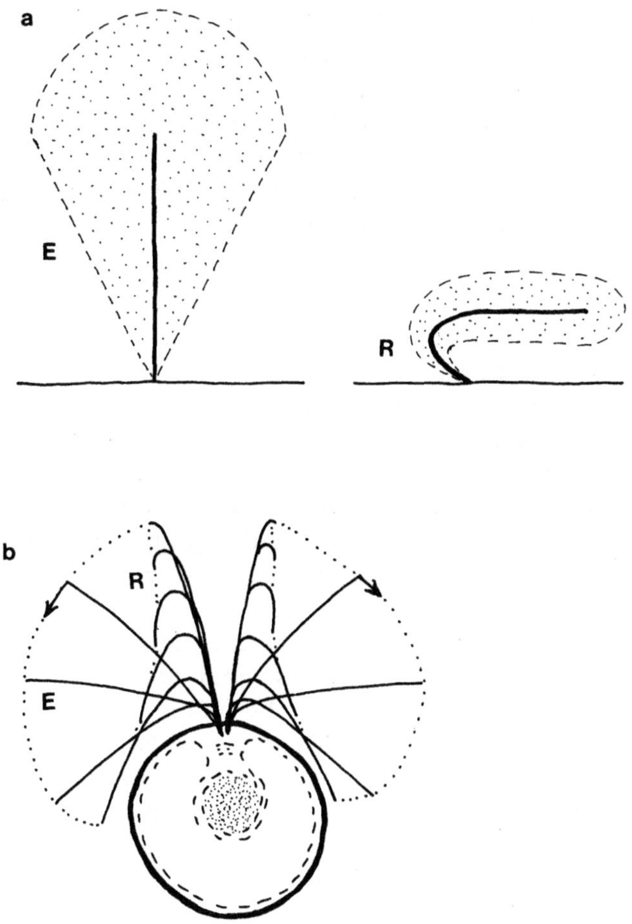

Fig. 5. Propulsion by ciliary beating. **a** The zone of water (stippled) carried around a cilium during its effective stroke E is much greater than that carried in the recovery stroke R. **b** When *Chlamydomonas* swims forward its two flagella perform ciliary beats with extended effective strokes E and recovery strokes R in which the flagella move close to the body

smaller C_T in the recovery stroke, and the differences in velocities in the two strokes, mentioned earlier.

A cell possessing a single ciliary organelle would surely merely rotate unless it was stabilised by some other structure. Normally organelles functioning as cilia are present in pairs or in larger numbers. *Chlamydomonas* is propelled forwards by approximately synchronous ciliary beats of its two flagella (Ringo 1967) (Fig. 5 b), the effective strokes swinging well out from the cell and achieving a large effect in propelling the cell forward, whilst in the recovery stroke the cilia move close to the cell surface, within the surrounding boundary layer, so that the backward movement of the cell caused by this stroke is thereby reduced. While the biflagellate *Chlamydomonas* moves in a jerky fashion, with longer forward effective steps alternating with shorter backward ones, the quadriflagellate *Polytomella*, which uses two opposite pairs of ciliary organelles in alternation, shows a smoother forward progression because the recovery strokes of one pair of flagella coincide with the effective strokes of the other pair. Where large numbers of ciliary organelles occur together, they beat in a close metachronal sequence which maintains an even smoother continuous forward movement (Sleigh 1989). A number of flagellates (e.g., some bodonids and euglenids) appear to be propelled by a single flagellum with a latero-posterior ciliary beat, but these forms also have a trailing (gliding?) flagellum and the relative roles of the two flagella are not known.

Although the ciliary beat pattern is most simply visualized as a cycle in which the cilium moves in one plane, it very seldom does so. During the recovery stroke, particularly, the cilium swings to one side or the other, gaining an advantage in propulsion because it can move even closer to the cell surface during this part of the beat and reduce the volume of fluid carried around the cilium in this stroke (Sleigh 1974, 1989). In ciliates, for example, Machemer and Sugino (1986) have shown by three-dimensional reconstructions that the cilia follow the outline of a cone, albeit sometimes rather compressed, particularly in the effective stroke; the frequency of cycling and the amplitude and inclination of the cone may be varied under membrane voltage control. The three-dimensional component of the motion has implications for the propulsion of water by the organelle, which is normally complicated by the fact that cilia operate in coordinated groups.

The propulsion of water around stationary cells

The flagella or cilia of cells that are anchored to a substratum circulate water around the cell. This water circulation is commonly used to bring food particles within reach of some food capturing structure. The forces generated by flagellar or ciliary action to produce these circulations are balanced by tensile or compressive forces at the anchorage point which determine the nature of the anchorage. The circulation of water created by an anchored cell is necessarily much more extensive than that produced around a swimming cell; for example, the local flows around a swimming flagellate (Fig. 6 a) usually extend less than a wavelength from the flagellum (velocity falls off exponentially with the distance) (Lighthill 1976), whilst attached flagellates (Fig. 6 b) or ciliates create flows extending for 10 or 20 body diameters around the organism (velocity falling off inversely with distance) (e.g., a collar fla-

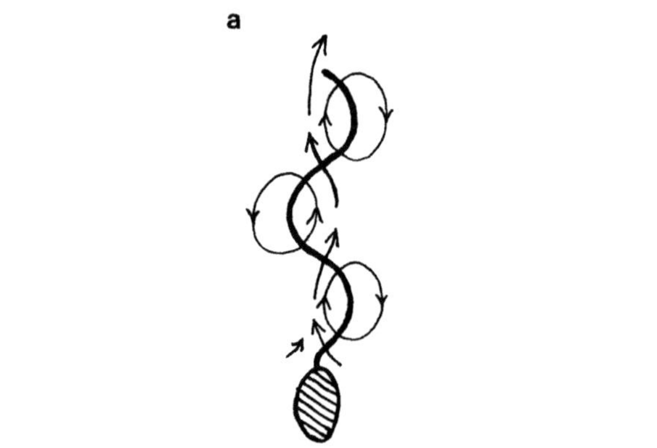

Fig. 6. The extent of water flow generated by flagella on **a** a swimming flagellate, and **b** a flagellate attached to the substratum (no attempt has been made to portray the local flows around individual flagellar waves in **b**)

gellate, Lapage 1925, Higdon 1979 b; ciliates, Sleigh and Aiello 1972, Barlow and Sleigh 1976). Attached flagellates always seem to show a more or less planar beat; the torsional resistance required in the attachment of a helically beating flagellum might be substantial. While slender strands are adequate to resist the tensions on attachments of flagellates bearing mastigonemes (e.g., the bicosoecid attached by its posterior flagellum), cylindrical stalks are necessary to resist compressive forces generated by the activity of smooth flagella (e.g., choanoflagellates). Functionally the use of the flagellum is somewhat modified in comparison with a swimming cell, with a lower optimum wavelength/amplitude ratio of 2.5 to 3 and a rather lower optimum flagellar length/body radius ratio of between 5 and 10; efficiency is improved if the flagellate cell is raised above the substratum to a height at least equal to the flagellar length (Higdon 1979 b). Ciliates are often also stalked or elongated to hold the cilia that create water flows well above the surface, although this depends on the orientation of the flow (Fenchel 1987).

Conclusions

There is a great variety in the detailed pattern of movement of the flagella of different protistan species; in fact they are so diverse that in some flagellate groups it is often more practicable to distinguish species from one another by their motility than by other features visible in the light microscope. This results partly from the diversities of body shape and flagellar appendages, and partly from specialisations of behaviour concerned not only with locomotion but also with the collection of food and other activities of the species concerned.

Although three basic patterns of movement have been described, the beat pattern observed may commonly include elements of more than one of the basic patterns; it may also vary with the rate of beating, and may therefore be regarded as being under cellular control. Clear examples are the fact that the beat patterns of many sperm tails are planar near the base and have a helical component near the tip, generating intermediate propulsive results (Holwill 1974), and the change in beat pattern of *Chlamydomonas* flagella from a ciliary form when swimming forward to an apparently planar flagellar form when swimming in reverse after appropriate stimulation. Minor changes in some of these features of the beat can produce substantial changes in movement. Changes in beat shape are understood to be produced by changes in the timing of active sliding within the axoneme, but their effects are modulated by the structure of the flagellum and features of the cell, such as size, shape, flagellar number and position on the cell.

Flagella are complex and highly adaptable organelles. Features of the structure and function of both the axoneme and the membrane can be varied to determine the specific features of motile behaviour displayed by a flagellated cell.

References

Baba SA, Hiramoto Y (1970) A quantitative analysis of ciliary movement by means of high speed microcinematography. J Exp Biol 52: 675–690

Blake JR, Sleigh MA (1974) Mechanics of ciliary locomotion. Biol Rev 49: 85–125

Bloodgood RA (1990) Gliding motility and flagellar glycoprotein dynamics in *Chlamydomonas*. In: Bloodgood RA (ed) Ciliary and flagellar membranes. Plenum, New York, pp 91–128

Brokaw CJ (1965) Non-sinusoidal bending waves of sperm flagella. J Exp Biol 43: 155–169

Chwang AT, Wu TY (1971) A note on the helical movement of micro-organisms. Proc Roy Soc Lond [Biol] 178: 327–346

Fenchel T (1987) Ecology of protozoa. Science Tech, Madison, Wisconsin/Springer, Berlin Heidelberg New York Tokyo

Gaines G, Taylor FJR (1985) Form and function of the dinoflagellate transverse flagellum. J Protozool 32: 290–296

Gray J, Hancock GJ (1955) The propulsion of sea-urchin spermatozoa. J Exp Biol 32: 802–814

Hancock GJ (1953) The self-propulsion of microscopic organisms through liquids. Proc Roy Soc Lond [A] 217: 96–121

Higdon JJL (1979 a) A hydrodynamic analysis of flagellar propulsion. J Fluid Mech 90: 685–711

– (1979 b) The generation of feeding currents by flagellar motions. J Fluid Mech 94: 305–330

– (1979 c) The hydrodynamics of flagellar propulsion: helical waves. J Fluid Mech 94: 331–351

Holwill MEJ (1966) Physical aspects of flagellar movement. Physiol Rev 46: 695–785

– (1974) Hydrodynamic aspects of ciliary and flagellar movement. In: Sleigh MA (ed) Cilia and flagella. Academic Press, London, pp 143–175

– Sleigh MA (1967) Propulsion by hispid flagella. J Exp Biol 47: 267–276

Lapage G (1925) Notes on the choanoflagellate *Codosiga* Ehrbg. Q J Microsc Sci 69: 471–508

Lighthill J (1976) Flagellar hydrodynamics. SIAM Rev 18: 161–230

Machemer H, Sugino K (1986) Parameters of the ciliary cycle under membrane voltage control. Cell Motil Cytoskeleton 6: 89–95

Ringo DL (1967) Flagellar motion and fine structure of the flagellar apparatus in *Chlamydomonas*. J Cell Biol 33: 543–571

Roberts AM (1981) Hydrodynamics of protozoan swimming. In: Levandowsky M, Hutner SH (eds) Biochemistry and physiology of protozoa, vol 4, 2nd edn. Academic Press, New York, pp 6–66

Satir P, Sleigh MA (1990) The physiology of cilia and mucociliary interactions. Annu Rev Physiol 52: 137–155

Shapiro AH (1961) Shape and flow: the fluid dynamics of drag. Doubleday, Garden City, NY

Silvester NR, Holwill MEJ (1972) An analysis of hypothetical flagellar waveforms. J Theor Biol 35: 505–523

Sleigh MA (1974) Patterns of movement of cilia and flagella. In: Sleigh MA (ed) Cilia and flagella. Academic Press, London, pp 79–92

Sleigh MA (1981) Flagellar beat patterns and their possible evolution. BioSystems 14: 423–431

– (1989) Ciliary propulsion in protozoa. Sci Prog (Oxf) 73: 317–332

– Aiello E (1972) The movement of water by cilia. Acta Protozool 11: 265–277

– Barlow D (1976) Collection of food by *Vorticella*. Trans Am Microsc Soc 95: 482–486

– Barlow D (1982) How are different ciliary beat patterns produced? Symp Soc Exp Biol 35: 139–157

Sugino K, Naitoh Y (1982) Simulated cross-bridge patterns corresponding to ciliary beating in *Paramecium*. Nature 295: 609–611

Taylor FJR (1987) The biology of dinoflagellates. Blackwell, Oxford

Taylor GI (1952) The action of waving cylindrical tails in propelling microscopic organisms. Proc Roy Soc Lond [A] 211: 225–239

Protoplasma (1991) 164: 54–69

PROTOPLASMA

High-speed video analysis of the flagellar beat and swimming patterns of algae: possible evolutionary trends in green algae

I. Inouye* and **T. Hori**

Institute of Biological Sciences, University of Tsukuba, Tsukuba, Ibaraki

Received September 25, 1990
Accepted April 23, 1991

Summary. The flagellar beat and swimming patterns of flagellated cells of 22 green plants, including 17 green flagellates (volvocalean and prasinophyte algae), motile cells of three seaweeds, *Bryopsis*, *Caulerpa*, and *Ulva*, sperms of a liverwort, *Marchantia*, and a fern, *Athyrium*, were examined using a high-speed video system. So-called breast-stroke is widely distributed in green plants, and occurs rarely in prasinophyte flagellates and ulvophycean algae; in these algal groups flagellar beat similar to that found in animal sperm is common, both during forward and backward swimming. Different types of swimming patterns were observed in prasinophytes. The results indicate evolutionary trends of flagellar beat and swimming patterns in green plants such as change from backward to forward swimming, from flagellar to ciliary beating and from uni-directional (parallel) to radial-directional (cruciate) beating. Such trends are shown in two prasinophyte groups, the *Pyramimonas*-lineage and *Tetraselmis*-lineage.

Keywords: Green algae; Flagellar beat; Video analysis; Evolution; Swimming patterns; Prasinophyceae.

Introduction

Flagellar movement has been studied mainly using flagella and/or cilia of animal cells, with the aim of understanding the molecular mechanism behind flagellar movement. Comparatively little is known about flagellar movement in phytoflagellates and reproductive plant cells (zoospores and gametes), except in one genus of the Chlorophyceae, *Chlamydomonas* (e.g., Ringo 1967; Hyams and Borisy 1978; Brokaw et al. 1982; Brokaw and Luck 1983; Omoto and Brokaw 1985; Kamiya and Hasegawa 1987; Rüffer and Nultsch 1985, 1987) and one genus of the Chrysophyceae, *Ochromonas* (Sleigh 1964, Holwill and Sleigh 1967). Based on the studies of *Chlamydomonas*, flagellar movement of green algae is generally thought to be what is described as "breast stroke", viz, cells swim forward with flagella beating like cilia (e.g., Sleigh 1981). We now know that the green plants consist of at least four different taxonomic groups at class rank, the Charophyceae (for convenience, land plants are included in this group), Chlorophyceae, Ulvophyceae, and Prasinophyceae (e.g., Stewart and Mattox 1978; Moestrup 1978; Moestrup and Ettl 1979; Melkonian 1982 a, 1984; O'Kelly and Floyd 1984). Further, many green flagellates do not exhibit breast stroke: some green flagellates swim with flagella forming undulatory waves and beating behind (e.g., *Scourfieldia*, Belcher 1964, Melkonian and Preisig 1982; *Pseudoscourfieldia*, Moestrup and Throndsen 1988; *Pedinomonas*, Melkonian 1990 b; genera of the Mamiellales and *Nephroselmis* (long flagellum), Melkonian 1990 a; *Pterosperma* and *Pachysphaera*, Parke and Adams 1961, Parke 1966; for review, see Melkonian and Robenek 1984, Melkonian 1990 a). The prasinophyte algae are believed to be an ancestral stock of green plants and the algae involved are morphologically very diverse compared to other groups of green plants (e.g., Norris 1980; Melkonian 1984, 1990 a; Moestrup and Throndsen 1988). Fossil records and a study of molecular phylogeny supports this view: fossils assignable to phycomata of the Pterospermataceae have been found in Precambrian sediments (Boalch and Parke 1971, Tappan 1980), and a recent analysis of 5S rRNA sequence of *Mamiella* indicates that prasinophytes branched off near the base of the evolutionary tree of green plants (Hori et al. 1990). Possible primitiveness of undulatory-type flagellar movement in

* Correspondence and reprints: Institute of Biological Sciences, University of Tsukuba, Tsukuba, Ibaraki 305, Japan.

green flagellates has been suggested (Melkonian 1983, Melkonian and Robenek 1984), and since most "non-breast stroke"-using green algae belong to the Prasinophyceae, evolutionary changes of swimming patterns must have taken place in this algal group. Prasinophytes may therefore have retained primitive or ancestral swimming and flagellar movement patterns. In this paper we present results of a general survey of various green plants examined with high-speed video, with particular emphasis on prasinophyte flagellates.

The prasinophytes belong to the "*Pyramimonas*-lineage" and "*Tetraselmis*-lineage". The *Pyramimonas*-lineage includes scale-covered quadriflagellate prasinophytes, *Pyramimonas, Halosphaera, Cymbomonas, Pachysphaera*, and *Pterosperma*. Homologous configuration of flagellar apparatus components has been demonstrated for *Pyramimonas, Halosphaera*, and *Pterosperma* (Inouye et al. 1990), indicating that these genera have a common ancestor. The recently rediscovered prasinophyte *Cymbomonas tetramitiformis* (Throndsen 1988) and a new prasinophyte genus *Prasinopapilla* will also be included in the *Pyramimonas*-lineage because they share many features with *Pyramimonas* and/or *Pterosperma*. For example, *Cymbomonas* (Throndsen 1988, Moestrup and Throndsen 1988) and *Prasinopapilla* (Inouye et al. in prep.) have limuloid scales and a scale duct system associated with the 1d flagellar rootlet sensu Moestrup and Hori (1989). A formal description of *Prasinopapilla* will be published separately (presently nomen nudum). Three genera of the *Tetraselmis*-lineage have been examined, *Pseudoscourfieldia, Nephroselmis*, and *Tetraselmis*.

Materials and methods

The following algae were used: *Brachiomonas submarina* (NIES: National Institute for Environmental Studies), *Bryopsis maxima* (Choshi, Chiba), *Carteria radiosa* (NIES), *Caulerpa racemosa* var. *clavife* f. *microphysa* (Okinawa), *Chlamydomonas* spp. (natural samples and culture strains collected from various localities), *Chlorogonium* sp. (Tsukuba, Ibaraki), *Cymbomonas tetramitiformis* (Shodoshima Isl. Kagawa), *Hafniomonas montana* (NIES), *Nephroselmis astigmatica* (Tokyo Bay, Tokyo), *N. olivacea* (NIES), *N.* aff. *rotunda* (Seto Inland Sea, Osaka), *Pedinomonas minor* (NIES), *Prasinopapilla vacuolata* nomen nudum (Tokyo Bay, Tokyo), *Pterosperma cristatum* (Tokyo Bay, Tokyo), *Pyramimonas parkeae* (Hachijo Isl. Tokyo), *Sphaerellopsis* sp. (Tsukuba, Ibaraki), *Spermatozopsis* sp. (Tsukuba, Ibaraki), *Tetraselmis* sp. (Iriomonte Isl. Okinawa), and *Ulva pertusa* (Shimoda, Shizuoka).

Sperms of a fern, *Athyrium nipponicum*, a liverwort, *Marchantia polymorpha*, and two gymnosperms, *Cycas revoluta* and *Gingko biloba*, were also examined. Spores collected from a sporophyte of *Athyrium nipponicum* collected in Tsukuba were inoculated on 0.8% agar plates containing modified Knop's medium (Pringsheim 1946) and prothalli were obtained one month after inoculation. Prothalli were transferred to a drop of distilled water and released sperms were studied. *Marchantia polymorpha* thalli possessing male receptacles were collected in Tsukuba, Ibaraki. The male receptacles were removed and placed upside down on a drop of distilled water. Sperms were then released.

A high-speed video system (Nac HMS-200) attached to a Nikon Optiphoto light microscope was used. This is capable of recording 200 frames per second. The video recorder was used in the negative mode and ordinary VHS video tapes used for recording. Still images of sequential frames on the monitor display were photographed with a single-lens-reflex camera with a black hood attachment. Neopan F (ISO 32) or Neopan 400 (ISO 400) films were used.

Results

In this paper, we use "ciliary beat" to indicate flagellar movement consisting of two different beat phases (effective and recovery strokes), and "flagellar beat" as flagellar movement forming undulatory waves that passes from base to tip.

The species examined fall into six different categories, based on the swimming patterns and mode of flagellar movement (Fig. 1). All results obtained are summarized in Fig. 1, and only selected species are described below.

Pyramimonas parkeae

Cells of *Pyramimonas* swim forward (e.g., Belcher 1969) and the flagellar movement is "cilia-type" beat and oppositely beating flagella form an angle of ca. 90° to each other (cruciate) (Melkonian and Robenek 1984, Melkonian 1990 a). *Pyramimonas parkeae* swims with the flagella in front of the cell and the cell body rotates around its longitudinal axis. Four flagella arise from a flagellar pit and extend in a cruciate manner. Figure 2 shows sequential movie frames of a cell swimming straight and very fast. Opposed flagella beat symmetrically and there is a phase difference between two opposed flagellar pairs (e.g., Fig. 2 c). A single beat cycle is shown in Fig. 3. The cell in this figure was caught under the coverslip by evaporation of the water. Flagellar beat is slower than in normally swimming cells and the phase difference between the two pairs had been disturbed. The flagella were still beating fast and a biphasic beat cycle is clearly visible. Figure 24 shows superimposed sequential waveforms. It is apparent that flagellar beat consists of effective and recovery strokes ("ciliary beat"). During the effective stroke, the flagella bend near the base and in the middle part (Fig. 3 f). During recovery stroke a wave is produced near the base (Fig. 3 h) and propagated along the flagellum toward the tip, resulting in the flagellum extending forward (Fig. 3 a–e, see also Fig. 24). By comparing Fig. 3 and Fig. 2, the beat cycle can be assessed.

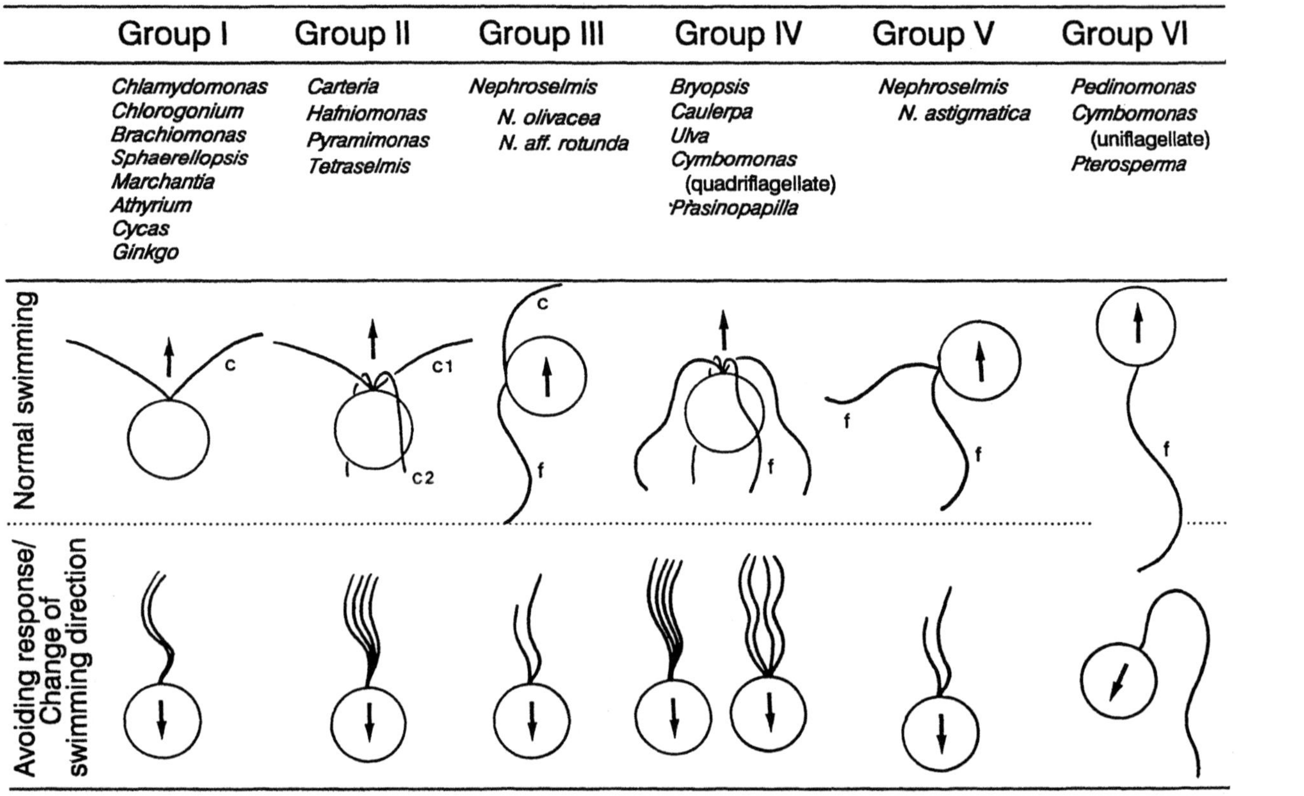

Fig. 1. Flagellar beat and swimming patterns of green plants may be categorized into six different groups. Arrows indicate direction of swimming. Ciliary beating is widely used for normal swimming in chlorophycean flagellates and sperms of land plants (Groups I and II). "Flagellar beat" is used in all the genera examined as normal swimming or an avoiding response. Two species of *Nephroselmis* (Group III) use both "ciliary" and "flagellar beat" for forward swimming. Algae placed in Groups IV, V, and VI use "flagellar beat" for normal swimming but differ in the direction of swimming. *c* "Ciliary beat"; *c 1* and *c 2* "ciliary beat" with ca. half beat cycle phase difference; *f* "flagellar beat"

It is clear that when one pair of flagella is in effective stroke, the other is in recovery stroke (Fig. 2 a, c, e, h, and i), indicating that phase difference between two opposed flagellar pairs is about half beat cycle (180°). Avoiding response was often observed, and occurred spontaneously. Cells suddenly stopped and threw the flagella forward, then swam backward beating flagella behind as has been observed in *Pyramimonas tetrarhynchus* (Belcher 1969) (Fig. 4). This phenomenon lasts very short. The flagella show undulate waveforms (Fig. 4), and beat apart from one another for several milliseconds. They gradually become located close to one another, resulting in a uni-directional ("parallel") beating. This behavior is similar to that of *Chlamydomonas* (Schmidt and Eckert 1976). *Pyramimonas* falls in Group II together with *Hafniomonas, Carteria*, and *Tetraselmis* (see below).

Pterosperma cristatum

This organism is characterized by four very long flagella. The swimming mode of *Pterosperma* and the allied genus *Pachysphaera* is backward swimming with undulatory beat of flagella (Parke and Adams 1961, Parke 1966, Melkonian 1990 a). The four flagella

Fig. 2. Forward swimming of *Pyramimonas parkae*. Note that opposed flagella show symmetrical wave forms. The two pairs are out of phase: when one flagellar pair is in recovery stroke the other is in effective stroke (e.g., **c**, **d**, **h**, **j**). Flagella in effective stroke (*E*) and those in recovery stroke (*R*). Arrow denotes direction of swimming. All movie frames in Figs. 2–23 were taken at 5 msec intervals unless otherwise noted

Figs. 3 and **4.** *Pyramimonas parkae*

Fig. 3. One beat cycle of a flagellum, consisting of recovery stroke (**a–e**) and effective stroke (**f–h**)

Fig. 4. Avoiding response. Arrow denotes swimming direction

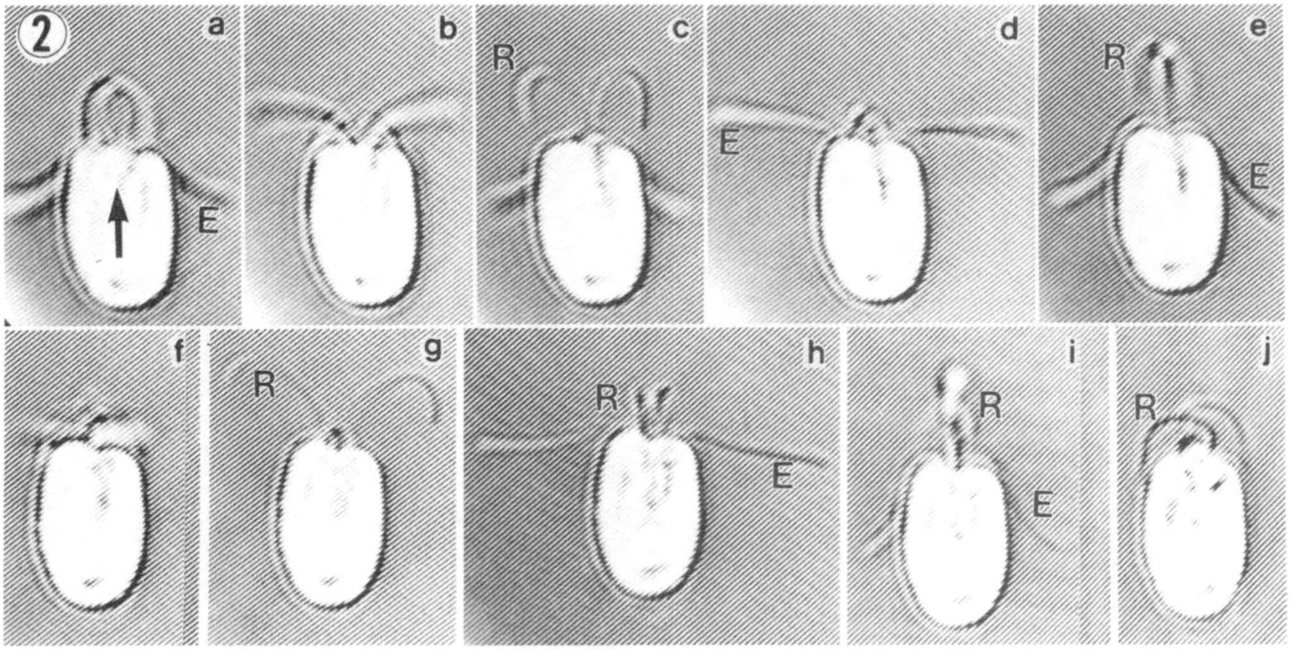
2
a
E
b
c
R
d
E
e
R
E
f
g
R
h
R
E
i
R
E
j
R

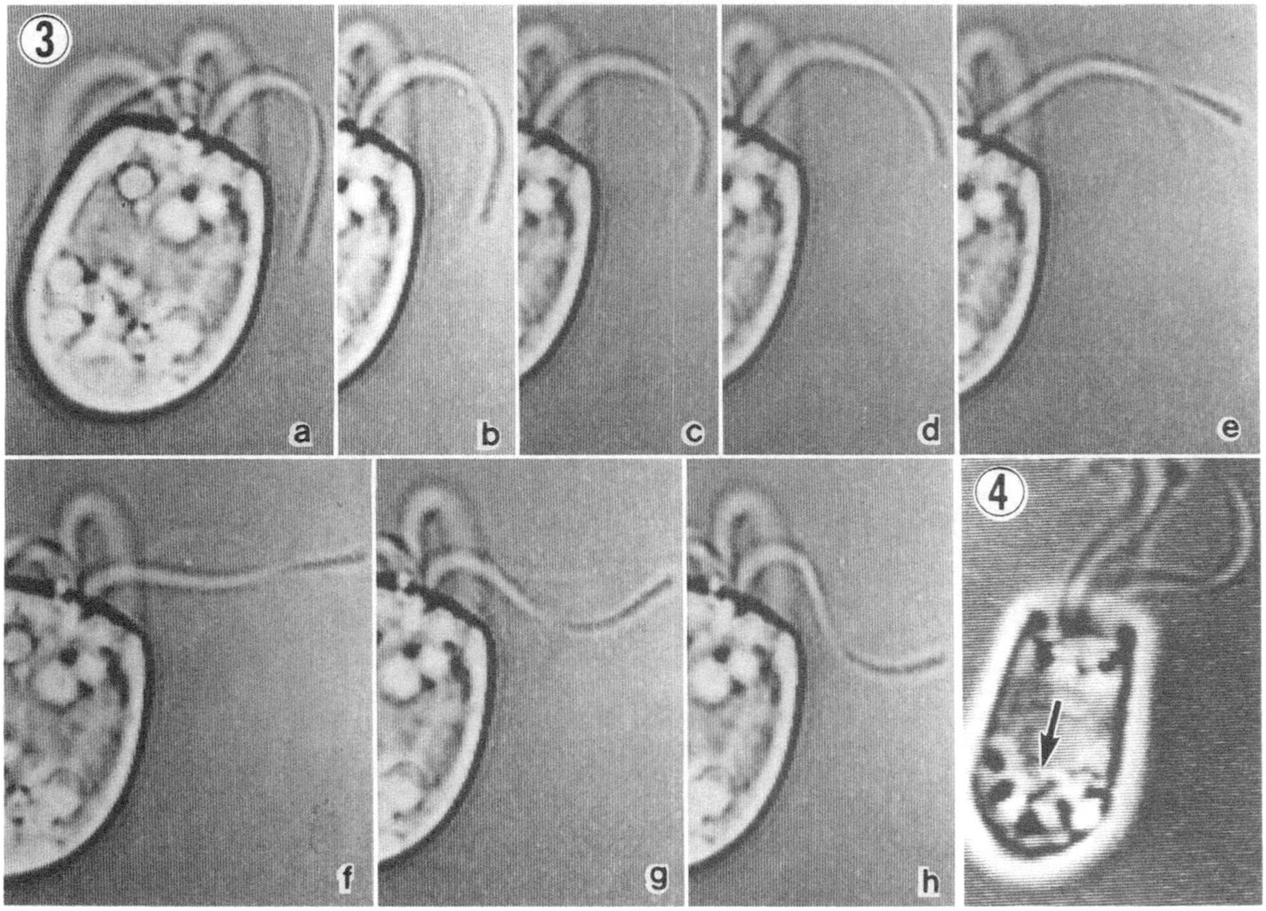
3
a
b
c
d
e
f
g
h
4

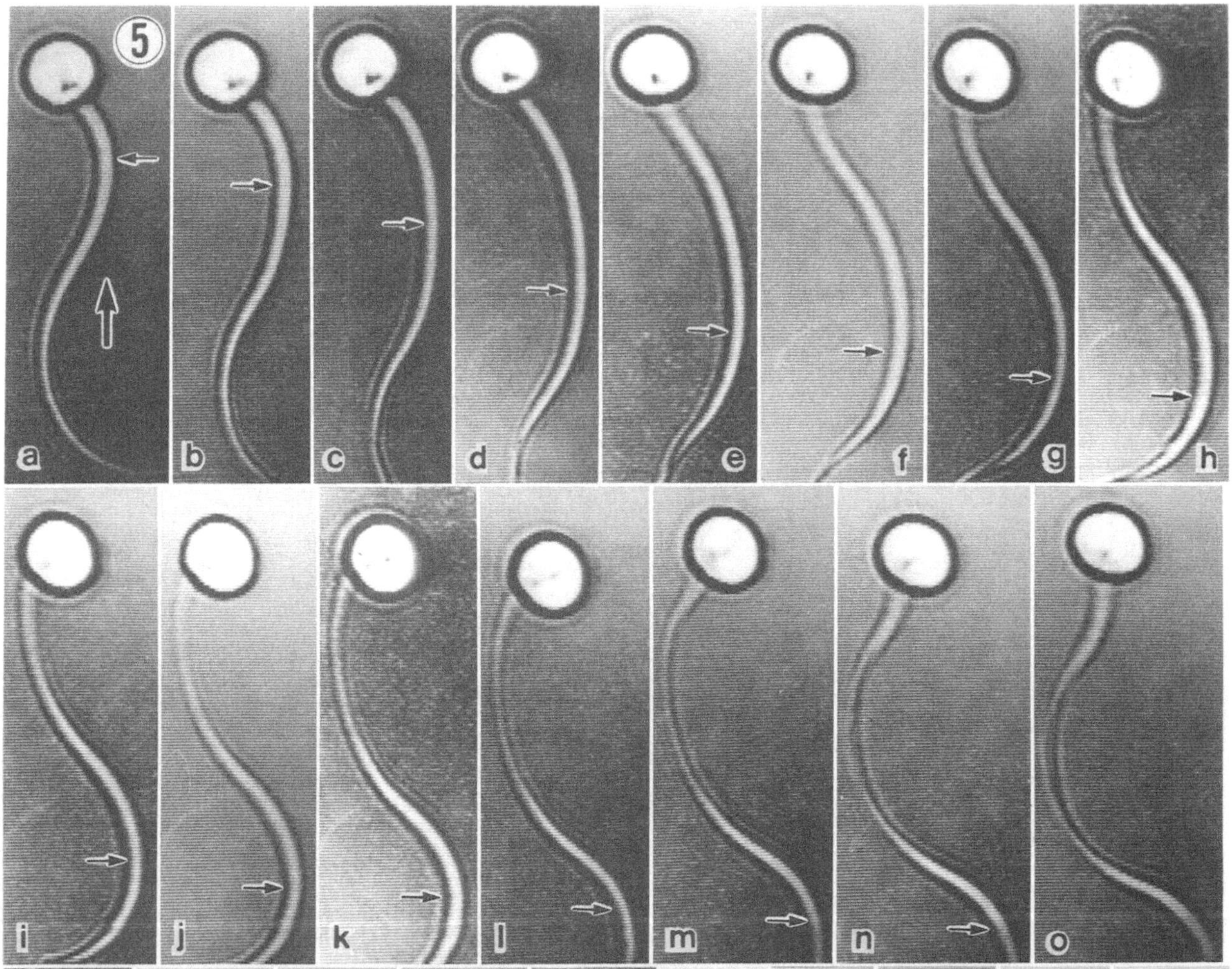

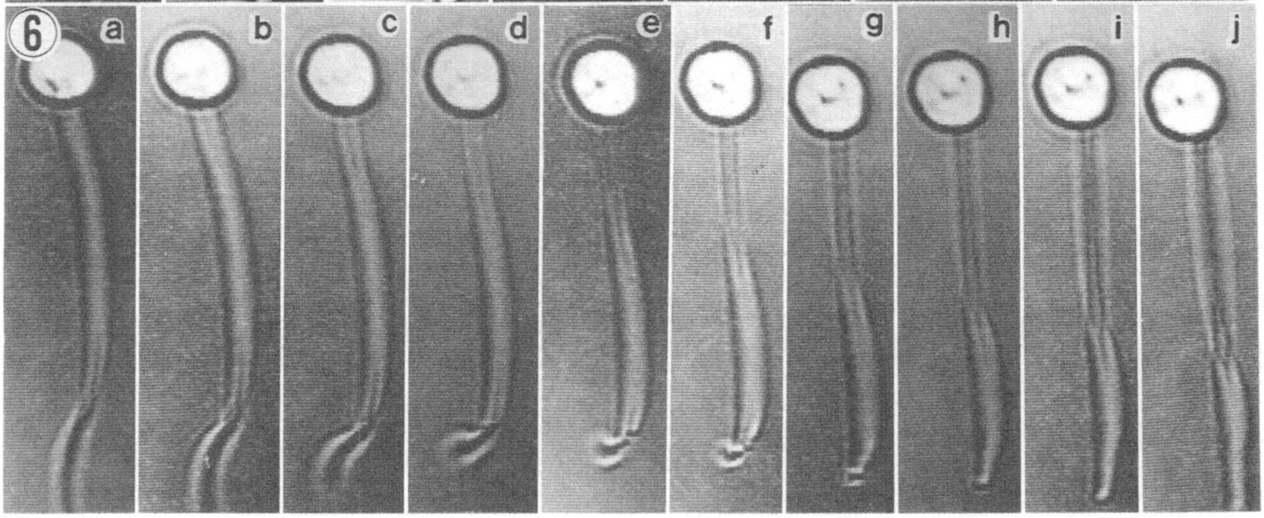

Figs. 5 and **6.** *Pterosperma cristatum* (same cell)

Fig. 5 a–o. Cell showing one beat cycle during normal backward swimming. The four flagella beat in exactly synchronized waves so that in micrographs four flagella are aligned on top of each other, resulting in an appearance like single flagellum. A wave generated at the base is propagated toward the tip (small arrows)

Fig. 6 a–j. Same cell as that in Fig. 5 but rotated approx. 90°. Plane of flagellar beat is almost perpendicular to the page. Profile of flagella in the micrographs is wider compared with Fig. 5, indicating that they are aligned side by side. Beat is uni-directional

emerge diagonally from the cell and are aligned in a row forming a compound flagellum (Figs. 5 and 6). There are, however, no visible structural connections between the flagella (Inouye et al. 1990). This is also clear in resting cells where the four flagella extend into different directions (Fig. 8). Figure 5 shows flagellar beat and swimming patterns. The four flagella act like one and beat synchronously in the same direction. Beat direction is perpendicular to the plane of the flagella (Fig. 6). The flagellar beat is "flagella-like" rather than "cilia-like". A wave is produced at the base and propagated along the flagella toward the tip (Fig. 5). No biphasic beat pattern was observed. A change in swimming direction was initiated by synchronized bend at the proximal part of the flagella (Fig. 7). This kind of motion often produces a circular movement of the cell. *Pterosperma* shows only backward swimming and thus falls into Group VI. *Pedinomonas* and the uniflagellate phase of *Cymbomonas tetramitiformis* (see below) also belong to this group (Fig. 1) probably together with *Scourfieldia* (Belcher 1964) and *Pseudoscourfieldia* (Moestrup and Throndsen 1988).

Prasinopapilla vacuolata

This organism swims by forward swimming motion (Fig. 9). There are, however, significant differences between forward swimming of *Prasinopapilla* and *Pyramimonas*. During forward swimming *Prasinopapilla* always swings its flagella back along the cell body with the flagellar tips pointing backward. The beat pattern is typical "flagellar beat" with undulatory waves produced at the base being propagated toward the tip of flagella (Figs. 9 and 24). Forward swimming generated by "flagellar beat" instead of "ciliary beat" has not been observed in green plants, but it is a common method of swimming in haptophyte algae (Sleigh 1981). Among the green algae, *Prasinopapilla, Cymbomonas*, and reproductive cells of ulvophycean algae (to be de-

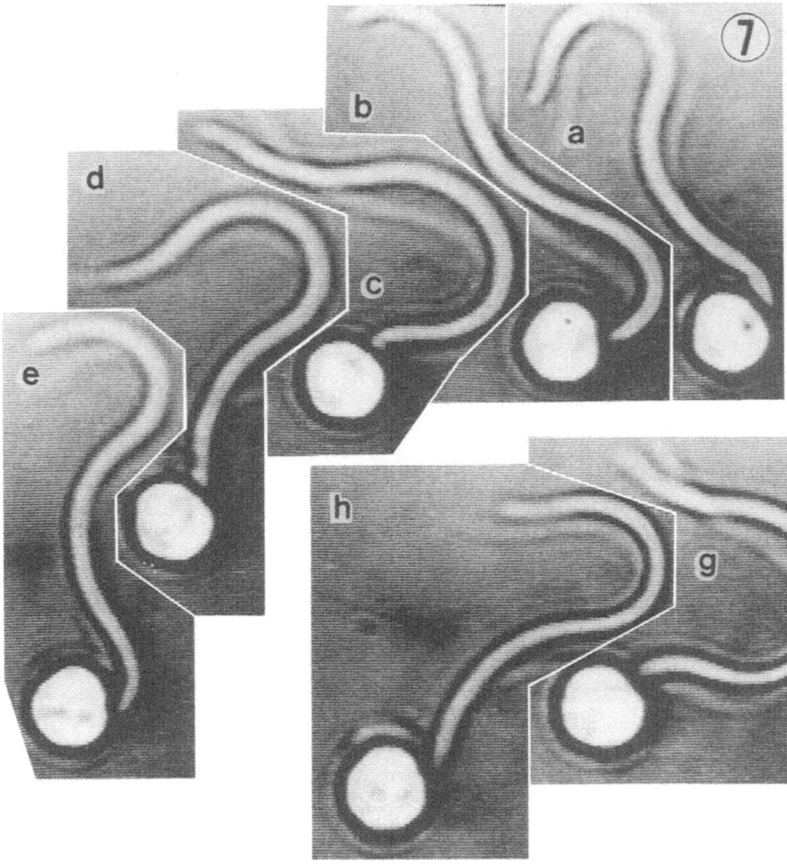

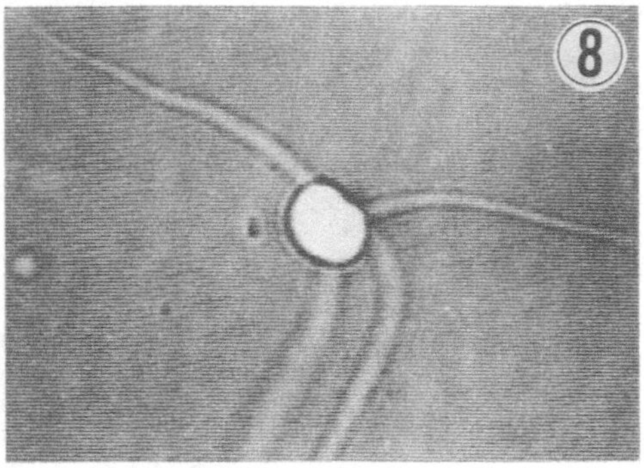

Figs. 7 and **8.** *Pterosperma cristatum*

Fig. 7 a–k. A cell changing swimming direction

Fig. 8. A cell drifting without beating flagella

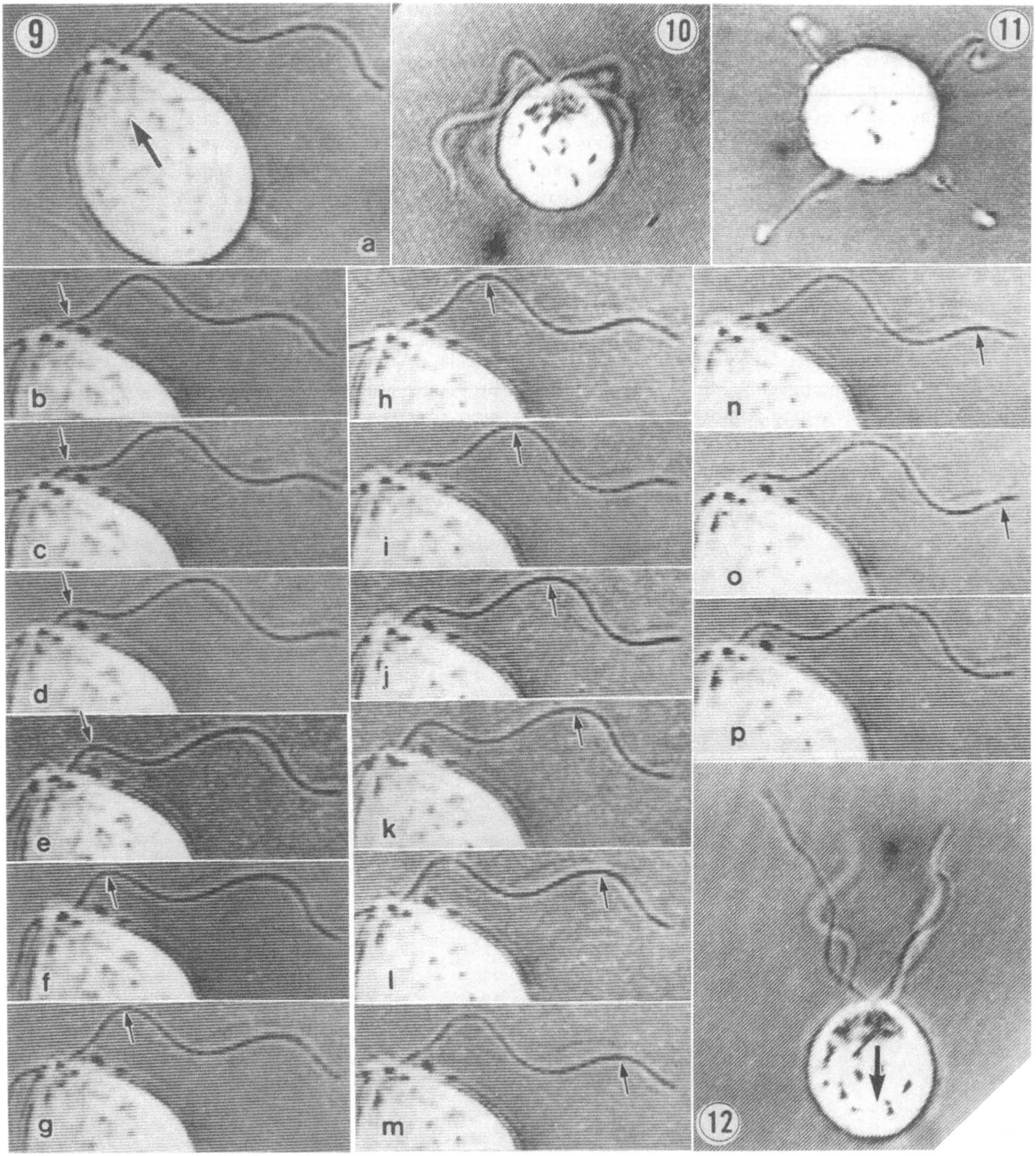

Figs. 9–12. *Prasinopapilla vacuolata*

Fig. 9 a–p. Overall appearance of a cell in normal forward swimming (**a**) with successive frames (**b–m**) showing "flagellar beat". Large arrow denotes direction of swimming. Small arrows indicate positions of propagated wave toward the tip

Fig. 10. Cell swimming forward. The four flagellar show symmetrical waveforms

Fig. 11. Cell viewed from the posterior side of the cell, showing cruciate profile of the flagella

Fig. 12. Cell in avoiding response. Arrow denotes direction of swimming

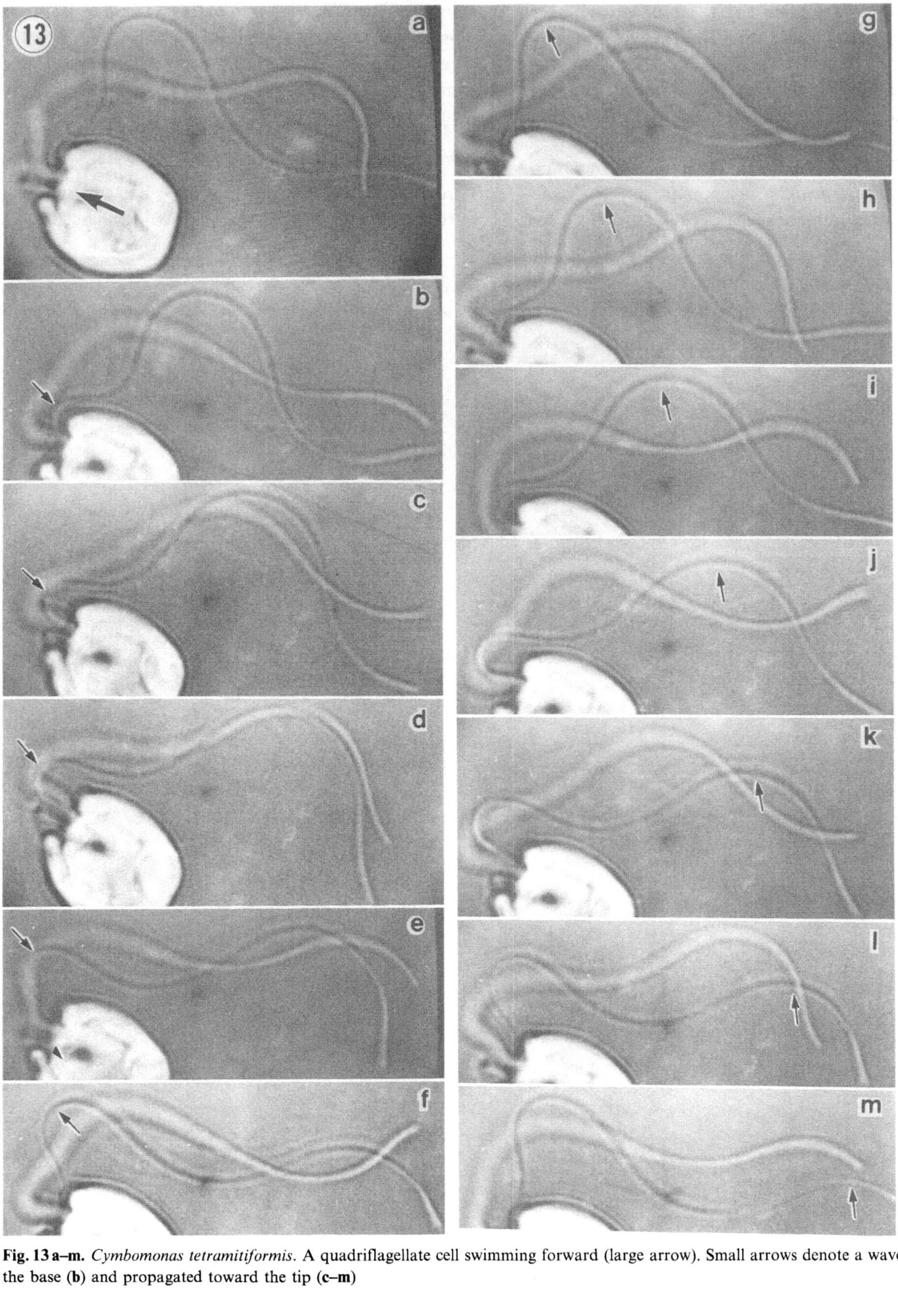

Fig. 13 a–m. *Cymbomonas tetramitiformis*. A quadriflagellate cell swimming forward (large arrow). Small arrows denote a wave produced at the base (**b**) and propagated toward the tip (**c–m**)

scribed below) are examples of this unusual swimming motion, and these algae form Group IV (Fig. 1). The direction of the flagellar beat is radial, that is, the four flagella extend in a cruciate pattern when viewed from anterior side of the cell (Fig. 11). There seems to be no phase difference between the beat of the four flagella (Fig. 10). This is another difference in the swimming patterns of *Prasinopapilla* in comparison to *Pyramimonas* and other quadriflagellates such as *Carteria* and *Hafniomonas* (Fig. 1). An avoiding response occurred spontaneously during our observations. The flagella were thrown forward and beat behind the cell (Fig. 12). During the avoiding response the flagella are out of synchrony.

Cymbomonas tetramitiformis

Throndsen (1988) reported that the cells swim by backward swimming, beating flagella behind the cell. Our

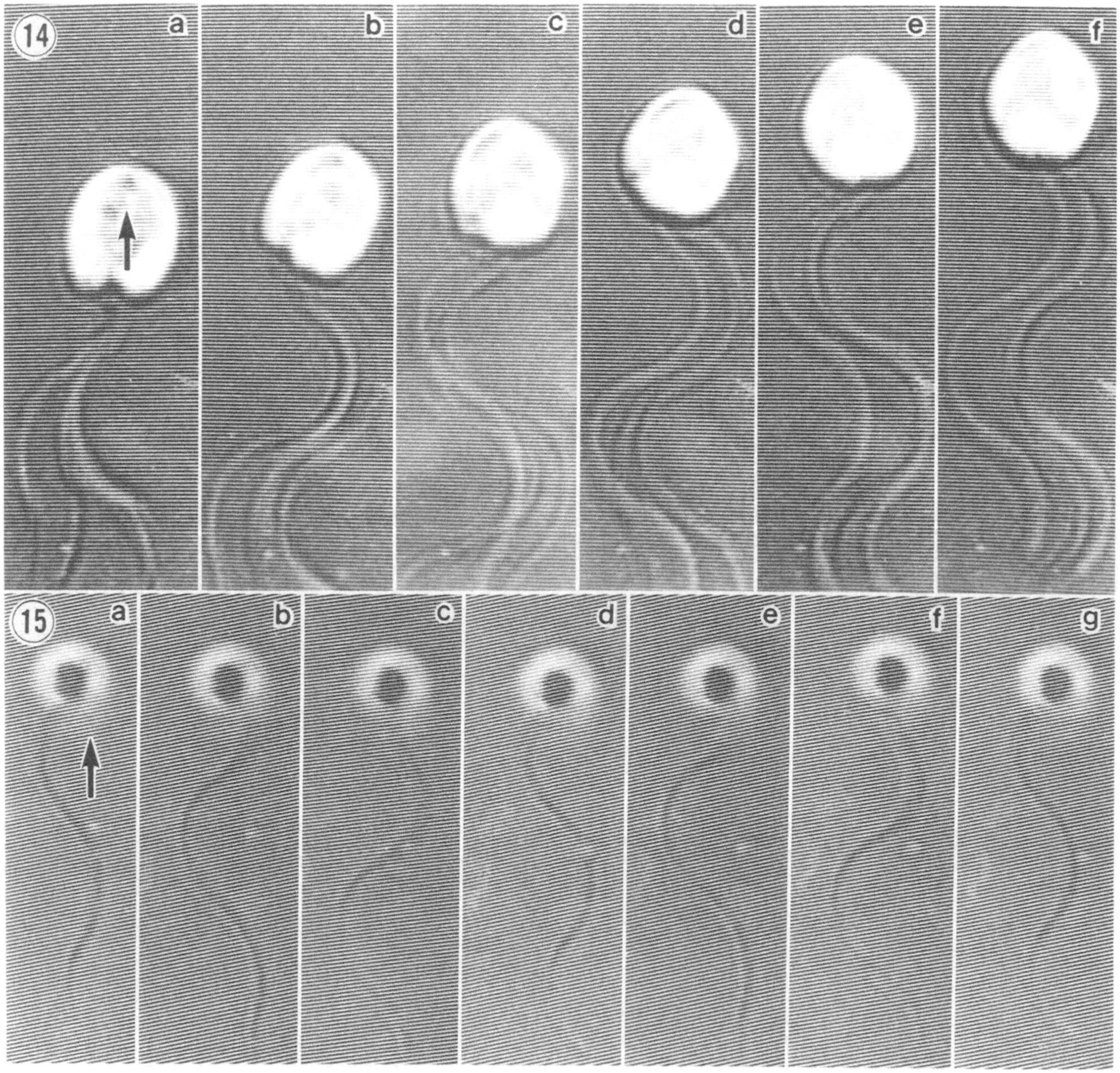

Figs. 14 and **15.** *Chlamydomonas tetramitiformis*

Fig. 14 a–f. A quadriflagellate cell swimming backward (arrow). The four flagella show essentially the same waveforms. Frames taken at 20 msec interval

Fig. 15 a–g. A uniflagellate cell swimming backward (arrow)

observations, however, revealed that the cells are capable of both backward and forward swimming, both being normal methods of swimming (Fig. 13). *Cymbomonas* is the only example of such an organism among the green algae so far examined. Forward and backward swimming are reversible spontaneously, but forward swimming seems to be more frequent. The movement of flagella during forward swimming is "flagellar beat" as in *Prasinopapilla* (Figs. 13 and 24). During the first 30 or 40 msec the wave is propagated forward in front of the cell (Fig. 13 b–f). The wave then moves outward and finally backwards along the cell (Fig. 13 g–m). This resembles the early stage of the recovery stroke of "ciliary beat" (e.g., see Fig. 2 a), but no biphasic cycle was observed. The beat cycle is synchronized in the four flagella and the direction of the flagella is radial (cruciate).

Backward swimming is by "flagellar beat" of the four flagella, which beat synchronously and uni-directionally (Fig. 14).

Cymbomonas may produce small uniflagellate cells in unialgal cultures. It is not clear what triggers formation of these cells which were produced in most strains we obtained. A detailed description of the uniflagellate cells will be given elsewhere. Only backward swimming was observed in the uniflagellate cells (Fig. 1) and it was generated by "flagellar beat" (Fig. 15). Change of swimming direction was done as in *Pterosperma* and animal sperms (Fig. 16).

Nephroselmis astigmatica, N. aff. rotunda, and N. olivacea

Three species of *Nephroselmis*, a genus of the *Tetraselmis*-lineage, were examined. This genus is characterized by the possession of two laterally inserted unequal flagella. In species so far examined, the short flagellum is directed forward and the other backward, and their heterodynamic movement was suggested (Manton et al. 1965; Moestrup and Ettl 1979; Melkonian and Robenek 1984; Moestrup 1983; Melkonian 1984, 1990 a). All three species examined here showed the long flagellum beating in "flagellar beat" (Figs. 17 and 18). By contrast, the beat pattern of the short flagellum is different between species, that is, both "flagellar beat" and "ciliary beat" were observed (Fig. 1). The short flagellum of *N. astigmatica* shows typical "flagellar beat" (Fig. 18) and the beat cycle and beat phase closely resembled those of the long flagellum (Fig. 18). The short flagellum of *N.* aff. *rotunda* and *N. olivacea* showed "ciliary beat" consisting of effective and recovery strokes (Fig. 17). Coexistence of "flagellar" and "ciliary beat" in a single organism is unusual and has not been reported before. All three species showed ordinary avoiding response, characterized by synchronized "flagellar beat" (Fig. 19).

Tetraselmis sp.

This genus may be considered one of the most advanced forms of the *Tetraselmis*-lineage together with *Scherf-*

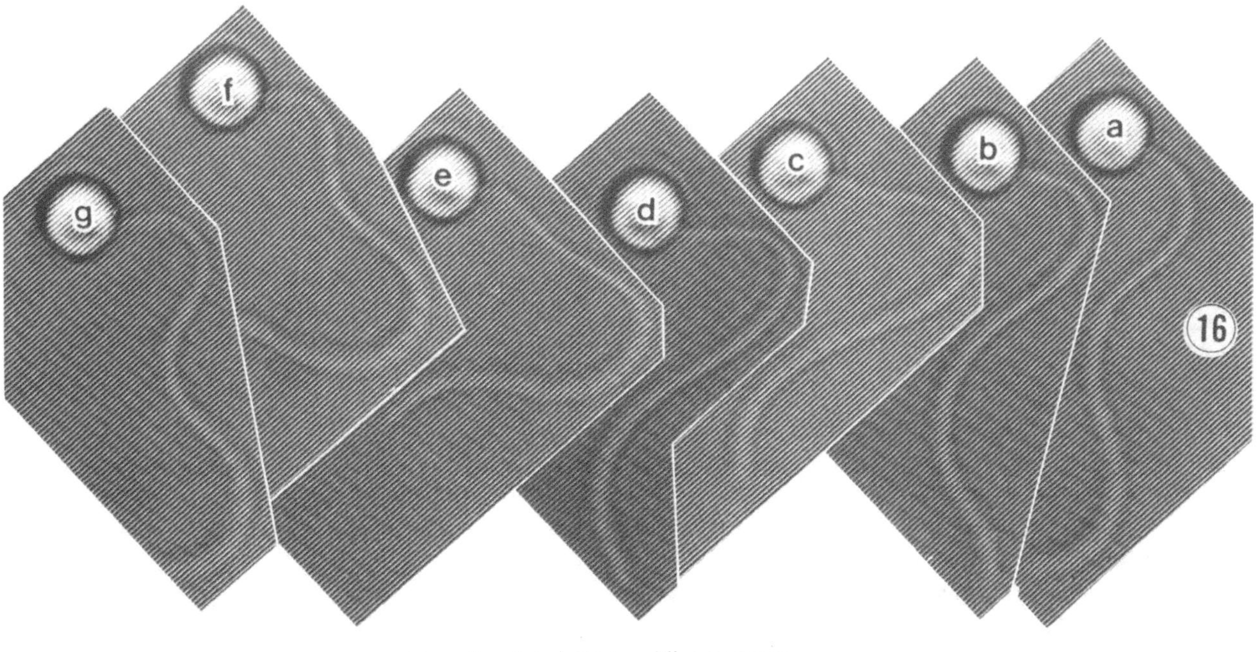

Fig. 16 a–g. *Cymbomonas tetramitiformis.* An uniflagellate cell in avoiding response

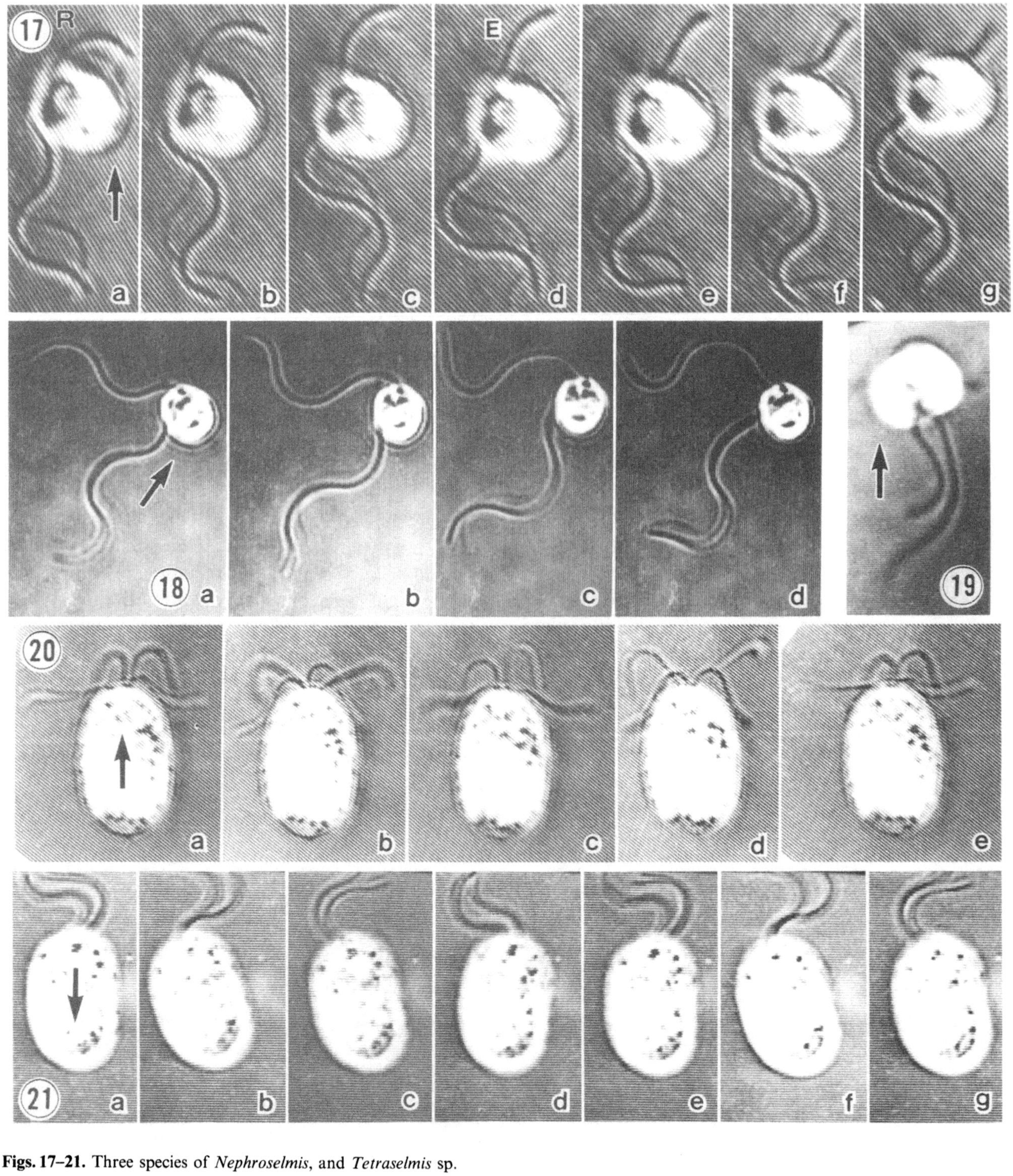

Figs. 17–21. Three species of *Nephroselmis*, and *Tetraselmis* sp.

Fig. 17 a–g. Sequential frames of *Nephroselmis* aff. *rotunda*. The short flagellum shows "ciliary beat" consisting of recovery (**a–c**) and effective (**d–g**) strokes, whereas the long flagellum shows "flagellar beat". *E* Flagellum in effective stroke. *R* flagellum in recovery stroke

Fig. 18 a–d. *Nephroselmis astigmatica*. Both flagella show "flagellar beat". The waveform of two flagella is essentially the same

Fig. 19. *Nephroselmis olivacea* in avoiding response

Fig. 20 a–e. Forward swimming of *Tetraselmis* sp. The two opposed flagella show symmetrical waveforms. The wave phase is different from the other flagellar pair (e.g., **a**, **c**, and **e**)

Fig. 21 a–g. Avoiding response of *Tetraselmis* sp., showing "flagellar beat". There is no wave phase difference between the flagella

felia (Melkonian and Preisig 1986). It differs from all other prasinophyte genera by possessing a thecate cell and a phycoplast-mediated cell division (e.g., Stewart et al. 1974). It differs from the two other genera of the *Tetraselmis*-lineage, *Nephroselmis* and *Pseudoscourfieldia*, by its quadriflagellate flagellation, however, their common origin is suggested by the presence, in all three genera, of rod-shaped flagellar scales (Melkonian 1982 b, 1990 a; Moestrup and Throndsen 1988). The flagella beat in four radial (cruciate) directions. Flagellar movement pattern of *Tetraselmis* has been described (Melkonian and Robenek 1984; Melkonian 1984, 1990 a). These authors observed that flagella exhibit "cilia-type" movement and two flagella act as a coupled pair beating in opposite directions. In the species examined here, flagella show "ciliary beat" in radial directions like in *Pyramimonas*, and do not act as coupled pairs (Fig. 20). The beat cycle is sychronized in two opposed flagellar pairs, and there is a phase difference of about half a beat cycle between the two flagellar pairs (Fig. 20). Avoiding response using unidirectional "flagellar beat" is similar to that of *Pyramimonas* and several other genera (Fig. 21). *Tetraselmis* resembles *Pyramimonas* and other green algal quadriflagellates in its swimming pattern (Group II; Fig. 1).

Remarks on some other green plants

The flagellated reproductive cells of three ulvophycean algae, *Bryopsis maxima* (Fig. 22), *Ulva pertusa*, and *Caulerpa racemosa* var. *clavifera* f. *microphysa* were studied. All these move by forward swimming (Fig. 22). When single frames are studied the flagellar beat appears like an ordinary breast stroke consisting of effective and recovery stroke. However, when the frames are arranged in order (Fig. 22) or superimposed on top of each other (Fig. 24), it is clear that the beat pattern is similar to "flagellar beat".

Beat pattern of sperms of *Marchantia polymorpha, Athyrium nipponicum, Cycas*, and *Ginkgo biloba* is "ciliary beat". In the latter two genera, sperms are multiflagellate and the flagella show metachronal waves, which makes tracing of single flagella difficult (Fig. 23). A biphasic beat cycle is visible, however (recovery stroke: Fig. 23 a–c; effective stroke: Fig. 23 d).

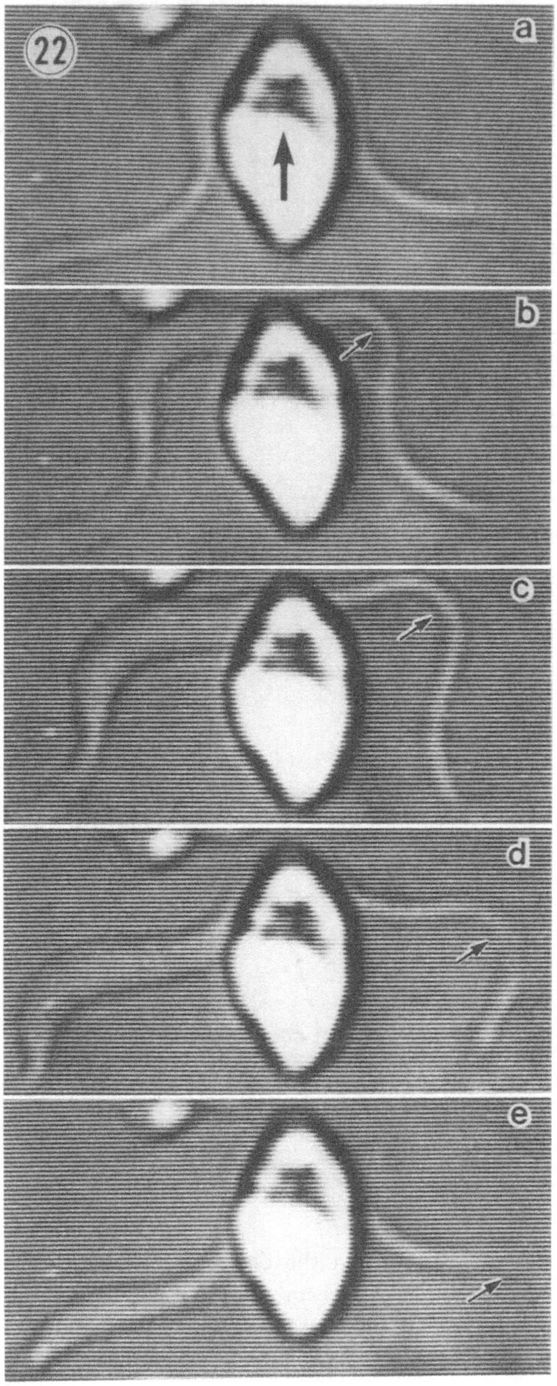

Fig. 22 a–e. Female gamete of *Bryopsis maxima.* Waves are produced at the bases and propagated toward the tips (small arrows). Large arrow denotes the direction of swimming

Discussion

It is clear, based on results presented here, that the swimming patterns of the green plants are very diverse and the "flagellar beat" is widely distributed in green plants (reviewed in Melkonian and Robenek 1984; Melkonian 1990 a, b) while the forward swimming mode using "flagellar beat" is described here for the first time in green plants. Studies on molecular mechanisms indicate that "ciliary beat" changes to the "flagellar beat" (from forward swimming to backward swimming as an

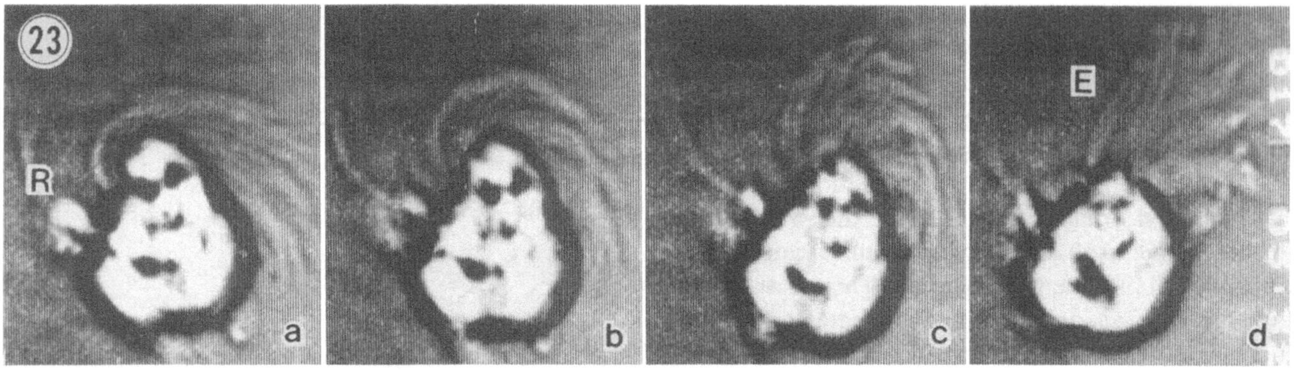

Fig. 23. Sperm cell of *Athyrium nipponicum*. Metachronal waves are visible, consisting of recovery stroke (*R*) (**a** and **b**) and effective stroke (*E*) (**c** and **d**)

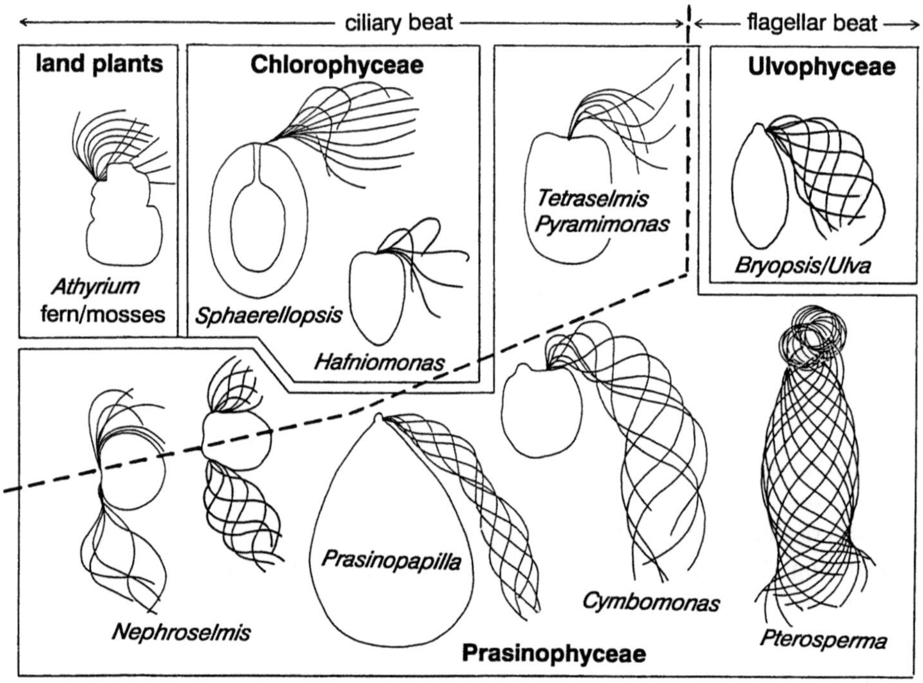

Fig. 24. Superimposed waveforms of the normal swimming mode of some representatives of green plants. Genera showing "ciliary beat" (top left) and those with "flagellar beat" (bottom right) are separated by a broken line through the genus *Nephroselmis*. Not drawn to scale

avoiding response) when the concentration of Ca^{2+} ions increases above a certain level (see Sleigh and Barlow 1982 for references). Hyams and Borisy (1978) showed this phenomenon in *Chlamydomonas* and found a threshold value for conversion of ca. 10^{-7} M Ca^{2+} (more detailed studies have been made later, e.g., Omoto and Brokaw 1985). This may reflect the evolution of the flagellar beat pattern. In the sea, where the concentration of Ca^{2+} is high, "flagellar beat" may have been established first and "ciliary beat" evolved in organisms which developed a mechanism to keep the concentration of Ca^{2+} inside flagella below the threshold value (Melkonian 1983, Moestrup and Throndsen 1988). This hypothesis may be tested by studying groups of organisms in which both flagellar and ciliary beating is present. The *Pyramimonas*-lineage is suitable for this purpose.

Because "flagellar beat"occurs in all the species examined as either an avoiding response or as normal swimming (e.g., *Pterosperma, Cymbomonas, Prasinopapilla*) it is suggested that "flagellar beat" is more primitive. Results support an idea that "flagellar beat" with backward swimming is the basic or primitive type (Sleigh 1981, Melkonian 1983). In the *Pyramimonas*-lineage, *Pryamimonas* may be the most advanced, being the only genus using the "ciliary beat" for forward swimming. Supporting evidence for this idea is that the limuloid flagellar scales in nearly all species are plate-like rather than spider's web-like. We have studied scale formation in *Cymbomonas tetramitiformis* and found that spider's web-like scales are formed in cisternae near the proximal side of the Golgi bodies and these are modified to plate-like limuloid scales in more distal cisternae (unpubl. data). If we consider *Pyramimonas* to be one of the most advanced genera in the lineage, *Pterosperma* may be considered one of the most primitive. *Cymbomonas* and *Prasinopapilla*, which use "flagellar beat" for forward swimming, may be inter-

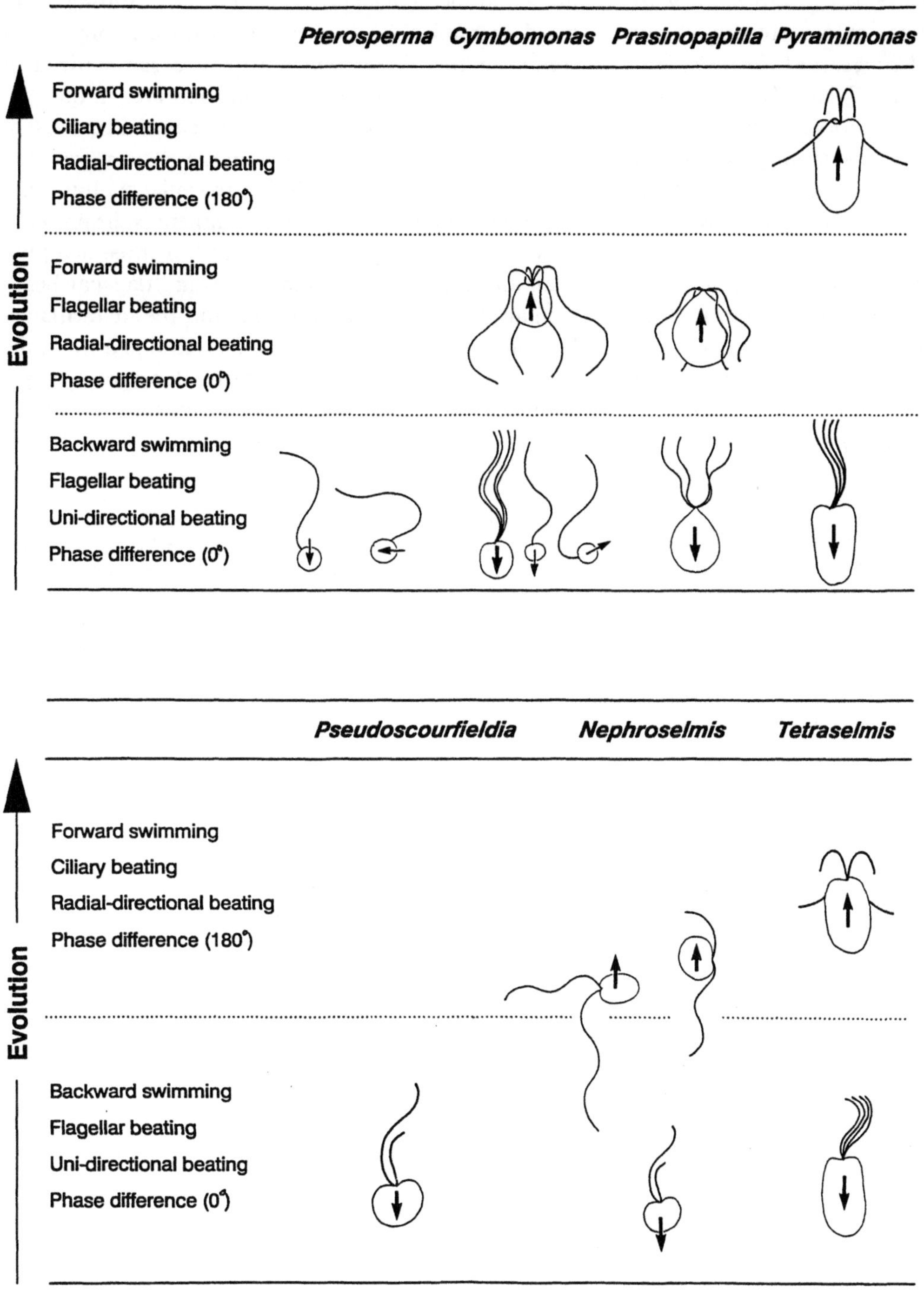

Fig. 25. Evolutionary trends in the *Pyramimonas*-lineage (**a**) and the *Tetraselmis*-lineage (**b**). The following characters are shared by all genera and considered to be evolutionary primitive: backward swimming, "flagellar beat", uni-directional beating, and no phase difference. Arrows indicate direction of swimming. Data for *Pseudoscourfieldia* are based on Moestrup and Throndsen (1988)

mediate. Generally speaking, the evolutionary trends of swimming patterns may be as follows. (1) A swimming pattern characterized by using both "ciliary beat" or breast stroke (as forward swimming) and "flagellar beat" (as avoiding response) is more advanced than that using only "flagellar beat" for both forward swimming and avoiding response (e.g., *Prasinopapilla*). Many bi- or quadriflagellate chlorophycean genera such as *Chlamydomonas, Chlorogonium, Carteria, Hafniomonas, Spermatozopsis,* and *Sphaerellopsis* use this swimming pattern (Fig. 1). (2) Forward swimming is more advanced than backward swimming as has been suggested by Melkonian (1983), because all green flagellates capable of forward swimming also swim by backward swimming (as an avoiding response). (3) Since uni-directional beating occurs in relation to the backward swimming, it is probably more primitive than radially-directed (cruciate) flagellar movement which is, in all cases except *Prasinopapilla*, associated with forward swimming. (4) The feature that flagella bend near the base when the cell changes swimming direction is primitive, because it was only observed in organisms which used backward swimming for normal swimming motion (e.g., *Pterosperma* and a uniflagellate phase of *Cymbomonas*). Flagellar beat of volvocalean colonial algae should be commented on. *Astrephomene gubernaculifera* exhibits ciliary type of motion with flagella beating in the same direction (unidirectional) (Hoops and Floyd 1983). Such a beat pattern is also known in *Volvox* (Hoops 1984) and *Gonium* (Greuel and Floyd 1985). This unusual flagellar movement is probably regarded as a secondarily adapted form in response to the colonial condition of these algae: in *Gonium*, arrangement and beat direction of the two flagella are different between centrally located cell and peripherally located cells in a colony, and flagella of the central cell beat oppositely like *Chlamydomonas* (Greuel and Floyd 1985). Possible evolutionary trends which may have occurred in the *Pyramimonas*-lineage are summarized in Fig. 25 a. These ideas also apply to other taxonomic groups, e.g., *Tetraselmis*-lineage.

Pseudoscourfieldia, Nephroselmis, and *Tetraselmis* form a distinct taxonomic group because of their characteristic rod-shaped flagellar scales. Such scales have not been observed in any other prasinophyte. *Pseudoscourfieldia marina* swims with two unequal flagella beating behind the cell showing the "flagellar beat" pattern (Throndsen 1969, Manton 1975, Moestrup and Throndsen 1988). Swimming patterns of *Nephroselmis* have been described (Manton et al. 1965, Moestrup and Ettl 1979, Moestrup 1983, Inouye and Pienaar 1984, Melkonian and Robenek 1984, Melkonian 1990 a). In all species so far examined, the posterior flagellum exhibits the "flagellar beat" while the anterior flagellum shows either the "ciliary beat" (*N. olivacea* and *N.* aff. *rotunda*) or the "flagellar beat" (*N. astigmatica*). *Tetraselmis* shows the "ciliary beat" during forward swimming and the "flagellar beat" occurs only during backward swimming (as avoiding response). Evolutionary trends in the *Tetraselmis*-lineage are summarized in Fig. 25 b. The evolutionary trends observed in the *Pyramimonas* and *Tetraselmis* lineages suggest that essentially the same changes in swimming pattern occurred independently during evolution (parallel evolution). Another point deserving some comment is the presence of swimming using "flagellar beat" in three ulvophycean algae. Forward swimming using "flagellar beat" similar to that of *Prasinopapilla* and the quadriflagellate phase of *Cymbomonas* is a common feature of the class Ulvophyceae. However, more ulvophycean algae should be examined before conclusions are drawn.

Acknowledgements

We are most grateful to Dr. Makoto, M. Watanabe and colleagues in the Laboratory of Plant Systematics and Phylogeny, University of Tsukuba for providing some of the species examined. We are also grateful to the editor and reviewers for their valuable comments. This study was supported by Grants-in-Aid for Scientific Researches #59480014 and #61440003 from Monbusyo, the Ministry of Education, Science and Culture, Japan.

References

Belcher JH (1964) Further notes on *Scourfieldia caeca*. Br Phycol Bull 2: 371–373

– (1969) Further observations on the type species of *Pyramimonas (P. tetrarhynchus* Schmarda) (Prasinophyceae): an examination by light microscopy, together with notes on its taxonomy. Bot J Linn Soc 62: 241–253

Boalch GT, Parke M (1971) The prasinophycean genera (Chlorophyta) possibly related to fossil genera, in particular the genus *Tasmanites*. In: Farinacci A (ed) Proceedings of the 2nd Planktonic Conference, Rome 1970, pp 99–105

Brokaw CJ, Luck DJL (1983) Bending patterns of *Chlamydomonas reinhardtii* flagella 1. Wild type bending patterns. Cell Motil 3: 131–150

– – Huang B (1982) Analysis of the movement of *Chlamydomonas* flagella: the function of the radial-spoke system is revealed by comparison of wild-type and mutant flagella. J Cell Biol 92: 722–732

Greuel BT, Floyd GL (1985) Development of the flagellar apparatus and flagellar orientation in the colonial green alga *Gonium pectorale* (Volvocales). J Phycol 21: 358–371

Holwill MEJ, Sleigh MA (1967) Propulsion by hispid flagella. J Exp Biol 47: 267–276

Hoops HJ (1984) Somatic cell flagellar apparatuses in two species of *Volvox* (Chlorophyceae). J Phycol 20: 20–27

– Floyd GL (1983) Ultrastructure and development of the flagellar apparatus and flagellar motion in the colonial green alga *Astrephomene gubernaculifera*. J Cell Sci 63: 21–41

Hori H, Satow Y, Inouye I, Chihara M (1990) Origins of organelles and algae evolution deduced from 5S ribosomal RNA sequence. In: Nardon P, Gianinazzi-Pearson V, Grenier AM, Margulis L, Smith DC (eds) Endocytobiology, IV. INRA Press, Paris, pp 557–559

Hyams JS, Borisy GG (1978) Isolated flagellar apparatus of *Chlamydomonas*: characterization of forward swimming and alteration of waveform and reversal of motion by calcium ions in vitro. J Cell Sci 33: 235–253

Inouye I, Pienaar RN (1984) Light and electron microscope observations on *Nephroselmis astigmatica* sp. nov. (class Prasinophyceae). Nord J Bot 4: 409–423

– Hori T, Chihara M (1990) Absolute configuration analysis of the flagellar apparatus of *Pterosperma cristatum* (Prasinophyceae) and consideration of its phylogenetic position. J Phycol 26: 329–344

Kamiya R, Hasegawa E (1987) Intrinsic difference in beat frequency between the two flagella of *Chlamydomonas reinhardtii*. Exp Cell Res 173: 299–304

Manton I (1975) Observations on the microanatomy of *Scourfieldia marina* Throndsen and *Scourfieldia caeca* (Korsch.) Belcher et Swale. Arch Protistenk 117: 358–368

– Rayns DG, Ettl H, Parke M (1965) Further observations on green flagellates with scaly flagella: the genus *Heteromastix* Korshikov. J Mar Biol Assoc UK 45: 241–255

Melkonian M (1982 a) Structural and evolutionary aspects of the flagellar apparatus in green algae and land plants. Taxon 3: 255–265

– (1982 b) Effect of divalent cations on flagellar scales in the green flagellate *Tetraselmis cordiformis*. Protoplasma 111: 221–233

– (1983) Functional and phylogenetic aspects of the basal apparatus in algal cells. J Submicrosc Cytol 15: 121–125

– (1984) Flagellar apparatus ultrastructure in relation to green algal classification. In: Irvine DEG, John DM (eds) Systematics of the green algae. Academic Press, London, pp 73–120

– (1990 a) Phylum Chlorophyta class Prasinophyceae. In: Margulis L, Chapman DJ, Corliss J, Melkonian M (eds) Handbook of Protoctista. Jones and Bartlett, Boston, pp 600–607

– (1990 b) Chlorophyte orders of uncertain affinities. Order Pedinomonadales. In: Margulis L, Chapman DJ, Corliss J, Melkonian M (eds) Handbook of Protoctista. Jones and Bartlett, Boston, pp 649–651

– Preisig HR (1982) Twist of central pair microtubules in the flagellum of the green flagellate *Scourfieldia caeca*. Cell Biol Int Rep 6: 269–277

– – (1986) A light and electron microscopic study of *Scherffelia dubia*, a new member of the scaly green flagellates (Prasinophyceae). Nord J Bot 6: 235–256

– Robenek H (1984) The eyespot apparatus of flagellated green algae: a critical review. In: Round FE, Chapman DJ (eds) Progress in phycological research, vol 3. Biopress, Bristol, pp 193–268

Moestrup Ø (1978) On the phylogenetic validity of the flagellar apparatus in green algae and other chlorophyll a and b containing plants. BioSystems 10: 117–144

– (1983) Further studies of *Nephroselmis* and its allies (Prasinophyceae). I. The question of the genus *Bipedinomonas*. Nord J Bot 3: 609–627

– Ettl H (1979) A light and electron microscopical study of *Nephroselmis olivacea* Stein (Prasinophyceae). Opera Bot 49: 2–39

– Hori T (1989) Ultrastructure of the flagellar apparatus in *Pyramimonas octopus* (Prasinophyceae). 2. Flagellar roots, connecting fibers, and numbering of individual flagella in green algae. Protoplasma 148: 41–56

– Throndsen J (1988) Light and electron microscopical studies on *Pseudoscourfieldia marina*, a primitive scaly green flagellate (Prasinophyceae) with posterior flagella. Can J Bot 66: 1415–1434

Norris RE (1980) Prasinophytes. In: Cox ER (ed) Phytoflagellates. Elsevier/North Holland, New York, pp 85–145

O'Kelly CJ, Floyd GL (1984) Flagellar apparatus absolute orientations and the phylogeny of the green algae. BioSystems 16: 227–251

Omoto CK, Brokaw CJ (1985) Bending patterns of *Chlamydomonas* flagella 2. Calcium effects of reactivated *Chlamydomonas* flagella. Cell Motil 5: 53–60

Parke M (1966) The genus *Pachysphaera* (Prasinophyceae). In: Barnes H (ed) Some contemporary studies in marine science. Allen and Unwin, London, pp 555–563

– Adams I (1961) The *Pyramimonas*-like motile stage of *Halosphaera viridis* Schmitz. Bull Res Counc Israel 10: 94–100

Pringsheim EG (1946) Pure culture of algae. Cambridge University Press, London

Ringo DL (1967) Flagellar motion and fine structure of the flagellar apparatus in *Chlamydomonas*. J Cell Biol 33: 543–571

Rüffer U, Nultsch W (1985) High-speed cinematographic analysis of the movement of *Chlamydomonas reinhardtii*. Cell Motil 5: 251–263

– – (1987) Comparison of the beating of cis and trans-flagella of *Chlamydomonas* cells held on micropipettes. Cell Motil Cytoskeleton 7: 87–93

Schmidt JA, Eckert R (1976) Calcium couples flagellar reversal to photostimulation in *Chlamydomonas*. Nature 262: 713–715

Sleigh MA (1964) Flagellar movement of the sessile flagellates *Actinomonas, Codonosiga, Monas* and *Poteriodendron*. Q J Microsc Sci 105: 405–414

– (1981) Flagellar beat patterns and their possible evolution. BioSystems 14: 423–431

– Barlow DI (1982) How are different ciliary beat patterns produced? Symp Soc Exp Biol 35: 139–157

Stewart KD, Mattox KR (1978) Structural evolution in the flagellated cells of green algae and land plants. BioSystems 10: 145–152

– – Chandler CD (1974) Mitosis and cytokinesis in *Platymonas subcordiformis*, a scaly green monad. J Phycol 10: 65–79

Tappan H (1980) The paleobiology of plant protists. WH Freeman, San Francisco

Throndsen J (1969) Flagellates of Norwegian coastal waters. Nytt Mag Bot 16: 161–216

– (1988) *Cymbomonas* Schiller (Prasinophyceae) reinvestigated by light and electron microscopy. Arch Protistenk 136: 327–336

Protoplasma (1991) 164: 70–90

PROTOPLASMA

Flagellar and cytoskeletal systems in amitochondrial flagellates: Archamoeba, Metamonada and Parabasala

G. Brugerolle*

Laboratoire de Zoologie et Protistologie, Université Blaise Pascal de Clermont-Ferrand, Aubiere

Received June 30, 1990
Accepted December 14, 1990

Summary. The hypothesis that protists without mitochondria, the so-called Archezoa of Cavalier-Smith, are primitive has received some support from rRNA sequence studies on Microsporidia and Diplomonadida. In spite of the lack of mitochondria the archezoan groups of protists show considerable differences in their organization: mastigont and cytoskeletal system, mitosis, Golgi apparatus, hydrogenosomes. This paper examines the characters of the flagellar apparatus and its associated cytoskeleton to obtain clues used for phylogenetic consideration on the three cited groups of flagellates.

Archamoebae of the Pelobiontida order comprising families such as Pelomyxidae and Mastigamoebidae share common features: a rudimentary mastigont system composed of only one basal body giving rise to a poorly motile flagellum and a basal body associated microtubular cone capping the nucleus. No Golgi apparatus has been detected.

Metamonada, comprising three orders: Retortamonadida, Diplomonadida, and Oxymonadida, have been tentatively assembled on the basis of the absence of mitochondria, Golgi apparatus, and basal body arrangement. They all have four basal bodies arranged in two pairs with always one recurrent flagellum generally included in a cytostomal depression. The recurrent basal body/flagellum is in relation to recurrent microtubular fibers. However, they display marked differences in their cytoskeletal system and fiber ultrastructure indicating a distant evolutionary relationship. The presence of a corset of microtubules in retortamonads and three microtubular fibers are distinguished in diplomonads, as well as a paracrystalline preaxostyle and axostyle in oxymonads are features that lend support to these groups being highly divergent.

Parabasala, comprising the orders Trichomonadida and Hypermastigida, is a monophyletic group with a set of homologous features such as the presence of the same arrangement of four basic basal bodies, the parabasal apparatus (striated fibre supporting Golgi), the microtubular pelta-axostyle complex, the external mitotic apparatus (crypto-pleuro-mitosis), the hydrogenosomes.

These three phyla appear distantly related, the Parabasala being a homogeneous group, perhaps also the Pelobiontida, while the Metamonada is heterogeneous and composed of three evolutionary lineages. Additional information such as rRNA and protein sequence data could contribute to a better understanding of the phylogenetic relationships among these groups.

Keywords: Amitochondrial flagellates; Archezoa; Archamoeba; *Chilomastix*; Cytoskeleton; Flagella; Metamonada; Microtubular root; Parabasalia; *Pelomyxa; Trichomonas; Trichonympha.*

Abbreviations: EM electron microscopy; MTOC microtubule organizing centre; PF parabasal fibre.

Introduction

The flagellar and cytoskeletal features have been constantly used for taxonomic and evolutionary purpose (Corliss 1986, 1987). In flagellates the mastigont system is usually a single unit comprised of the flagella with their basal bodies, the flagellar roots attached to the basal bodies, and the centrosomal structures associated with the basal bodies. The mastigont/cytoskeletal system is involved in several functions: cell motility, mostly achieved by the flagella, cell shape, and organelle positioning. In relation with centrosomal structures it participates in cell polarity and cell morphogenesis during division. It contains part of the history of the organism evolution since the appearance of the flagella and the cytoskeletal structures, and it is also one of the most accessible systems to obtain clues for phylogenetic hypotheses.

To use the characters of the mastigont system in evolutionary studies it is necessary to recognize homologies: in terms of structures and their organization. This is relatively easy when comparing genera of the

* Correspondence and reprints: Laboratoire de Zoologie et Protistologie, Université Blaise Pascal de Clermont-Ferrand, F-63177 Aubiere, France.

same lineage (short distance evolution) but more difficult between genera from different lineages (long distance evolution). However, the flagellar/cytoskeletal systems used with some precautions and in addition with other criteria provide a large set of identification traits for a group.

What are the features to be considered and what is their value? The structure of the basal body and axoneme is quasi universal. However, some interesting peculiarities have been recognized: the structure of the transitional zone, of the transitional helix, terminal plate, paraxial rod (Taylor 1978, Hibberd 1979, Melkonian 1980, Moestrup 1982, Grain et al. 1988) and flagellar necklace as revealed by freeze fracture (Bardele 1983). Another criterion which merits consideration is the basal body arrangement and the modalities of basal body duplication and partition during the division. Among the basal body associated roots, microtubular, periodic and microfibrillar structures have been recognized, but very often they represent convergent structures whose phylogenetic value is difficult to ascertain. Morphologically only close similarities or very peculiar organizations are useful. Often biochemical or immunological comparisons involving protein sequence or the results of gene hybridizations are necessary to demonstrate homologies. All of these features can be considered in association to other criteria such as mitochondria, plastids, life cycle, etc. in a phenetic or cladistic phylogeny (Corliss 1986, 1987; Lipscomb 1985; Smith and Patterson 1986).

The goal of this paper is to examine the typical organization and structures of the flagellar/cytoskeletal systems in the different groups of the amitochondrial flagellates to show their traits and to find clues to their phylogenetic relationships. These flagellate groups have been assembled by Cavalier-Smith (1983, 1987 a, b, 1989 a, b) in the subkingdom Archezoa, because they lack mitochondria. In opposition, the Protozoa kingdom, comprising all other eukaryotic protists, which have mitochondria except for organisms which have lost them secondarily. This concept has received some support from the results of rRNA sequencing which placed the diplomonad *Giardia* (Sogin et al. 1989) and the nonflagellate microsporidian *Vairimorpha* (Vossbrink et al. 1987) among the most ancient protists and eukaryotes. Among the so-called Archezoa, Archamoeba of the order Pelobiontida have been redefined recently (Griffin 1988) but no rRNA sequence study is available. In Metamonada (consisting of three orders: Retortamonadida, Diplomonadida, and Oxymonadida) only a single rRNA sequence from the diplomonad *Giardia*, has been obtained (Sogin et al. 1989). Ribosomal RNA sequences obtained from trichomonads species in the Parabasala group show a relatively less primitive position than the archezoan groups. These data suggest the parabasalid group has branched off either between the mitochondria-bearing Euglenozoa and *Dictyostelium* (Baroin et al. 1988, Perasso et al. 1989) or before the Euglenozoa (Sogin 1989).

From the published ultrastructure studies we know that all these groups of Archezoan flagellates are deprived of mitochondria; do they share any other common feature to be used in phylogenetic consideration? This will be discussed in the conclusion.

Mastigamoebids: phylum Archamoeba

Order Pelobiontida

This order, created by Page (1987), has been redefined after the discovery of flagella in *Pelomyxa* by Griffin

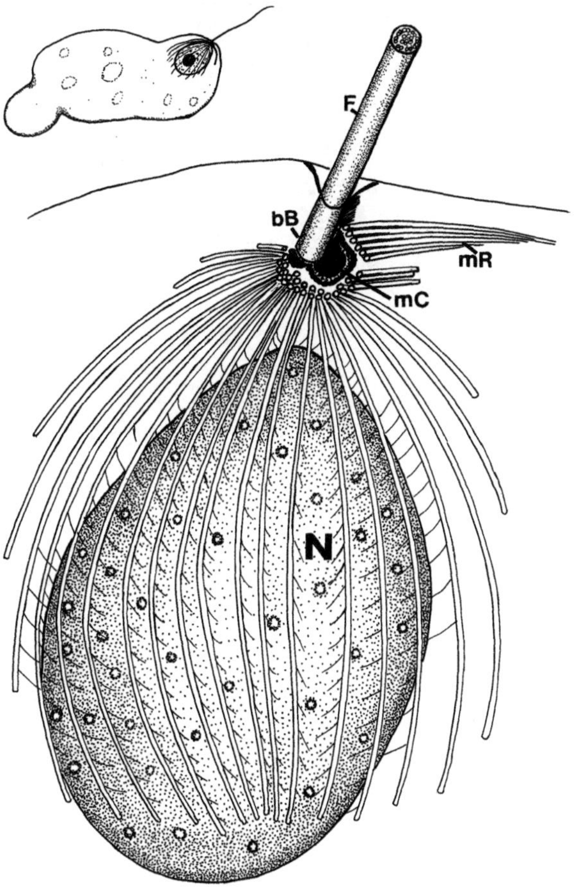

Fig. 1. Diagram of the flagellar/cytoskeletal system forming one karyomastigont in a mastigamoebid type: *Mastigina*. From a microtubule organizing centre (*mC*) associated with the basal body (*bB*) of the unique flagellum (*F*) arises a cone of microtubules which caps and attaches to the nucleus (*N*). A surface root (*mR*) is also differentiated

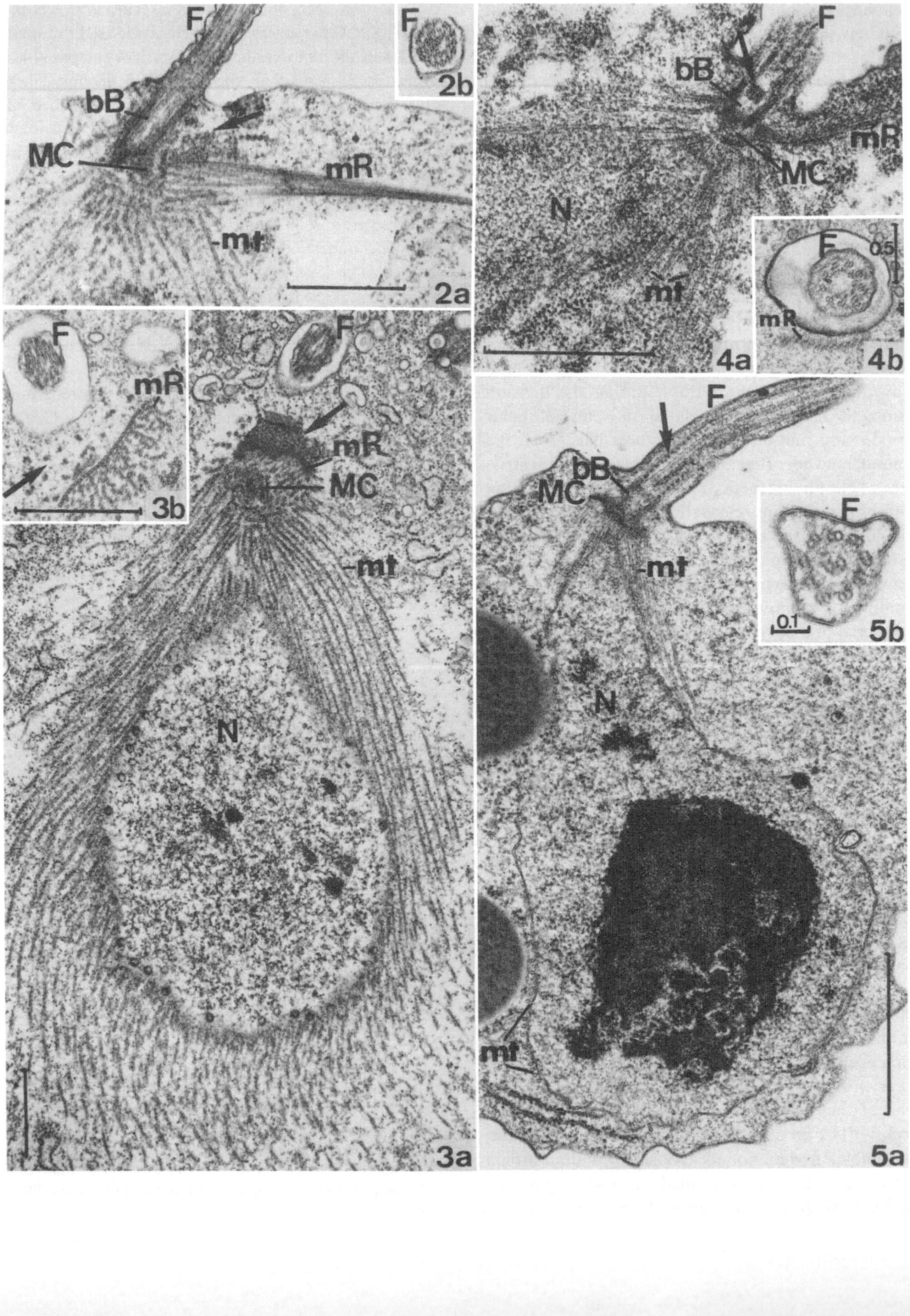
F
bB
MC
mR
mt
2a
2b
N
0.5
4a
4b
3a
3b
0.1
5a
5b

(1988). These organisms have a mastigont system composed of only one flagellum and one basal body attached to the nucleus by a cone of microtubules radiating from the basal body, that constitutes the karyomastigont unit. They have neither mitochondria nor Golgi apparatus and belong to Archamoeba defined by Cavalier-Smith (1987a, b, c). This order includes *Pelomyxa* which has several karyomastigonts, at least at certain stages of development (Griffin 1988, Goodkov 1989), and the genus *Mastigina* already studied by electron microscopy (Brugerolle 1982). Also the genera *Mastigamoeba* and *Mastigella* show characteristic features of the group from light microscopic studies, and this is confirmed by electron microscopic photographs of *Mastigamoeba* sp. (Brugerolle unpubl. data). The examination of the ultrastructure of a new genus *Phreatamoeba* (Chavez et al. 1986) and electron microscopy photographs of a very similar species (Brugerolle unpubl. data) has led to include this genus in the group.

The flagellar apparatus

In these organisms each karyomastigont comprises a unique flagellum (Fig. 1) but in genera such as *Mastigina* and *Pelomyxa* there are several karyomastigonts per cell (Griffin 1988). One of the peculiarities of the mastigamoebids is that they have only one flagellum and one basal body. Other flagellate organisms generally have at least two flagella and two basal bodies, although sometimes one basal body is barren in the interphasic cell. In the mastigamoebid genera examined to date, there is no trace of a second basal body in the interphasic cell. We suppose a daughter basal body appears during the mitotic process, but the division has not been observed in these poorly known organisms. In several groups of protists the flagella share peculiar structures such as the transitional helix (in heterokonts), which help to define the group; are there such structures in mastigamoebids? The axoneme observed in *Mastigamoeba* and in *Phreatamoeba* has the 9 + 2 standard fibers (Figs. 4 and 5) but the flagellum does not beat normally. There is no regular wave propagation along its length like in other flagella. Only some portions of the flagellum are motile, the movement is sometimes described as an "ineffectual flailing" (Chavez et al. 1986; and pers. obs.). In *Phreatamoeba*, the flagellum breaks very easily at its base and the central fibers do not reach an axosome or the basal plate (Fig. 4). In *Mastigamoeba*, the central fibers do not originate from the center of the basal plate but from the periphery and there is a transitional helix-like structure on the basal region of the axoneme (Fig. 5a). In *Mastigina* and *Pelomyxa*, the flagella are non motile and the axoneme looks more like a bundle of microtubules than the classical cylinder of 9 doublets with 2 central microtubules (Figs. 2 and 3). There are no described flagellar hairs on the flagellar surface. A structure assimilated to a basal body is always present at the base of the flagellum; it is very short in *Phreatamoeba* and *Mastigamoeba* (Figs. 4 and 5). In *Mastigina*, the secondary karyomastigonts do not develop a flagellum but a basal body like structure is present and is associated with the nucleus by a reduced cone of microtubules (Brugerolle 1982). However, in no genera has any clear transverse section been obtained to determine the fine structure of the basal body and of the transitional region. This latter zone does not show any common particular structure useful to define the group. We notice that the flagellar system is poorly developed, it shows abnormalities and it does not serve too much for the locomotion which is mainly realized by amoeboid movements. Is this a rudimentation of a normal mastigont system in an amoeboid organism or a stage in the completion of a functional system in very primitive cells? This is difficult to guess. However, an example of mastigont system regression occurs in the parabasalid lineage: *Histomonas* and *Dientamoeba* have a progressive lack of the flagella and cytoskeletal fibers while the amoeboism increases. In *Histomonas*, one of the four flagella remains, but all the basal bodies are retained (Rybicka et al. 1972). In *Dientamoeba*, the fla-

Figs. 2a, b and 3a, b. *Mastigina hylae*. The microtubule organizing center (*MC*) associated with the basal body (*bB*) of the unique flagellum (*F*) gives rise to a cone of microtubules (*mt*) capping the nucleus (*N*). A microtubular root (*mR*) originating from the microtubule center (*MC*) lies under the cell surface (Fig. 2a). Dense material is interposed between the microtubular root (*mR*) and the flagellar base (arrow). The axoneme of the flagellum is formed by a bundle of microtubules (Fig. 2b)

Fig. 4a, b. *Phreatamoeba*. The microtubular cone (*mt*) capping the nucleus (*N*) arises from a microtubular center (*MC*) associated with the basal body (*bB*) of the flagellum (*F*) as is the lateral microtubular root (*mR*). The central fibers of the axoneme do not contact the internal structure seen of the transition zone (arrow)

Fig. 5a, b. *Mastigamoeba*. The cone of microtubules (*mt*) arises from a center (*MC*) at the base of a short basal body (*bB*) and is associated with the nuclear envelope. The large endosomal structure is remarkable in the pear-shape nucleus (*N*). The flagellar axoneme (*F*) appears normal (**b**) but a helix-like structure is apparent in the basal region (arrow)

gella and the basal bodies have disappeared, only the parabasal fibers remain (Camp et al. 1974). In the Kinetoplastida group, the bodonids and cryptobiids have two fully developed flagella whereas in trypanosomatids the flagellar number is reduced to one but a second barren basal body remains (Vickerman 1969). However, different modes of regression of the flagellar/cytoskeletal system could occur in different lineages of organisms.

The basal body attached cytoskeleton

In all studied genera the basal body is surrounded by amorphous dense material from which radiating microtubules arise and this structure is therefore assimilated to a microtubule organizing centre (MTOC) (Figs. 1–5). Independent microtubules are mostly directed towards the top of the pear-shape nucleus forming a cone covering most of the nucleus surface in, e.g., *Mastigina hylae* (Fig. 3). In *Phreatamoeba* (Fig. 4 a) and in the only studied *Mastigamoeba* sp. (Fig. 5 a), the microtubules of the cone are not so numerous, they cap the conical extremity but some reach the posterior end of the nucleus. In *Mastigina hylae*, the secondary karyomastigonts only have a reduced number of microtubules originating from the base of the barren basal body (Brugerolle 1982). In *Pelomyxa palustris*, radiating microtubules around the basal bodies have been demonstrated as well as microtubules in contact to the nuclear envelope (Griffin 1988). The microtubules of the cone are probably linked to the nuclear surface by thin threads whose nature is unknown. This linkage could explain the close relation between the nucleus and the basal body noted in *Mastigina, Mastigamoeba*, and *Phreatamoeba*. The presence of a very long finger-like cone in front of the nucleus in *Mastigamoeba* is in favor to a solid linkage between the leading anterior flagellum and the nucleus by the intermediary of this microtubular nuclear connector. A nuclear connector apparatus was found in the *Chlamydomonas* and other chlorophytes (Salisbury 1988, Schulze et al. 1987). However, the fibers are microtubular in mastigamoebids whereas they are microfibrillar and striated in these algae and they probably differ by their composition (Salisbury et al. 1984).

An accessory microtubular fiber composed of about ten microtubules in *Phreatamoeba* and more than a hundred in *Mastigina hylae*, is directed laterally beneath the cell surface (Figs. 1–4). Microfilamentous zones are clearly seen in cytoplasmic areas forming pseudopods, they look like those found in other amoeba and probably are composed of actin and myosin.

Phylogenetic considerations

Several questions in relation to the evolution could be raised. Is this group a monophyletic evolutionary lineage? The original composition and organization of the mastigont system and associated skeletal apparatus favor this view, as well as the absence of mitochondria and Golgi bodies. However, the number of characters is small and the structure of the flagella base shows some differences which could indicate some heterogeneity in the group.

Is this group primitive as it was previously hypothesized (Cavalier-Smith 1987 a, b)? The absence of mitochondria and Golgi bodies has been noted in groups that branch very early in phylogenic trees based on rRNA sequence comparisons, such as Microsporidia (Vossbrink et al. 1987) and in Diplomonadida (Sogin et al. 1989). Organisms which have not acquired mitochondria and a complete respiratory chain could have taken refuge in anoxic sites as the atmosphere became oxygenated. Alternatively, mitochondria could have disappeared in relation to a mode of life in the low oxygen conditions that characterize the habitat of the mastigamoebid species. The apparently very simple flagellar/cytoskeletal apparatus can be considered as either a primitive system or a rudimentary system which is poorly developed because of the existence of a competent amoeboid form of locomotion.

The unique features described above fail to support any close relationship with other groups of flagellate protists, and any possible relationship with other less characterized genera of amoebas are not obvious (Page 1987, Chavez et al. 1986, Griffin 1988).

Metamonads: phylum Metamonada

The word metamonad was created by Grassé (1952) for grouping several parasitic flagellates now known under the names retortamonads, oxymonads, trichomonads, and hypermastigids (the diplomonads were set apart). After electron microscopic studies demonstrated the characteristic features of each group, the trichomonads and hypermastigids were grouped together in a separate lineage, Parabasalia (Honigberg 1973, Brugerolle and Taylor 1977), and they were excluded from the metamonadean phylum. The three orders Retortamonadida, Diplomonadida, and Oxymonadida were then tentatively grouped on the basis of their basal body arrangement and the absence of mitochondria and Golgi apparatus (Brugerolle and Taylor 1977) and subsequently they were classified in the phylum Metamonada (Cavalier-Smith 1987a).

Retortamonads: order Retortamonadida

Only two genera have been identified (*Chilomastix* and *Retortamonas*) and studied by electron microscopy (Brugerolle 1973, 1977; Brugerolle and Mignot 1990). They were considered to be intestinal parasites, however, recently free-living species have been isolated from salt marsh (Farmer unpubl. data) and recognized as real retortamonads on scanning EM photographs. This finding gives consistency to the overlooked description of the free living species *Chilomastix undulata* by Skuja (1956).

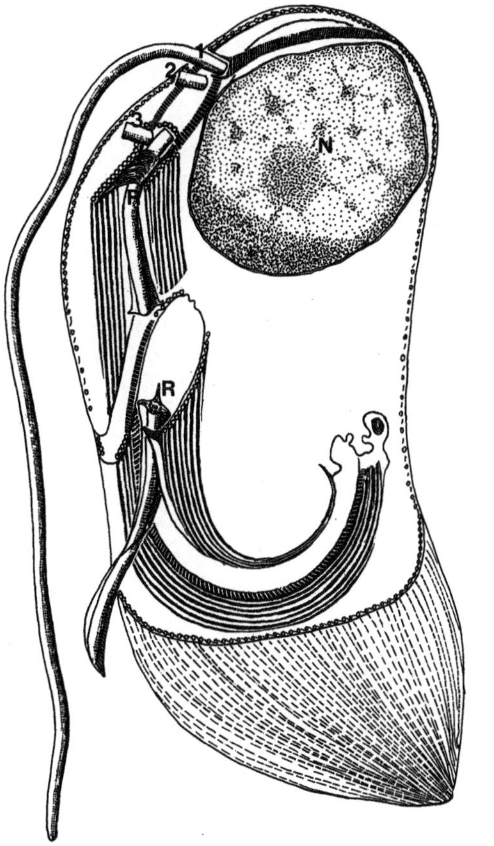

Fig. 6. Reconstruction diagram of a retortamonad cell skeleton. The basal bodies are arranged in two pairs (*1, 2* and *3, R*); the distal pair contains the basal body of the recurrent and modified flagellum (*R*). A corset of microtubules surrounds the cell. A large ventral cytostomal pocket is bordered by lips containing a row of microtubules and in the right lip an additional striated lamellar fiber. Note the nuclear connector at the top of the nucleus (*N*)

Flagellar apparatus

It is basically composed of 4 flagella arising from 4 basal bodies arranged in two pairs (Figs. 6 and 7). In *Retortamonas* one basal body of each pair is barren and the two pairs appear more distant than in *Chilomastix* (Fig. 8). This reduction of the flagellar number shows that basal body 1 of the first pair and basal body R of the second pair seem dominant. Basal bodies 1 and 2 are orthogonal and give rise to two anterior free flagella. Slightly posterior basal bodies 3 and R are orthogonal; basal body 3 gives rise to a third anterior flagellum and basal body R is typically directed backward and gives rise to the recurrent flagellum lodged in the cytostomal pocket. There are no peculiar structures in the axoneme and in the transitional zone of the flagella. The recurrent flagellum has lateral fin-like lamellae (2 in *Chilomastix* and 2 or 3 in *Retortamonas*) which contain a striated dense armature linked to the axoneme (Figs. 10, 11, and 13).

Cytoskeleton

A monolayer of microtubules underlying the plasma membrane surrounds the whole cell except in the deepest area of the cytopharynx where endocytosis takes place. The microtubules are linked to each other and to the plasma membrane by bridges (Fig. 9). They originate in the anterior part of the cell near, but not directly in contact with, the basal bodies (Fig. 7). The ventral cytostomal aperture is supported by a prominent right lip longer than the left lip. The lips are bordered by bridged microtubules and are linked on their summit by an arched microfibrillar bundle (Fig. 12). The right lip contains a flattened fiber that is very thin in the upper part of the cytostome and thicker in the distal part, where it forms a hook (Fig. 12). In the posterior half this fiber replaces the pellicular microtubules (Fig. 13). This fiber, lamellar in appearance, is striated both in longitudinal section (Fig. 13) and in transverse section (Fig. 14), and this is an unusual appearance for a basal body root. A grazing section shows the paracrystalline meshwork (Fig. 14). Nothing is sure about its possible contractility.

The nucleus is maintained close to the basal bodies by a periodic connector that is particularly apparent in *Retortamonas* (Fig. 6). The scarce mitotic figures obtained, show an intranuclear spindle in *Retortamonas* (Brugerolle 1977) indicating a close mitosis. A pair of basal bodies are situated near each spindle pole, suggesting that the two pairs of basal bodies separate first and then complete their number in each daughter cell;

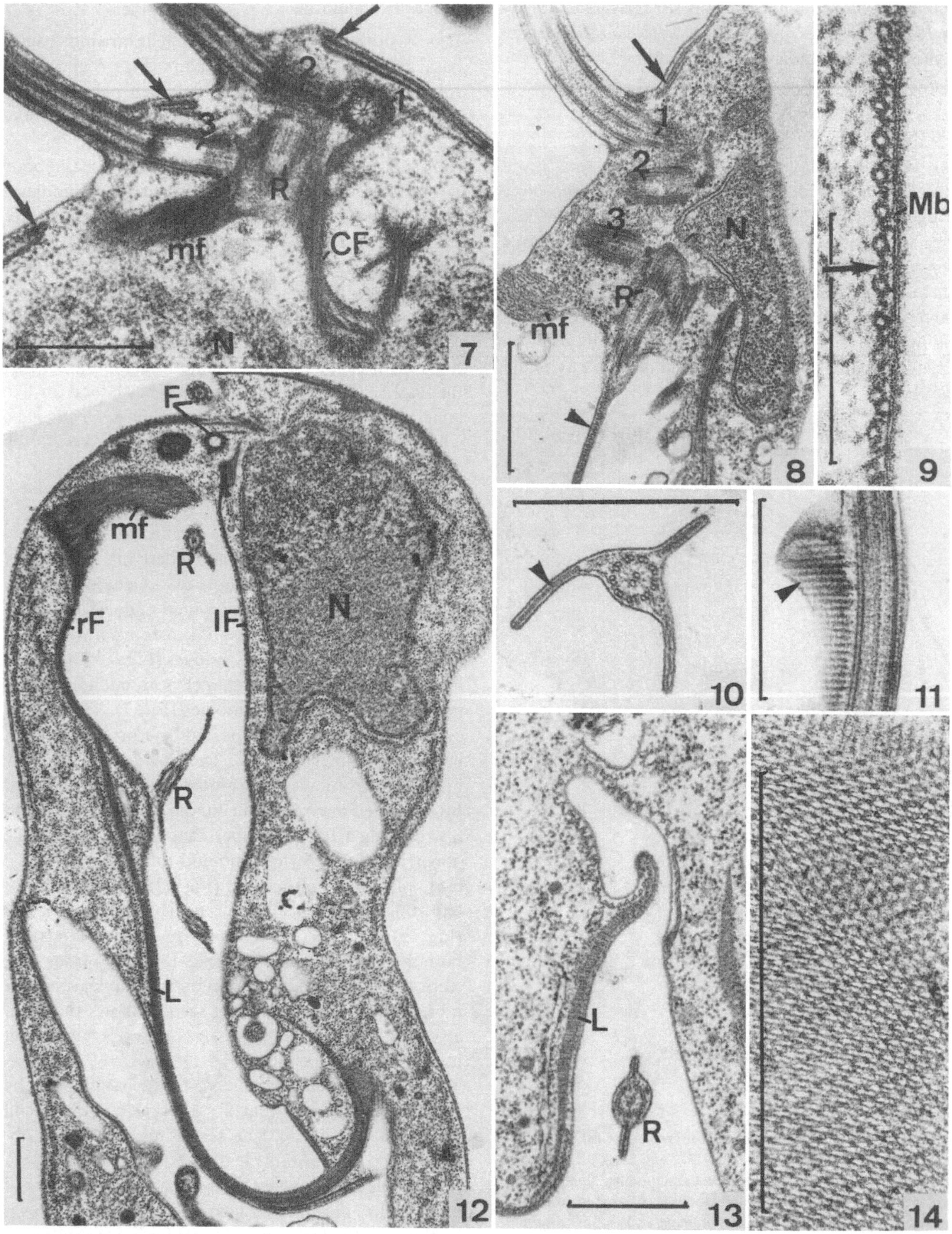
2
1
3
R
mf
CF
N
7
1
2
3
N
R
mf
8
Mb
9
10
11
F
mf
R
rF
IF
N
R
L
12
L
R
13
14

however, much more remains to be done on the mitosis and cell morphogenesis.

Phylogenetic considerations

This original group shows basic characters such as the two pairs of basal bodies including a recurrent one, and the absence of mitochondria and Golgi apparatus. The pellicular corset of microtubules not divided into several fibers has an ambiguous phylogenetic significance. The cytostomal striated fiber appears more specific and could be compared to other similar structures. The mitotic apparatus and the morphogenesis of the flagellar apparatus are incompletely known.

Diplomonads: order Diplomonadida

Most of the genera of this order have been studied by electron microscopy and we have a relatively complete idea of the mastigont system organization and of the structures of the cytoskeleton. The sub-order Diplomonadina are like double organisms, possessing two mastigont systems, two cytostomes, two nuclei (2 karyomastigonts) organized in two fold rotational symmetry. The genera that have been studied are: *Trepomonas* (Brugerolle et al. 1973 a, Eyden and Vickerman 1975), *Hexamita* (Brugerolle 1974), *Spironucleus* (Brugerolle et al. 1973 b), *Octomitus* (Brugerolle et al. 1974), *Giardia* (Brugerolle 1975 c). But the probable stem of the order Diplomonadida is a cell bearing one mastigont system and one nucleus (1 karyomastigont) like the genera *Enteromonas* and *Trimitus* of the sub-order Enteromonadina (Brugerolle 1975 a, d; Brugerolle and Taylor 1977; Kulda and Nohynkova 1978; Vickerman 1990).

Flagellar apparatus

The basic number of flagella is four and they originate from two pairs of basal bodies (Fig. 15). In each pair the two basal bodies are linked and show different configurations according to the genera (Figs. 17–20, and 25). In *Hexamita*, for example, the anterior pair, basal bodies 1 and 2, gives rise to anterior flagella and the second pair, basal bodies 3 and R, gives rise to one anterior flagellum and to the recurrent one R (Fig. 15). Axonemes of the flagella are quite normal and the transition zone does not show any peculiarity. Only the ventral flagellum of *Giardia* presents a fin-like lamella (not shown) attached to the axoneme which is reminiscent of the recurrent flagella of the retortamonads (Brugerolle 1975 c). In most of the genera the two pairs of basal bodies are connected by dense material to the surface of the nuclear envelope where it forms a cup-like depression (Fig. 19).

Cytoskeletal system

There are typically three microtubular fibers attached to the base of the basal bodies (Figs. 15 and 16). The microtubular fiber linked to basal body 1 is directed anteriorly and covers a part of the nucleus. It has been named the supra-nuclear fiber (Figs. 15 and 16). In double organisms (e.g., *Hexamita*) the two opposed supra-nuclear fibers cross on the top of the cell. These fibers are diversely developed according to the genera and species. In *Giardia* the microtubules of the striated disk arise in relation to basal bodies 1 and could be homologous to the supra-nuclear fiber (Brugerolle 1975 c) (Fig. 25). The infra-nuclear fiber passes under the nucleus, and in double organisms such as *Trepomonas* and *Hexamita* it crosses with the opposite fiber and continues its course along the symmetrical cytopharynx (Fig. 16). The direct microtubular fiber follows the recurrent flagellum and constitutes the main part of the microtubular armature of the cytopharynx. When the cytopharynx is closed and has disappeared

Figs. 7 and **8.** Basal body arrangement in *Chilomastix* (Fig. 7) and *Retortamonas* (Fig. 8). Basal bodies 1 and 2 are from the anterior pair and 3 and R are the posterior pair. One basal body (*R*) is turned backward and gives rise to the recurrent flagellum. In *Retortamonas* basal bodies 2 and 3 are barren. The pellicular microtubules originate at some distance from the basal bodies (arrows). The cytostomal microtubular fiber (*CF*) is connected to the basal bodies R and 1. Note the arched microfibrils (*mf*) and nucleus (*N*)

Fig. 9. *Chilomastix* bridged microtubules (arrow) form a corset under the plasma membrane (*Mb*)

Figs. 10 and **11.** *Retortamonas* transverse (Fig. 10) and longitudinal (Fig. 11) sections of the recurrent flagellum (R) showing 3 fin-like lamellae containing a striated structure (arrowheads) linked to the axoneme (also seen in Fig. 8)

Fig. 12. *Chilomastix* lips of the cytostomal pocket are bordered by a left fiber (*lF*) and by a right fiber (*rF*), the latter containing a striated hooked lamina (*L*) in addition to the pellicular microtubules. Both right and left fibers are joined on their summit by arched microfibrils (*mf*). Sections of the modified recurrent flagellum (*R*) are seen in the cytostomal aperture. Note the anterior nucleus (*N*) and basal bodies of flagella (*F*)

Figs. 13 and **14.** *Chilomastix* transverse section (Fig. 13) and grazing section (Fig. 14) of the striated right cytostomal lamina (*L*) replacing the microtubular corset. A clear periodicity is seen in Fig. 13 and the paracrystalline meshwork in Fig. 14. Recurrent modified flagellum (*R*)

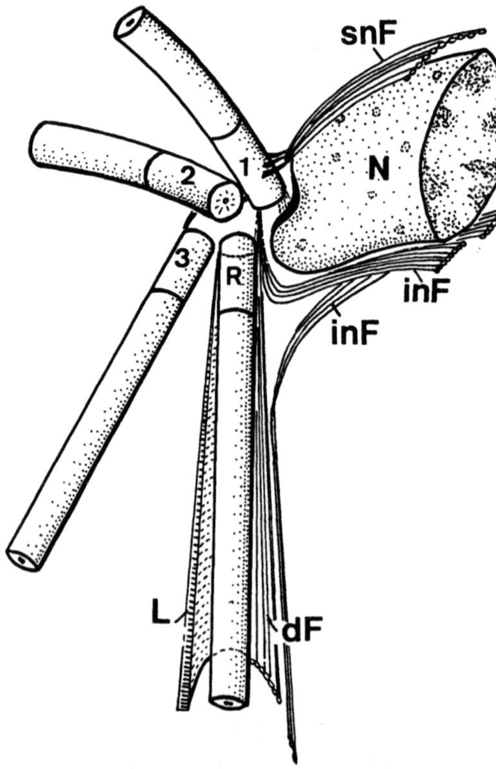

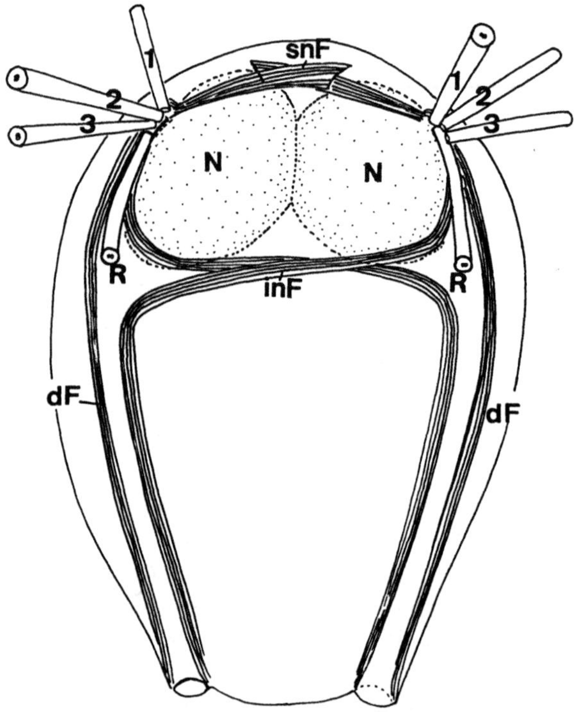

Figs. 15 and **16.** Schematic reconstruction of a diplomonad cell skeleton to show the basic arrangement of the basal bodies in two pairs 1, 2 and 3, R in relation to the nucleus. Three microtubular cytoskeletal fibers are associated to the basal bodies and comprise a supra-nuclear fiber (*snF*), an infra-nuclear fiber (*inF*) and a direct fiber (*dF*). The latter forms the cytopharyngeal tube armature in association with a dense lamina (*L*). In organisms with two karyomastigonts the two sets of basal bodies and their associated fibers are arranged in a two-fold rotational symmetry (Fig. 16). The supra-nuclear fibers could overlap over the nuclei and the infra-nuclear fibers cross each other and integrate in the armature of the opposed cytopharynx

in *Octomitus* and *Giardia* this fiber follows the axoneme of the two recurrent flagella (Fig. 21). The cytostomal pockets in *Trepomonas* or tubes in *Hexamita* (Figs. 16, 22, and 24) are reinforced by a lamina (Fig. 24) layered under the membrane in *Trepomonas* and *Hexamita* (Figs. 23 and 24) and by diverse periodic fibers which have been identified in *Trepomonas, Hexamita*, and *Octomitus*. These striated fibers are diversely developed according to the genera and species.

The flagellar apparatus/cytoskeleton of *Giardia* merits a particular mention. The cell presents a flattened face consecutive to the development of a large ventral disk and therefore flagella seem arranged in a bilateral symmetry. A precise study of the arrangement of the basal bodies shows them to be in a two-fold rotational symmetry and very similar to that of *Octomitus* (Figs. 25 and 28). The ventral disk is composed of a spiral of microribbons supported by the submembranal micro-

---

Figs. 17 and **18.** Basal body arrangement in *Trepomonas*: anterior pair 1, 2 and distal pair 3, R of basal bodies. The supra-nuclear fiber (*snF*) is linked to basal body 1, the direct fiber (*dF*) is connected to basal body R. Base of the nucleus (*N*)

Fig. 19. Dense material (arrow) interposed between basal bodies 1, 2, R and the nucleus (*N*) in *Hexamita*. The recurrent flagellum (*R*) penetrates in the cytopharynx (*Cp*)

Fig. 20. Basal body arrangement in *Octomitus*, the basal bodies of the first pair 1, 2 are in opposition and well separated from the second pair of basal bodies 3, R

Fig. 21. In *Octomitus* the axonemes of the two recurrent flagella (*R* and *R′*) arise at the posterior end and are accompanied by the two direct microtubular fibers (*dF*) lined by a dense lamina (*L*)

Figs. 22–24. Longitudinal sections (Figs. 22 and 23) and transversal section (Fig. 24) of one cytopharynx (*Cp*) in *Hexamita* where the two microtubular fibers (*mtF*), a striated fiber (*sF*) and a dense lamina (*L*) compose the armature of the cytopharyngeal tube containing the recurrent flagellum (*R*)

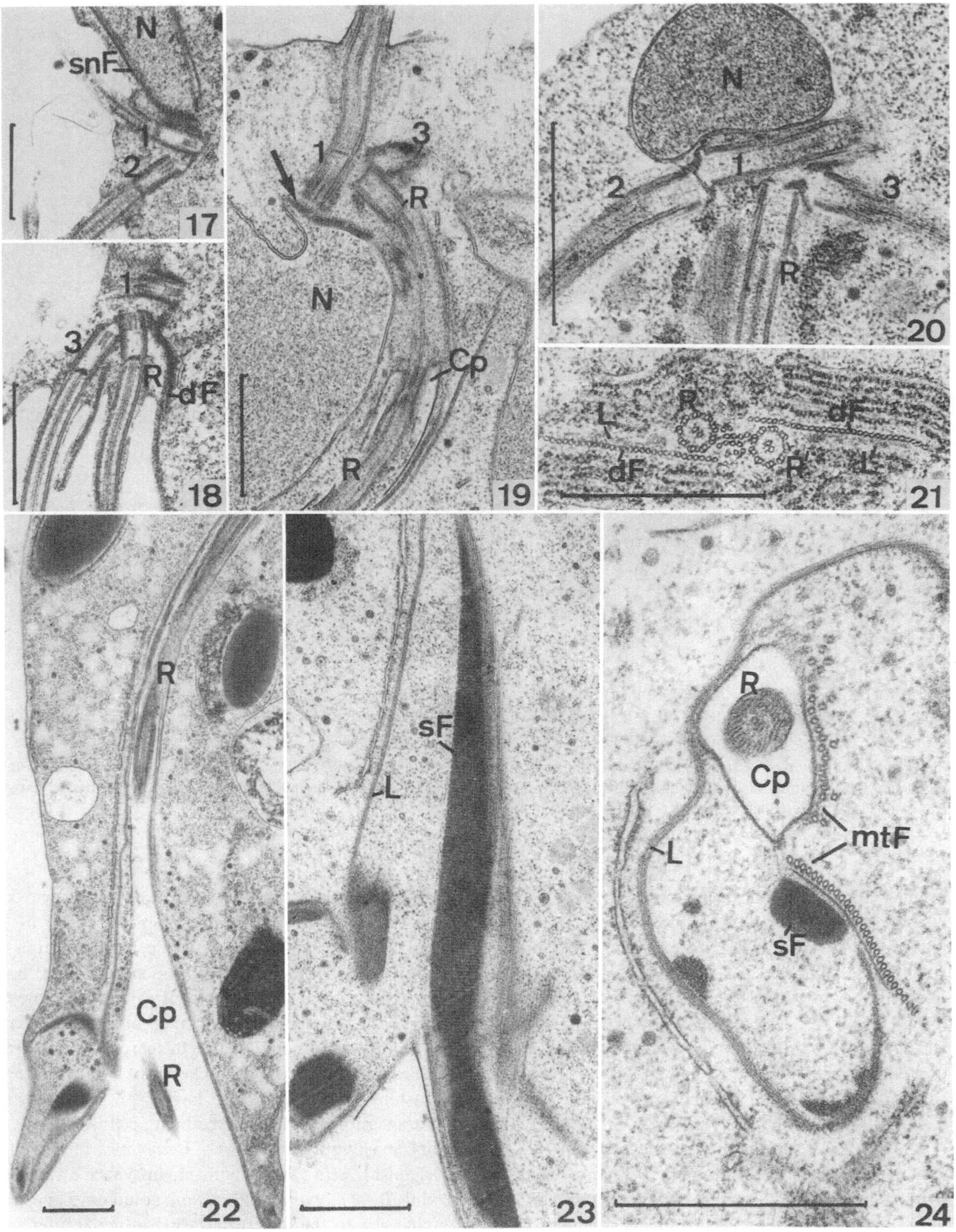
N
snF
1
2
17
1
3
R
dF
18
3
1
R
N
Cp
R
19
N
1
2
3
R
20
L
R
dF
dF
R
L
21
R
Cp
R
22
sF
L
23
R
Cp
mtF
L
sF
24

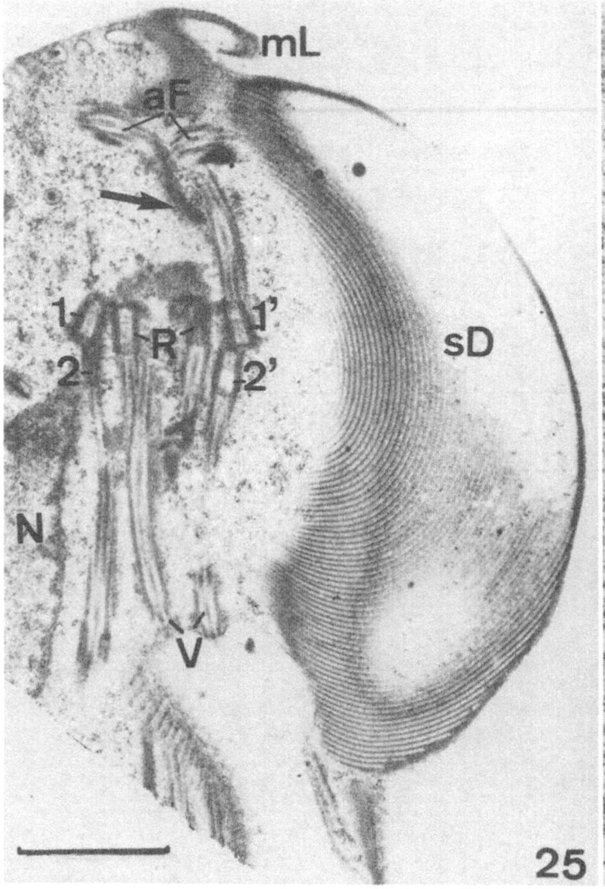

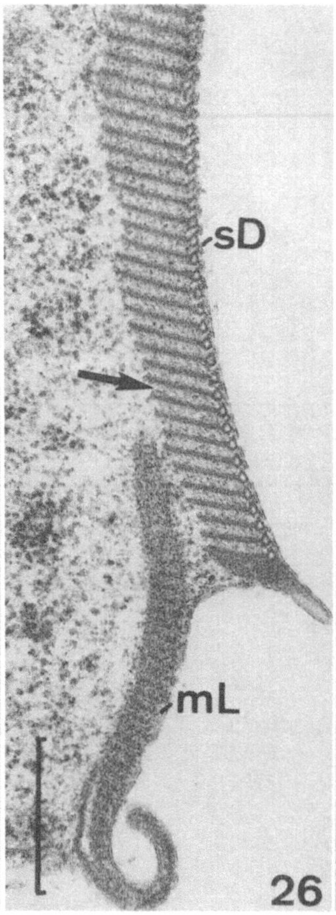

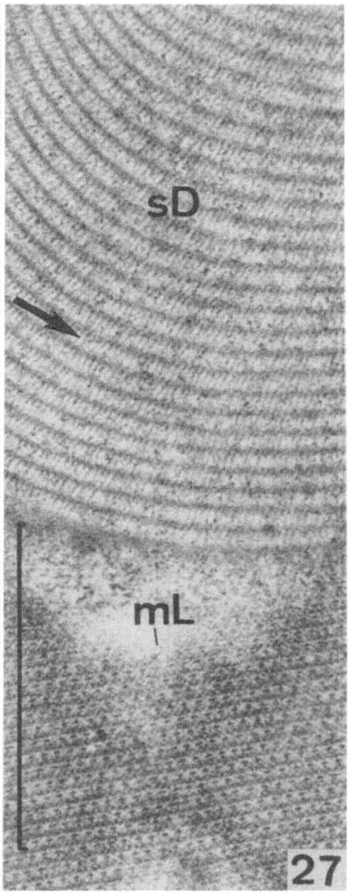

Fig. 25. Oblique section of *Giardia* showing the two sets of basal bodies side by side between the nuclei (*N*). The basal bodies 1 and 1′ of each set give rise to the antero-lateral flagella (*aF*) which cross anteriorly. The micro-ribbons of the striated disk (*sD*) originate (arrow) near the anterior basal bodies 1, 1′. Axoneme of ventral flagella (*V*) and of the marginal lamina (*mL*)

Figs. 26 and **27.** Transverse section (Fig. 26) and grazing section (Fig. 27) in the striated disk (*sD*) of *Giardia* showing the microtubules under the membrane supporting their dorsal micro-ribbons. Thin threads are linking neighbouring micro-ribbons (arrow). The striated marginal lamina (*mL*) has a crystalline meshwork

tubules (Figs. 25 and 26). The microtubules originate near the basal body 1 and follow the axoneme of the antero-lateral flagella (Fig. 25) and could be homologous to the supra-nuclear fiber of other genera. The microribbons are linked to each other by cross-bridges (Holberton 1973, 1981; Brugerolle 1975 c; Holberton and Ward 1981) (Figs. 25 and 26) forming a solid structure which has been isolated and analyzed. The major components are tubulin and giardin, a protein of about 25 to 30 kDa (Crossley and Holberton 1983, Holberton et al. 1988, Baker et al. 1988, Peattie et al. 1989).

The margin of the disk contains a lamina which is striated in transverse section (Figs. 25 and 26) and shows a paracrystalline meshwork in grazing section (Fig. 27). The ventral disk of *Giardia* is an interesting example of cytoskeletal differentiation within an evolutionary lineage. The over-development of cytoskeletal structures in relation to the mode of life of the organisms can greatly modify the cell organization except for the basal body arrangement and basal body fiber connections. The genera of this order show the same basic flagellar/cytoskeletal organization and consitute an evolutionary lineage (Brugerolle 1975 d, Brugerolle and Taylor 1977, Vickerman 1990). The relationships within the group have been suggested, the first ancestor could be a free-living cell with only one karyomastigont similar to *Enteromonas* or *Trimitus*. Organisms with two karyomastigonts such as the free-living *Trepomonas* and *Hexamita* could have appeared giving rise to the genera parasitic in vertebrates such as *Spironucleus*, *Octomitus*, and *Giardia*. The evolution from free-living to parasitic genera is marked by a

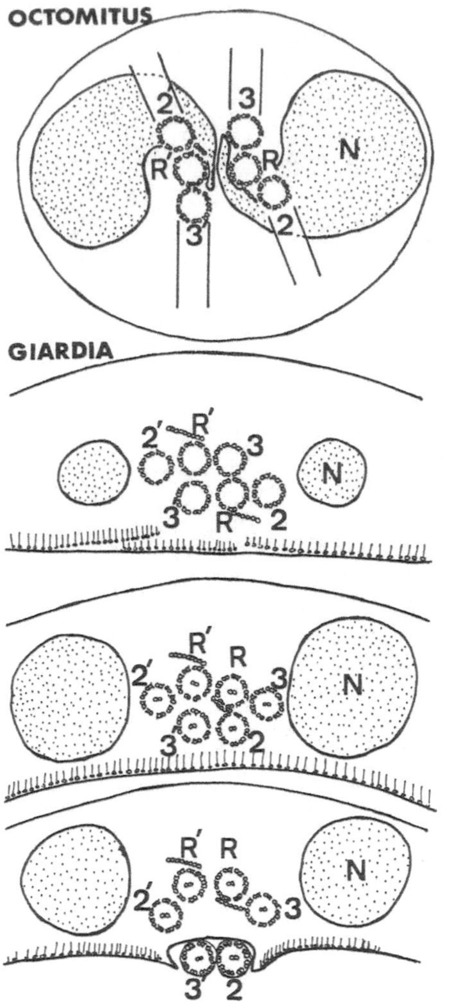

Fig. 28. Scheme showing the two-fold rotational symmetry at the basal body level in *Octomitus* and in *Giardia*

change in the site of flagellar insertion, and of the type of swimming, a progressive reduction of the cytostome which is finally lacking in *Octomitus* and in *Giardia*. The most pronounced adaptation is the development of a specific sucking apparatus in *Giardia*.

Phylogenetic relationships

This group shows the same basal body arrangement as the retortamonads, i.e., two pairs of flagella with a recurrent one. Rare photographs on division in *Hexamita* indicate that there is a pair of basal bodies on each pole of the nucleus; mitosis is of the semi-open type and synchronous in the two nuclei (Brugerolle et al. 1974). The fibers forming the cytoskeleton are difficult to compare in the two groups, and both lack mitochondria and Golgi apparatus.

Oxymonads: order Oxymonadida

This order comprises many parasitic genera, mostly living in symbiosis in termites and roaches. Polymastigidae with the genera *Polymastix* and *Monocercomonoides* are cosmopolitan and represent the basic state of organization (Brugerolle and Joyon 1973, Brugerolle 1980), Oxymonadidae and Pyrsonymphidae have developed modifications such as a holdfast in relation to their life-cycle in the host.

Flagellar apparatus

Organisms have four flagella arising from basal bodies separated into two pairs which are maintained distant by the preaxostylar structure (Figs. 29 and 30). In the basic Polymastigidae one pair, basal bodies 1 and 2, is anteriorly directed and gives rise to free anterior flagella; in the second pair, the basal body R of the recurrent flagellum is directed posteriorly and basal body 3 is directed anteriorly. There is no particularity on axoneme or on the transitional zone of the flagella.
In *Pyrsonympha* all the flagella are recurrent and adhere to the cell body. Striated fibers line the axoneme of the flagella (paraxial rods) (Brugerolle 1970, Hollande and Carruette-Valentin 1970) (Fig. 34).

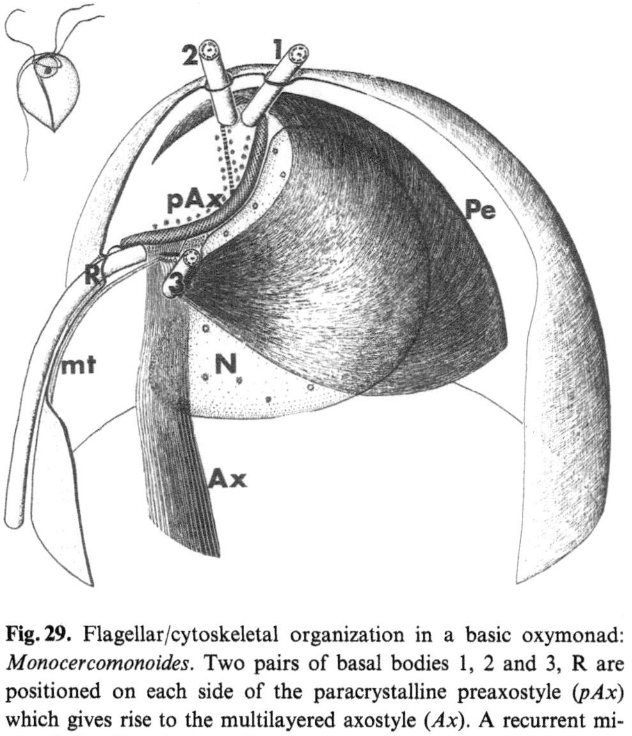

Fig. 29. Flagellar/cytoskeletal organization in a basic oxymonad: *Monocercomonoides*. Two pairs of basal bodies 1, 2 and 3, R are positioned on each side of the paracrystalline preaxostyle (*pAx*) which gives rise to the multilayered axostyle (*Ax*). A recurrent microtubular fiber (*mt*) is adjacent to the recurrent flagellum (*R*); the microtubules of the pelta (*Pe*) cap the anterior part of the nucleus (*N*)

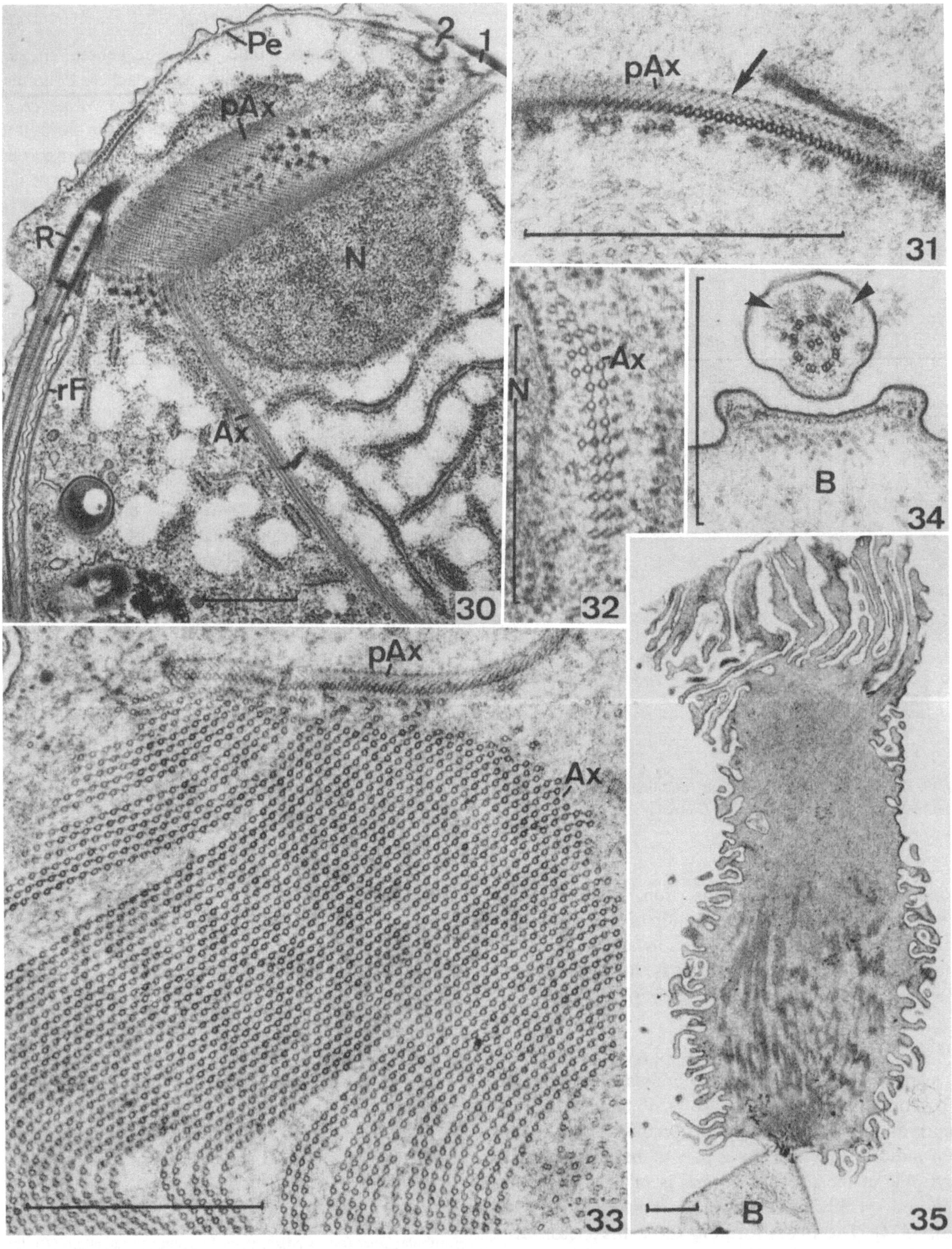
Pe
2
1
pAx
R
N
rF
Ax
30
pAx
31
N
Ax
32
B
34
pAx
Ax
33
B
35

Cytoskeletal system

The major trait is the presence of a preaxostylar fiber which gives rise to the paracrystalline axostyle (Figs. 29, 30, 32, and 33). The preaxostyle is a lamina positioned between the two pairs of basal bodies very close to the anterior nucleus in Polymastigidae. It shows an original composite structure: the face adjacent the nucleus consists of a row of microtubules; the microtubules bear intercrossed projections that form a paracrystalline structure on the other face (Fig. 31). The preaxostyle is the site of microtubule nucleation for axostylar microtubules and could be compared to a MTOC. The paracrystalline axostyle is a multilayer of microtubules linked by permanent intra-row bridges and by labile inter-row links (Figs. 32 and 33). It is relatively simple and non-contractile in Polymastigidae (3 or 4 rows of microtubules) (Fig. 32), but it is more developed in Oxymonadidae and Pyrsonymphidae (more than 50 rows of microtubules) (Fig. 33) where it is able to move in the cytoplasm (McIntosh et al. 1973; McIntosh 1973, 1974; Mooseker and Tilney 1973; Langford and Inoue 1979; Bloodgood and Fitzharris 1979; Woodrum and Linck 1980).

Two other microtubular fibers arise from the basal bodies: one is composed of few microtubules which follow the course of the recurrent flagellum under the membrane (Figs. 29 and 30), the other forms a cap on the nucleus and has been named the pelta (Figs. 29 and 30). In *Pyrsonympha* and in several Oxymonadidae genera, a specific holdfast that modifies the cell shape has developed on the front of the cell, but the basic organization described for Polymastigidae is always retained (Brugerolle 1970, Hollande and Carruette-Valentin 1970, Cochrane et al. 1979) (Fig. 35). The holdfast of *Pyrsonympha* is filled by microfilaments which react with actin antibodies (Brugerolle unpubl. data). This apparatus appears to be an adaptation to the parasitic life and a secondary character similar to the development of the ventral disk in *Giardia*. The mitosis is of the closed type with an internal spindle; a pair of basal bodies are seen at the poles (Brugerolle and Joyon 1973; and unpubl. obs.), however, the complete mitosis and morphogenesis have not been observed at the ultrastructural scale.

Phylogenetic relationships

This group constitutes a monophyletic lineage, the genera show homologous characters. The most apparent are the paracrystalline preaxostyle and axostyle and the same 2 + 2 basal body arrangement with a recurrent flagellum. This arrangement is similar to that of retortamonads and diplomonads but the paracrystalline axostyle and preaxostyle reveals differences and suggests probable long evolutionary distance between this group and the two others.

Evolutionary considerations on the metamonad group

The three orders (Retortamonadida, Diplomonadida, Oxymonadida) of metamonads belong to three different lineages and show few common features: they share the same basic basal body arrangement and they lack mitochondria and Golgi apparatus. The types of mitosis and morphogenesis are incompletely known. Among the cytoskeletal fibers only the microtubular fiber regularly associated to the recurrent basal body could indicate a homology. Other cytoskeletal structures are specific to one lineage or difficult to compare. The metamonads appear heterogeneous and cytological/ultrastructural studies are unable to appreciate the distance between these 3 lineages. Nuclear encoded rRNA sequence comparisons may be able to resolve evolutionary relationships where ultrastructural comparisons meet difficulties.

Parabasalids: phylum Parabasala

Light and electron microscopy have brought evidence that the two orders Trichomonadida and Hypermas-

Fig. 30. The paracrystalline preaxostyle (*pAx*) which generates the axostyle (*Ax*) is shown in *Monocercomonoides*. A pair of basal bodies 1, 2 is positioned on one side of the preaxostyle, on the other side only the basal body (*R*) of the recurrent flagellum is seen. The recurrent microtubular fiber (*rF*) and the microtubules of the pelta (*Pe*) are also shown; *N* nucleus

Fig. 31. Cross section showing the composite structure of the preaxostyle (*pAx*) in *Pyrsonympha*; a row of joined microtubules bears intercrossed projections which form an adjacent paracrystalline layer (arrow)

Figs. 32 and **33.** Cross section of the multilayered axostyle (*Ax*) in the basic *Monocercomonoides* (Fig. 32) and in the contractile axostyle of *Pyrsonympha* (Fig. 33) where intra-row bridges are seen as well as a portion of the preaxostyle (*pAx*)

Fig. 34. Cross section of a recurrent flagellum adhering to the cell body (*B*) in *Pyrsonympha* showing the additional fibers around the axoneme (arrowheads)

Fig. 35. Anterior holdfast of *Pyrsonympha* containing microfilaments arising on the top of the cell body (*B*)

tigida belong to the same evolutionary lineage and can be named Parabasalia (Honigberg 1963, 1973; Brugerolle 1976; Brugerolle and Taylor 1977). Most of the phylogenetic features used include the mastigont system and associated cytoskeleton, but the mitotic process, the presence of hydrogenosomes and the lack of mitochondria are also important features.

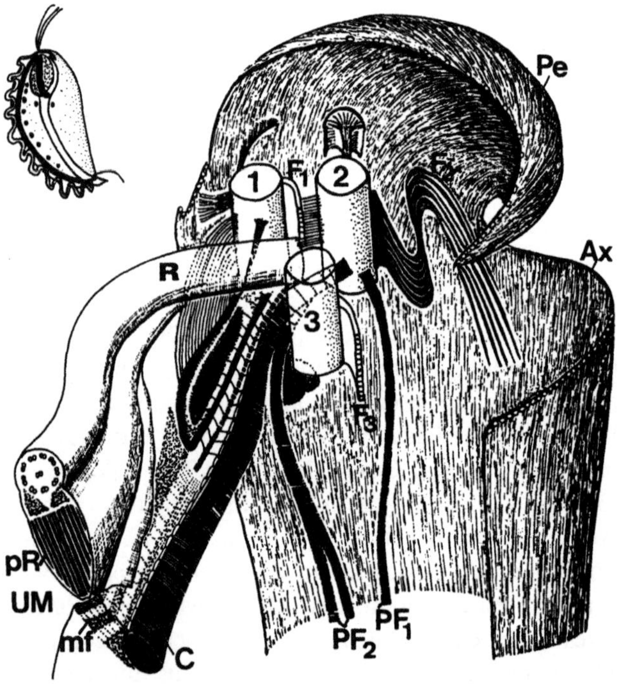

Fig. 36. Reconstruction of the flagellar/cytoskeletal system in *Tritrichomonas muris*; basal bodies 1, 2, 3 of the anterior flagella are arranged in a fixed position around the basal body R of the recurrent flagellum. The periodic roots connected to the basal bodies are: the two parabasal fibers (PF_1 and PF_2), the costa (*C*), the sigmoid fibers (F_2), the hooked lamina (F_1 and F_3). Notice the paraxial rod of the recurrent flagellum (*pR*), the microfibrillar layers (*mf*) forming the cell body portion of the undulating membrane (*UM*) which are linked to the subjacent costa, and the microtubular fiber of the pelta (*Pe*) axostyle (*Ax*) complex

Organization of the mastigont system

The basic flagellar system is found in the more simple genus *Monocercomonas*. The cell has 4 flagella, 3 are anteriorly directed and one is recurrent (Figs. 36 and 37). The basal bodies are always arranged in the same way: basal body R is orthogonal to the three others which have a fixed position. Basal body 1 and 3 bear a hooked lamina (F 1 and F 3) and basal body 2 bears a set of sigmoid fibers (F 2) which come in contact with the pelta-axostyle junction (Figs. 36, 38, and 39). There is no particularity in the transitional region of the flagella. The recurrent flagellum which is adherent to the cell body in Trichomonadidae and Devescovinidae develops paraxonemal fibers (paraxial rods) whose ultrastructure is species or genus specific. An example is shown with *Tritrichomonas muris* (Figs. 36 and 45) where the axoneme of the flagellum is doubled by a large paracrystalline fiber. When the number of flagella increases the basic organization of the basal bodies 1, 2, 3, R does not change and additional basal bodies,

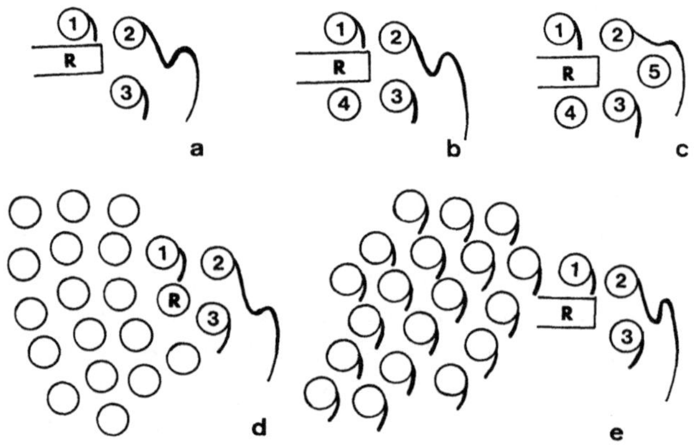

Fig. 37. Scheme showing the arrangement of the additional basal bodies around the privileged basal bodies 1, 2, 3, R in **a** *Monocercomonas*, **b** *Tetratrichomonas* or *Trichomonas*, **c** *Hexamastix*, **d** *Lophomonas*, and **e** *Joenia*; notice that in this latter genus all the additional basal bodies forming the flagellar area bear a hooked lamina

Figs. 38 and **39.** Basal bodies arrangement in *Monocercomonas* (Fig. 38) and *Tritrichomonas* (Fig. 39). Basal bodies 1, 2, 3 of the anterior flagella are orthogonal to basal body R of the recurrent flagellum. Hooked lamina (F_1 and F_3), sigmoid fibers (F_2) applied against the pelta (*Pe*) axostylar (*Ax*) junction, parabasal fibers (*PF*) and costa (*C*) are shown

Fig. 40. Periodic structure of the costa (*C*) connected to the basal bodies in the genus *Tritrichomonas*

Figs. 41 and **42.** Parabasal fibers (*PF*) supporting the Golgi cisternae (*G*) in *Monocercomonas* (Fig. 41) and some branches of the parabasal fiber complex in the hypermastigid *Trichonympha* (Fig. 42)

Fig. 43. Longitudinal section of the pelta (*Pe*) axostyle (*Ax*) complex in *Tritrichomonas*. Anterior flagella (*aF*), nucleus (*N*)

Fig. 44. Internal fibrous structure (arrow) in the lamellar undulating membrane (*uM*) in Trichomonadinae (*Trichomitus*)

Fig. 45. Paraxonemal structures (arrows) in the recurrent flagellum (*R*) of *Tritrichomonas muris* and microfibrillar connections (*mf*) with the costa (*C*) in the cell body portion of the undulating membrane

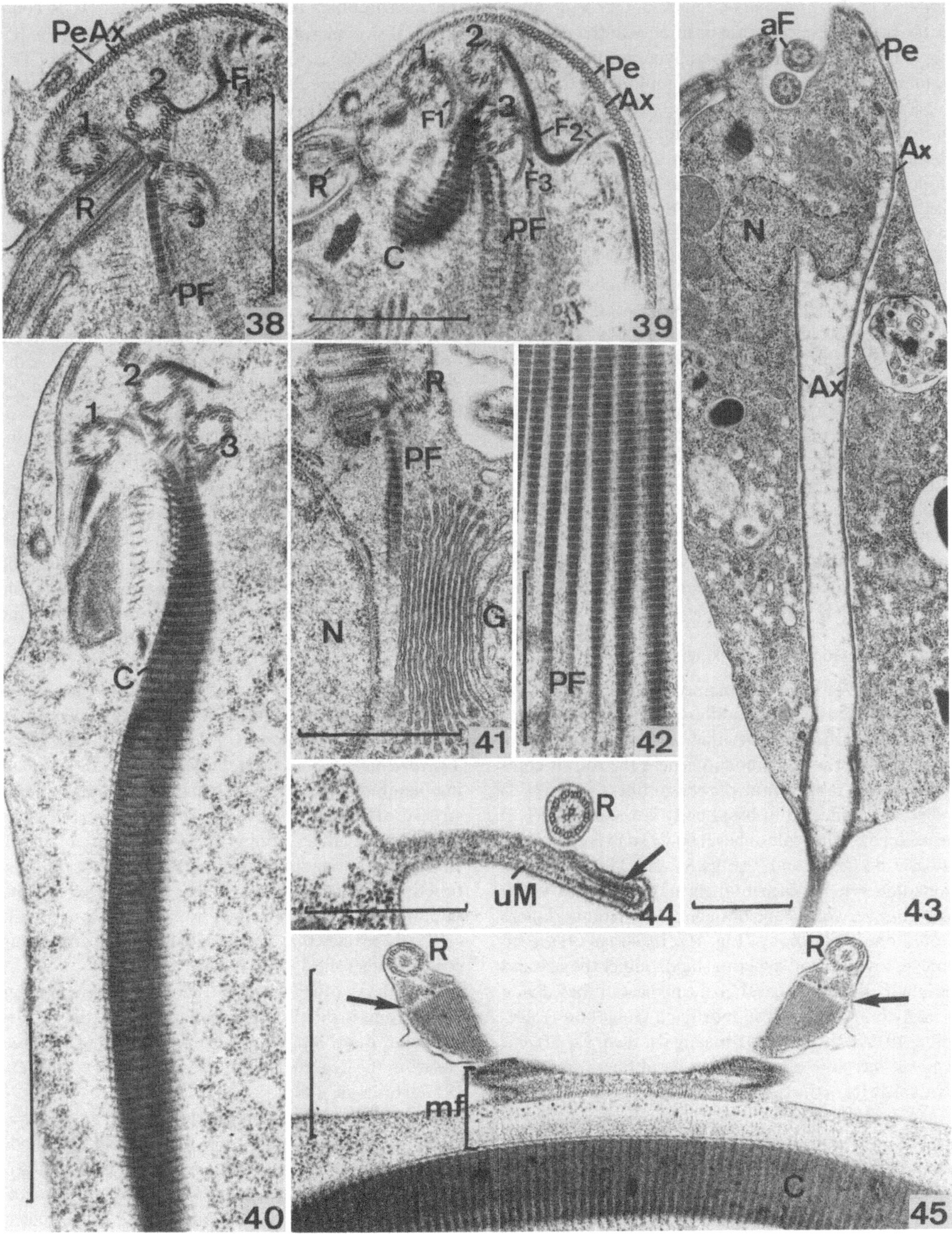
PeAx
F1
2
1
R
3
PF
38
2
1
Pe
Ax
F1
3
F2
R
F3
PF
C
39
aF
Pe
Ax
N
Ax
43
2
1
3
C
40
R
PF
N
G
41
PF
42
R
uM
44
R
R
mf
C
45

number 4 in *Tetratrichomonas* and number 5 in *Hexamastix*, occupy a very precise position near the previous ones (Fig. 37). In highly evolved hypermastigids where the number of flagella has greatly multiplied, the basic or privileged basal bodies (1, 2, 3, R) could be recognized in most species and particularly in *Lophomonas* (Hollande and Carruette-Valentin 1972) and in *Joenia* (Hollande 1979) (Fig. 37). In the latter, each additional basal body bears a hooked fiber that is similar to the F 1 and F 3 attached to basal bodies 1 and 3, suggesting these basal bodies have formed replicas. When the number of flagella is reduced, the basic barren basal bodies are retained in *Histomonas* (Rybicka et al. 1972) but disappear in *Dientamoeba* (Camp et al. 1974). In trichomonads prior to mitosis a new basal body appears opposite to R and becomes the new recurrent basal body (R'). Then the partition of the parental basal bodies occurs, one parental basal body migrates towards R' forming the first pair of basal bodies of the daughter cell. Finally each sister cell reconstitutes its number of basal bodies (Brugerolle 1975 b). Although the mitosis has been observed in trichomonads, several points concerning basal body partitioning and fibre reconstitution remain to be cleared.

Basal body associated cytoskeleton

Certainly the striated root named the parabasal fiber, which polarizes the Golgi bodies forming the parabasal apparatus, is the most original feature of the group. In the basic monocercomonad genera the root is composed of two branches or parabasal fibers, one (PF 1) is directly connected to basal body 2, the other (PF 2) passes on the other side of basal body 3 and is connected to basal bodies R and 2 (Figs. 36 and 38). In Trichomonadida genera, each branch may support a Golgi body, but very often one branch is dominant and there is only one Golgi body (Fig. 41). In larger cells such as devescovinids and hypermastigids, either the size and the length of the parabasal fibers increase or they divide in many branches each supporting a Golgi body (Hollande 1979). In the hypermastigids such as *Trichonympha*, there are about 50 parabasal fibers which individualize from the two semi-circular rostral structures (Fig. 42). In all cases parabasal fibers are connected to the basic or privileged basal bodies. In *Dientamoeba* where the basal bodies are lacking, the two parabasal branches converge on a point, named here the attractophore, which is also the centrosomal center of the cell (Camp et al. 1974).

Another character common to all the parabasalids is the microtubular pelta-axostyle complex. It is composed of two microtubular sheets which overlap on the pelta-axostyle junction (Figs. 36, 38, 39, and 43). The anterior semi-circular pelta surrounds the basal body area and the axostyle forms the longitudinal axis of the cell. Microtubules are maintained side by side by microfibrillar bridges. In the very simple and basic Monocercomonadinae and Trichomonadinae, the row of microtubules forms a groove. But in larger cells of Devescovinidae and Joenidae, the row enrolls to give a very broad structure (Brugerolle 1976, Hollande 1979). In *Dientamoeba*, where basal bodies are lacking, the pelta-axostyle complex is absent (Camp et al. 1974). Evidence shows that microtubules begin to assemble on the surface of the F 2 sigmoid fiber which plays the role of a MTOC. Several other fibers connected to the basal bodies are more specific (Fig. 36). The costa occurs only in Trichomonadidae which have an undulating membrane. It is a very broad striated root connected to basal body R of the recurrent flagellum and courses under the undulating membrane (Figs. 36 and 40). Several types of costa have been distinguished, mainly by the band pattern (Honigberg and Brugerolle 1990), and the costa of *Trichomitopsis* is contractile (Amos et al. 1979). In the genus *Tritrichomonas* the costa is linked to the undulating membrane by microfibrillar layers (Figs. 36 and 45).

The undulating membranes contain fibrous structures which are species-, genus-, or sub-family specific. In Trichomonadinae such as *Trichomitus, Trichomonas, Tetratrichomonas* and *Pentatrichomonas*, the undulating membrane has a lamellar shape containing a dense structure (Fig. 44) which looks like a woven tissue when seen after negative staining (Brugerolle unpubl. data). In Tritrichomonadinae, the undulating membrane is a functional unit composed of a modified flagellum containing a paraxial rod and a rail-like swelling of the cell body surface that is associated with the underlying costa (Brugerolle 1976) (Figs. 36 and 45).

Among many other specific structures described in this lineage one is interesting to mention: it is the furcate striated root prolongating the proximal end of the basal bodies in the hypermastigid *Deltotrichonympha* (Tamm 1972, Hollande 1986).

The mitotic spindle is completely external and the nuclear envelope persists during the division (Hollande 1971, Hollande and Carruette-Valentin 1972, Brugerolle 1975 b). Structures, named battachio or attractophores, are appended at the base of the basal bodies; they polarize the spindle and are equivalent to a centrosome.

Evolutionary considerations

The presence of a set of homologous characters indicates this group is a monophyletic lineage. It constitutes a very interesting example of flagellar and cytoskeletal structure evolution and of protein evolution in relation to species differentiation. This study has been undertaken (Viscogliosi and Brugerolle 1990) but the major drawback is the difficulty to cultivate many of these parasitic species that restricts the study to a few genera.

Summary of the taxonomic characters of the flagellates without mitochondria: Archamoeba, Metamonada, Parabasala

Phylum Archamoeba

Order Pelobiontida

- basically one karyomastigont composed of one basal body bearing a poorly motile flagellum and,
- a cape of microtubules attaching the basal body to the nucleus;
- no Golgi apparatus, no microbodies;
- type of mitosis unknown.

Phylum Metamonada

- 4 basal bodies and flagella arranged in two pairs;
- recurrent basal body associated with at least one recurrent microtubular fiber;
- no Golgi apparatus, no microbodies;
- close or semi-open mitosis.

Order Retortamonadida

- the two pairs of basal bodies are very close;
- complete corset of bridged microtubules under the plasma membrane;
- recurrent flagellum included in a ventral cystostomal-cytopharyngeal depression bordered by 2 microtubular fibers and by a paracrystalline lamina;
- close mitosis with an internal spindle (incompletely known).

Order Diplomonadida

- basically one karyomastigont in the Enteromonadina sub-order; two karyomastigonts in the Diplomonadina sub-order;
- the two pairs of basal bodies are very close;
- 3 microtubular fibers: supranuclear, infranuclear and cytostomal fibers are connected to the basal bodies;
- ventral cytostomal-cytopharyngeal depression bordered by 2 microtubular fibers, by a striated lamina and striated roots.

Order Oxymonadida

- the two pairs of basal bodies are maintained apart by a preaxostylar paracrystalline lamina;
- internal crystalline axostyle originating from the preaxostylar lamina;
- anterior microtubular fiber or pelta and recurrent microtubular fiber connected to the basal body pair comprising the recurrent flagellum;
- no cytostomal depression associated with the recurrent microtubular fiber;
- close mitosis with an internal spindle.

Phylum Parabasala

The two orders Trichomonadida and Hypermastigida have the same basic characters.

- basically 4 basal bodies typically arranged, the basal body R of the recurrent flagellum being orthogonal to the 3 basal bodies 1, 2, 3 of the anteriorly directed flagella. Additional basal bodies forming a flagellar area in Hypermastigida;
- basal bodies 1 and 3 bearing a hooked lamina, basal body 2 a set of sigmoid fibers connected to the pelta-axostyle complex;
- two striated fibers connected to the basal bodies and supporting the Golgi cisternae forming the parabasal apparatus; the number of parabasal fibers multiplies in Hypermastigida;
- a pelta-axostyle complex composed of a row of microtubules; the row enrolls giving a multilayered axostyle in some hypermastigids;
- hydrogenosomes organelles always present;
- typical close mitosis (pleuromitosis) with an internal spindle composed of pole to pole microtubules forming a paradesmosis and of chromosomal microtubules attached to the kinetochores inserted in the nuclear envelope.

Conclusion

These three groups of flagellates (Archamoeba, Metamonada, Parabasala) do not show recognizable homologous characters and they constitute several lines of descent in spite of the fact that they all lack mitochondria. Groups which lack mitochondria are considered more primitive than others (Stewart and Mattox 1980; Cavalier-Smith 1987a, 1989). However, they could have secondarily lost their mitochondria as has

probably occurred in rumen Ciliates (Müller 1988, Cavalier-Smith 1987 b). The lack of a Golgi apparatus in Archamoeba and Metamonada could indicate a very simple mode of compartmentalization and secretion in these two groups, and that differentiates them a little more from Parabasala. Archamoeba have a very unique and simple mastigont/cytoskeletal system which could be considered either primitive or regressive. A similar basal body arrangement suggests the metamonad groups are probably paraphyletic or highly divergent, but we have only a few cytological and ultrastructural characters. The use of an anti-ciliary tubulin antibody which reacts with oxymonad flagella but not with diplomonad ones could be an indication of this heterogeneity (Brugerolle and Adoutte 1988).
Considerable clarification of the phylogenetic position of these groups and of their interrelationships is expected from ongoing studies of their ribosomal RNA sequences. Comparison of conserved portions of rRNA sequences has already proven very efficient for deriving broad phylogenies of protists (Sogin et al. 1986, 1989; Gunderson et al. 1986, 1987; Vossbrink et al. 1987; Baroin et al. 1988; Wolters and Erdman 1988; Qu et al. 1988; Perasso et al. 1989; Sogin 1989). At this time these methods have been tested on many phylogenetic examples and checked by other data on fossils (Le et al. 1989). If these methods are valuable for the short and medium evolutionary distances, they could be extrapolated to long distance evolution which we particularly need in protistan phylogeny. At this moment, only two representatives of the three phyla discussed in this paper have been analyzed: *Giardia lamblia* (Diplomonadida) and *Trichomonas vaginalis* (Trichomonadida) (Baroin et al. 1988, Perasso et al. 1989, Sogin et al. 1989, Sogin 1989). Both emerge very deeply in the evolutionary tree of eukaryotes, *Giardia* so far constitutes the earliest known eukaryotic branch (Sogin et al. 1989, Sogin 1989). The microsporidian *Vairimorpha* branches between *Giardia lamblia* and *Trichomonas vaginalis* confirming the fact that diplomonads and trichomonads are indeed in two distinct phyla. Recent sequence data (unpubl.) from our group concerning five other parabasalids shows them all branching from a common stem, therefore confirming the monophyly of this group. Further data on Pelobiontida and other Metamonada are eagerly awaited. Because these organisms are difficult to grow, their sequences will be accessible only through the PCR (polymerase chain reaction) approach. The coming period will probably witness the solution of the phylogenetic problems discussed in this paper.

References

Amos WD, Grimstone AV, Rothschild LJ, Allen RD (1979) Structure, protein composition and birefringence of the root fiber in the flagellate *Trichomonas*. J Cell Sci 35: 139–164

Baker DA, Holberton DV, Marshall J (1988) Sequence of a *Giardia* subunit cDNA from *Giardia lamblia*. Nucleic Acids Res 16: 71–77

Bardele CF (1983) Comparative freeze-fracture study of the ciliary membrane of protists and invertebrates in relation to phylogeny. J Submicrosc Cytol 15: 263–267

Baroin A, Perasso R, Qu LH, Brugerolle G, Bachellerie JP, Adoutte A (1988) Partial phylogeny of the unicellular eukaryotes based on rapid sequencing of a portion of 28 S ribosomal RNA. Proc Natl Acad Sci USA 85: 3474–3478

Bloodgood RA, Fitzharris TP (1979) Initiation of bends in the microtubular axostyle of *Pyrsonympha*. Cytobios 23: 109–117

Brugerolle G (1970) Sur l'ultrastructure et la position systématique de *Pyrsonympha vertens* (Zooflagellata Pyrsonymphina). CR Acad Sci Paris 270: 2558–2560

– (1973) Etude ultrastructurale du trophozoite et du kyste chez le genre *Chilomastix* Alexeieff 1910 (Zoomastigophorea, Retortamonadida Grassé 1952). J Protozool 20: 574–575

– (1974) Contribution à l'étude cytologique et phylétique des Diplozoaires (Zoomastigophorea, Diplozoa, Dangeard 1910). III Etude ultrastructurale du genre *Hexamita* (Dujardin 1838). Protistologica 10: 83–90

– (1975 a) Etude ultrastructurale du genre *Enteromonas* da Fonseca (Zoomastigophorea) et revision de l'ordre Diplomonadida Wenyon. J Protozool 22: 468–475

– (1975 b) Etude de la cryptopleuromitose et de la morphogenèse de division chez *Trichomonas vaginalis* et chez plusieurs trichomonadines primitives. Protistologica 11: 457–468

– (1975 c) Contribution à l'étude des Diplozoaires (Zoomastigophorea, Diplozoa, Dangeard 1910). V Nouvelle interprétation de l'organisation cellulaire de *Giardia*. Prostistologica 11: 99–109

– (1975 d) Contribution à l'étude des Diplozoaires (Zoomastigophorea, Diplozoa, Dangeard 1910). VI Caractères généraux des Diplozoaires. Protistologica 11: 111–118

– (1976) Cytologie ultrastructurale, systématique et évolution des Trichomonadida. Ann Stat Biol Besse en Chandesse Univ Clermont-Ferrand 10: 1–92

– (1977) Ultrastructure du genre *Retortamonas* Grassi 1879 (Zoomastigophorea, Retortamonadida, Wenrich 1932). Protistologica 13: 233–240

– (1980) Etude ultrastructurale du flagellé parasite *Polymastix melolonthae* (Oxymonadida). Protistologica 17: 139–145

– (1982) Caractères ultrastructuraux d'une Mastigamibe: *Mastigina hylae* (Frenzel). Protistologica 18: 227–235

– (1986) Structural diversity of trichomonads as the basis for systematic and evolutionary considerations. Acta Univ Carolinae Biol Prague 30: 199–210

– Adoutte A (1988) Probing protistan phylogenies with an antitubulin antibody. BioSystems 21: 255–268

– Joyon L (1973) Ultrastructure du genre *Monocercomonoides* (Travis). Zooflagellata, Oxymonadida. Protistologica 9: 71–80

– Mignot JP (1990) Phylum Zoomastigina class Retortamonadida. In: Margulis L, Corliss JO, Melkonian M, Chapman DJ (eds) Handbook of Protoctista. Jones and Bartlett, Boston, pp 259–265

– Taylor FJR (1977) Taxonomy, cytology and evolution of the

Mastigophora. In: Hutner SH (ed) Proceedings Vth International Congress of Protozoology, New York. Pace University, New York, pp 14–28
– Joyon L, Oktem N (1973 a) Contribution à l'étude cytologique et phylétique des Diplozoaires (Zoomastigophorea, Diplozoa, Dangeard 1910). I Etude ultrastructurale du genre *Trepomonas* (Dujardin). Protistologica 9: 339–344
– – – (1973 b) Contribution à l'étude cytologique et phylétique des Diplozoaires (Zoomastigophorea, Diplozoa, Dangeard 1910). II Etude ultrastructurale du genre *Spironucleus* (Lavier 1936). Protistologica 9: 495–502
– – – (1974) Contribution à l'étude cytologique et phylétique des Diplozoaires (Zoomastigophorea, Diplozoa, Dangeard 1910). IV Etude ultrastructurale du genre *Octomitus* (Prowazek 1904). Protistologica 10: 457–463
Camp RR, Mattern CFT, Honigberg BM (1974) Study of *Dientamoeba fragilis* Jepps et Dobell. I Electron microscopic observations of the binucleate stages. II Taxonomic position and revision of the genus. J Protozool 21: 69–82
Cavalier-Smith T (1983) A 6-kingdom classification and a unified phlyogeny. In: Schenk HEA, Schemmler WS (eds) Endocytobiology II: intracellular space as oligogenetic, vol 2. De Gruyter, Berlin, pp 1027–1034
– (1987 a) Eukaryotes with no mitochondria. Nature 326: 332–333
– (1987 b) The simultaneous symbiotic origin of mitochondria, chloroplasts, and microbodies. Ann NY Acad Sci 503: 17–54
– (1987 c) The origin of eukaryotes and archaebacterial cells. Ann NY Acad Sci 503: 55–72
– (1989 a) Archaebacteria and Archezoa. Nature 339: 100–144
– (1989 b) Systems of kingdoms. Yearbook of Science and Technology. McGraw-Hill, New York, pp 175–179
Chavez LA, Balamuth W, Gong T (1986) A light and electron microscopical study of a new, polymorphic free-living amoeba, *Phreatamoeba balamuthi* n. g., n. sp. J Protozool 33: 397–404
Cochrane M, Harriett E, Bushe HE, Scammel JG (1979) Structure of the attached stage of *Pyrsonympha* in the termite *Reticulitermes flavipes* Kollar. Protistologica 15: 259–270
Corliss JO (1986) Progress in protistology during the first decade following reemergence of the field as a respectable interdisciplinary area in modern biological research. Prog Protozool 1: 11–63
– (1987) Protistan phylogeny and eukaryogenesis. Int Rev Cytol 100: 319–370
Crossley R, Holberton DV (1983) Characterization of proteins from the cytoskeleton of *Giardia lamblia*. J Cell Sci 59: 81–103
Eyden BP, Vickerman K (1975) Ultrastructure and vacuolar movements in the free-living diplomonad *Trepomonas agilis* Klebs. J Protozool 22: 54–66
Grain J, Mignot JP, De Puytorac P (1988) Ultrastructure and evolutionary modalities of flagellar and ciliary systems in protists. Biol Cell 63: 219–237
Grassé PP (1952) Classe des Zooflagellés, Zooflagellata ou Zoomastigina. In: Grassé PP (ed) Traité de zoologie, vol 1, fasc 1. Masson, Paris, pp 963–982
Griffin JO (1988) Fine structure and taxonomic position of the giant amoeboid flagellate *Pelomyxa palustris*. J Protozool 35: 300–315
Goodkov AV (1989) Ultrastructure of the giant amoeba *Pelomyxa palustris*. 1 Cytoplasmic microtubules, subcentrioles and flagella. A comparative morphological analysis of organization. Tsitologiya 31: 371–379
Gunderson JH, Elwood H, Ingold A, Kinde K, Sogin ML (1987) Phylogenetic relationships between Chlorophytes, Chrysophytes, and Oomycetes. Proc Natl Acad Sci USA 84: 5823–5827
– Mc Cutchan TF, Sogin ML (1986) Sequence of the small subunit ribosomal RNA gene expressed in the bloodstream stages of *Plasmodium berghei*: evolutionary implications. J Protozool 33: 525–529
Hibberd DJ (1979) The structure and phylogenetic significance of the flagellar transition region in the chlorophyll *c* containing algae. BioSystems 11: 243–261
Holberton DV (1973) Fine structure of the ventral disk apparatus and the mechanism of attachment in the flagellate *Giardia muris*. J Cell Sci 13: 11–41
– (1981) Arrangement of subunits in microribbons from *Giardia*. J Cell Sci 77: 167–185
– Ward AP (1981) Isolation of the cytoskeleton from *Giardia*. J Cell Sci 47: 139–166
– Baker DA, Marshall J (1988) Segmented alphahelical coiled-coil structure of the protein giardin from *Giardia* cytoskeleton. J Mol Biol 204: 789–795
Hollande A (1971) Les atractophores, l'induction du fuseau et la division cellulaire chez les Hypermastigines. Etude ultrastructurale et révision systématique des Trichonymphines et des Spirotrichonymphines. Protistologica 7: 2–100
– (1979) Complexes parabasaux et symétrie bilatérale du rostre chez *Joenia annectens*. Protistologica 15: 407–426
– (1986) Morphologie, ultrastructure et affinités de *Deltotrichonympha operculata* Sutherland (1953) symbionte du termite australien *Mastotermes darwinensis* Frogatt. Protistologica 22: 415–439
– Carruette-Valentin J (1970) La lignée des Pyrsonymphines et les caractères infrastructuraux communs aux genres *Opisthomitus, Oxymonas, Saccinobacculus, Pyrsonympha* et *Streblomastix*. CR Acad Sci Paris 270: 1587–1590
– – (1972) Le problème du centrosome et la cryptopleuromitose atractophorienne chez *Lophomonas striata*. Protistologica 8: 267–278
Honigberg BM (1963) Evolutionary and systematic relationships in the flagellate order Trichomonadida Kirby. J Protozool 10: 20–63
– (1973) Remarks upon trichomonad affinities of certain parasitic protozoa. In: De Puytorac P, Grain J (eds) Progress in protozoology. Université de Clermont-Ferrand, Clermont-Ferrand, p 187
– Brugerolle G (1990) Structure. In: Honigberg BM (ed) Trichomonads parasitic in humans. Springer, Berlin Heidelberg New York Tokyo, pp 2–35
Kulda J, Nohynkova E (1978) Flagellates of the human intestine and of the intestines of other species. In: Kreier JP (ed) Parasitic protozoa, vol 2. Academic Press, New York, pp 2–138
Langford GM, Inoue S (1979) Motility of the microtubular axostyle in *Pyrsonympha*. J Cell Biol 80: 521–538
Le HLV, Perasso R, Billard R (1989) Phylogénie moléculaire des "poissons" basée sur l'analyse des séquences d'ARN ribosomique 28S. CR Acad Sci Paris Sér III 309: 493–538
Lipscomb DL (1985) The eukaryotic kingdoms. Cladistics 1: 127–140
McIntosh JR (1973) The axostyle of *Saccinobaculus*. II Motion of the microtubule bundle and a structural comparison of straight and bent axostyles. J Cell Biol 56: 324–339
– (1974) Bridges between microtubules. J Cell Biol 61: 166–187

McIntosh JR, Ogata ES, Landis SC (1973) The axostyle of *Saccinobaculus*. I Structure of the organism and its microtubule bundle. J Cell Biol 56: 304–323

Melkonian M (1980) Ultrastructural aspects of basal body associated fibrous structures in green algae, critical review. BioSystems 12: 85–104

Moestrup Ø (1982) Flagellar structure in algae: a review with new observations on the Chrysophyceae, Phaeophyceae (Fucophyceae), Euglenophyceae, and *Reckertia*. Phycologia 21: 427–528

Mooseker MS, Tilney LG (1973) Isolation and reactivation of the axostyle, evidence for a dynein-like ATPase in the axostyle. J Cell Biol 56: 13–26

Müller M (1988) Energy metabolism of protozoa without mitochondria. Annu Rev Microbiol 42: 465–488

Page FC (1987) The classification of "naked" amoebae (phylum Rhizopoda). Arch Protistenk 133: 199–217

Peattie DA, Alonso RA, Hein A, Caulfield JP (1989) Ultrastructural-localization of giardins to the edges of disk microribbons of *Giardia lamblia* and the nucleotide and deduced protein sequence of alpha giardin. J Cell Biol 109: 2323–2335

Perasso R, Baroin A, Qu LH, Bachellerie JP, Adoutte A (1989) Origin of the algae. Nature 339: 142–144

Qu LH, Perasso R, Baroin A, Brugerolle G, Bachellerie JP, Adoutte A (1988) Molecular evolution of the 5′-terminal domain of large-subunit rRNA from lower eukaryotes. A broad phylogeny covering photosynthetic and non-photosynthetic protists. BioSystems 21: 203–208

Rybicka K, Honigberg BM, Holt SC (1972) Fine structure of the mastigont system in culture forms of *Histomonas meleagridis* (Smith). Protistologica 8: 107–120

Salisbury JL (1988) The lost neuromotor apparatus of *Chlamydomonas*, rediscovered. J Protozool 35: 574–577

– Baron A, Surek B, Melkonian M (1984) Striated flagellar roots: isolation and partial characterization of a calcium-modulated contractile organelle. J Cell Biol 99: 962–970

Schulze D, Robenek H, Mc Fadden GI, Melkonian M (1987) Immunolocalization of a Ca^{++}-modulated contractile protein in the flagellar apparatus of green algae: the nucleus-basal body connector. Eur J Cell Biol 45: 51–61

Skuja H (1956) Taxonomische und biologische Studien über das Phytoplankton schwedischer Binnengewässer. Nova Acta Reg Soc Sci Upsaliensis Ser IV 16: 104–105

Smith R Mc K, Patterson DJ (1986) Analysis of heliozoan interrelationships. Example of the potentials and limitations of ultrastructural approaches to the study of protistan phylogeny. Proc R Soc Lond [Biol] 227: 325–366

Sogin ML (1989) Evolution of eukaryotic microorganisms and their small subunit ribosomal RNAs. Amer Zool 29: 487–499

– Elwood HJ, Gunderson JH (1986) Evolutionary diversity of eukaryotic small-subunit rRNA genes. Proc Natl Acad Sci USA 83: 1383–1387

– Gunderson JH, Elwood HJ, Alonso RA, Peattie DA (1989) Phylogenic meaning of the kingdom concept: An unusual ribosomal RNA from *Giardia lamblia*. Science 243: 75–77

Stewart KD, Mattox K (1980) Phylogeny of phytoflagellates. In: Cox ER (ed) Phytoflagellates, development in marine biology, vol 2. Elsevier/North Holland, Amsterdam, pp 433–462

Tamm SL (1972) Free kinetosomes in Australian flagellates. I Types and spatial arrangement. J Cell Biol 54: 39–55

Taylor FJR (1978) Problems in the development of an explicit hypothetical phylogeny of the lower eukaryotes. BioSystems 10: 67–89

Vickerman K (1969) The fine structure of *Trypanosoma congolense* in its bloodstream phase. J Protozool 16: 54–69

– (1990) Phylum Zoomastigina class Diplomonadida. In: Margulis J, Corliss JO, Melkonian M, Chapman DL (eds) Handbook of Protoctista. Jones and Bartlett, Boston, pp 200–210

Viscogliosi E, Brugerolle G (1991) Biochemical and immunological characterization of cytoskeletal proteins in trichomonads: *Tritrichomonas foetus, Tritrichomonas muris, Trichomonas vaginalis*. J Protozool 38: 32 A

Vossbrinck CR, Maddox JV, Friedman S, Debrunner-Vossbrinck BA, Woese CR (1987) Ribosomal RNA sequence suggests microsporidia are extremely ancient eukaryotes. Nature 326: 411–414

Wolters J, Erdmann VA (1988) Cladistic analysis of ribosomal RNAs. The phylogeny of eukaryotes with respect to the endosymbiotic theory. BioSystems 21: 209–214

Woodrum DT, Linck RW (1980) Structural basis of motility in the microtubular axostyle: implications for cytoplasmic microtubule structure and function. J Cell Biol 87: 404–414

Protoplasma (1991) 164: 91–104

PROTOPLASMA

An ultrastructural comparison of the mitotic apparatus, feeding apparatus, flagellar apparatus and cytoskeleton in euglenoids and kinetoplastids

R. E. Triemer[1,]* and **M. A. Farmer**

[1] Department of Biological Sciences, Rutgers University, Piscataway, New Jersey, and [2] Center for Advanced Ultrastructural Research, Barrow Hall, University of Georgia, Athens, Georgia

Received June 30, 1990
Accepted December 14, 1990

Summary. The euglenoids and kinetoplastids form a diverse assemblage of organisms which show no obvious phylogenetic relationship with other flagellates. An ultrastructural examination and comparison of the flagellar apparatus, the feeding apparatus, and mitotic nucleus indicate a number of shared morphological features which support a common ancestry for the two groups. Of particular interest is the euglenoid, *Petalomonas cantuscygni*, which shares many of the ultrastructural features common to both groups. Based on the data presented, we hypothesize that a euglenoid with features similar to those now present in *P. cantuscygni* was ancestral to both the euglenoid and kinetoplastid lines.

Keywords: Bodonid; Crithidia; Cytoskeleton; Cytostome; *Diplonema; Euglena*; Feeding apparatus; Flagellar apparatus; Microtubular root; Mitosis; Pellicle; *Rhynchomonas; Trypanoplasma.*

Abbrevation: MTR complex of reinforcing microtubules.

Introduction

It is well accepted that higher plants and animals are descended from ancestors in the kingdom Protoctista. Thus, the earliest branches of the evolutionary "tree" leading to higher eukaryotes are occupied by representative protists. This paper will focus on two specific groups of protists with a long, but uncertain evolutionary history, the kinetoplastid and euglenoid flagellates. At the molecular level, phylogenies inferred from 18 S-like rRNA sequences place the euglenoid flagellates closer to the base of the evolutionary tree than most other eukaryotes with the next branch of the tree being occupied by the trypanosomes (Sogin et al. 1986). It must be noted, however, that the molecular data also suggest that trypanosomes and euglenoids, though more closely related to each other than to other eukaryotes, have had a long and separate evolutionary history since the point of divergence. Sogin et al. (1986) note that "the genetic diversity in this collection of eukaryotes is seen to exceed that displayed within either the eubacterial or the archebacterial lines of descent".

The euglenoids and kinetoplastids form an extremely diversified, yet isolated group of protists occupying a variety of habitats and exhibiting numerous lifestyles. Our limited understanding of the phylogeny and relationships of these organisms with other algae and protists is evidenced by their past taxonomic histories. The kinetoplastids were at one time classified in the order Protomonadina along with other flagellates having from 1–4 flagella. In this classification, the bodonids were separated from the trypanosomes on the basis of flagellar number and arrangement and were grouped with other biflagellates. Hollande (1952) also recognized the trypanosomes and bodonids as two separate orders and included in his Bodonidea genera which did not possess kinetoplasts. The bodonids and trypanosomes were brought together into the single order Kinetoplastida by virtue of their possession of a kinetoplast by Honigberg (1963).

The taxonomic position of the euglenoids is similarly colored. In past phylogenetic schemes the euglenoids have had the distinction of occupying many different branches of the phylogenetic tree due to some char-

* Correspondence and reprints: Department of Biological Sciences, Rutgers University, Piscataway, NJ 08855-1059, U.S.A.

acteristics they share with other groups. They have been placed near the green algae based on the presence of chlorophyll *b* in their plastids (Klebs 1883, Dougherty 1955). Flagellar hairs, although structurally different (Bouck et al. 1978) are present on euglenoid flagella and on the flagella of chlorophyll *a* and *c* containing organisms (Bouck 1971, 1972). The eyespot is free of the chloroplast as in the eustigmatophytes (Hibberd and Leedale 1970, 1971, 1972). A paraxial rod exists in the flagella of dinoflagellates and euglenoids (Cachon et al. 1988). None of these features by itself is sufficient to give us a clear indication as to the relationship of the Euglenophyta with other algae, a fact which has long been recognized (Klebs 1983; Senn 1900; Fritsch 1935; Dodge 1975; Leedale 1967, 1978). Since many euglenoids are phagotrophic they were claimed by protozoologists as well as phycologists. Their protozoological history is likewise confused with the euglenoids having a status ranging from an order, the Euglenida, in the class Phytomastigophorea (Honigberg et al. 1964) to a phylum Euglenida in the kingdom Protoctista (Margulis et al. 1989).

A number of investigators have suggested a common ancestry for the euglenoids and kinetoplastids (e.g., Mignot 1964; Schuster et al. 1968; Leedale 1970; Porter 1973; Vickerman and Preston 1976; Taylor 1976, 1980) and some have gone as far as to erect a new taxon the "Euglenozoa" (Cavalier-Smith 1981, Corliss 1984). This merger is supported in hypotheses presented by Kivic and Walne (1984) and Willey et al. (1988). These hypotheses focus on four primary ultrastructural features which are used to link the ancestry of euglenoids and kinetoplastids: (1) mitosis, (2) the feeding apparatus, (3) the flagellar apparatus, (4) the cytoskeleton. Suspected homologies in each of these four features are described and analyzed based on the available data. The addition of a substantial amount of new data concerning mitosis, the flagellar apparatus and feeding apparatus in the Euglenozoa warrant a review of the current hypotheses.

Mitosis

Prior to 1985, ultrastructural features of mitosis had been described for only four genera of euglenoids: *Euglena* (Leedale 1968, 1982; Pickett-Heaps and Weik 1977; Gillott and Triemer 1978), *Astasia* (Sommer and Blum 1965, Chaley et al. 1977), *Phacus* (Pickett-Heaps and Weik 1977), and *Colacium* (Kugrens and Rosowski 1972) all belonging to the same taxonomic order, the Euglenales (sensu Leedale 1967). If ultrastructural features of mitosis are to be of any phylogenetic value they must be examined in a wide variety of euglenoids, including the colorless phagotrophic forms which are more likely to be closely related to the kinetoplastids than the photosynthetic forms. Mitosis has now been examined in four colorless, phagotrophic euglenoids: *Anisonema* sp. (Triemer 1985), *Ploeotia costata* (Triemer 1986, Triemer and Fritz 1988), *Entosiphon sulcatum* (Triemer 1988), and *Petalomonas cantuscygni* (Triemer and Farmer 1988). Furthermore, the fine structural features of nuclear division have been examined in *Diplonema ambulator* (syn. = *Isonema*; Triemer and Ott 1990) an unusual protist included in the larger taxon, Euglenozoa (Corliss 1984) and believed to be closely related to the euglenoids and kinetoplastids (Kivic and Walne 1984, Willey et al. 1988).

The fine structural aspects of mitosis in kinetoplastids have been largely restricted to the trypanosomes (*Leishmania*, Bianchi et al. 1969, Croft 1979, Triemer et al. 1986; *Herpetomonas*, De Souza et al. 1976; *Leptomonas*, Souto-Padron et al. 1980; *Blastocrithidia*, Solari 1983) and focus on the genus *Trypanosoma* (Inoki and Ozeki 1969; Vickerman and Preston 1970; De Souza and Meyer 1974; Heywood and Weinman 1978; Solari 1980 a,b; Paterson and Woo 1984). As in the euglenoids, little information is available on mitosis in the organisms which are believed to be at the base of the evolutionary line, the bodonids. There is but a single short report on mitosis in *Trypanoplasma borelli* (Skarlato 1987). Despite the limitations on the available data noted above, some observations can be made.

Kivic and Walne (1984: 270) state that "Nuclear architecture and division in euglenoids... are essentially the same as in bodonids, trypanosomatids, and *Isonema*" (syn. *Diplonema*, Triemer and Ott 1990) " ... which have a totally closed mitosis, an intranuclear spindle, and no metaphase plate formation." Recent evidence contradicts this statement on a number of

Fig. 1. Dividing nucleus in *Diplonema ambulator* with persistent elongate nucleolus (*Nu*) and metaphase plate (▶). *C* Chromosome. Bar: 1 μm

Fig. 2. *Petalomonas cantuscygni.* Early division showing central spindle (*Sp*) and persistent nucleolus (*Nu*). Bar: 1 μm

Fig. 3. *Petalomonas cantuscygni.* Mid division nucleus with chromosomes (*C*) on subspindles (▶). Bar: 1 μm

Fig. 4. *Petalomonas cantuscygni.* Late anaphase nucleus showing nucleolus (*Nu*) partially dispersed (▶) over interzonal spindle (*IZS*). Bar: 1 μm

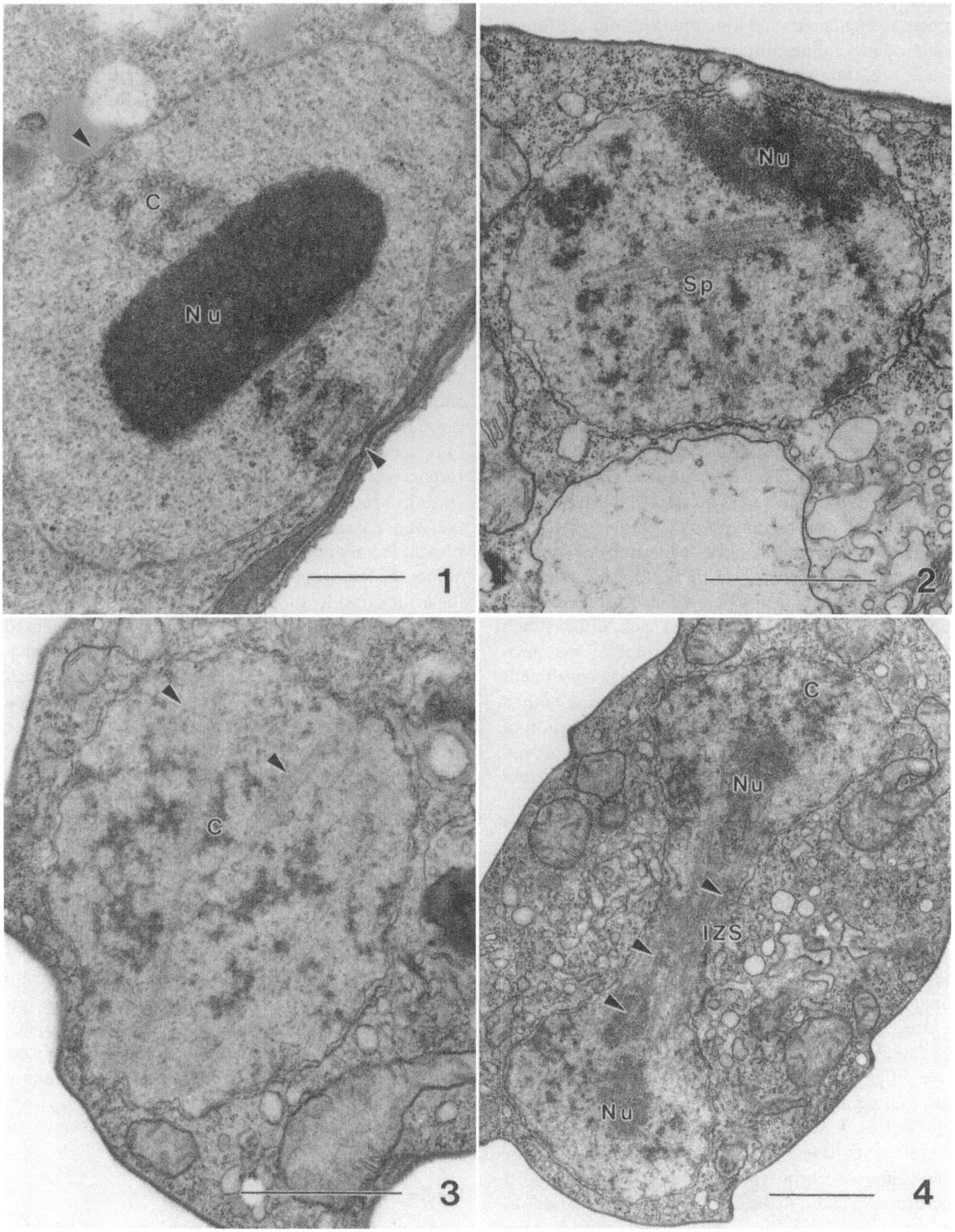
C
Nu
1
Nu
Sp
2
C
C
3
C
Nu
IZS
Nu
4

points. First, closed intranuclear spindles are not restricted to euglenoids and kinetoplastids but are found in some fungi, other protozoans, and some algae (for review, see Raikow 1982). Second, *Diplonema* has a distinct metaphase plate (Fig. 1). In euglenoids, a distinct metaphase plate with compacted chromosomes is not present. However, the chromosomes can assemble in a loose equatorial plate at metaphase in some genera (Triemer 1988, Triemer and Fritz 1988). Likewise, kinetoplastid nuclei contain a number of dense plaques arranged loosely about the equatorial plane during mitosis (Solari 1980 a, b, 1983; Triemer et al. 1986; Skarlato 1987). The exact nature of these plaques is unknown but it has been suggested that they function as kinetochores (Solari 1980 a, b). In any case, *Diplonema* has a definite metaphase plate and euglenoids and kinetoplastids do align kinetochores or plaques at the equatorial plane at some point in mitosis. Third, the types of spindles formed differ. The spindle of the trypanosomes consists of a large central bundle of microtubules surrounded by radial bundles which apparently attach to the chromosomal plaques. During mitosis the plaques first move to the poles followed by elongation of the central spindle (Solari 1980 a, b). The spindle in the bodonid, *Trypanoplasma*, consists of three to four bundles of about forty microtubules each, which attach to kinetochore-like plaques. Since some of the chromatin remains associated with the nuclear envelope it has been hypothesized that mitosis includes both a primitive division mechanism involving segregation of chromosomes attached to the nuclear membrane and a microtubular spindle utilizing kinetochore-like structures (Skarlato 1987). The euglenoid spindle is composed of a number of independent subspindles each of which contains chromosomal and non-chromosomal microtubules. Chromosomal segregation is concomitant with nuclear elongation and does not appear to rely on shortening of the chromosomal microtubules as occurs in the kinetoplastids. The distance between the chromosomes and the nuclear envelope remains about the same until late in anaphase (Triemer 1985). In effect, anaphase A and B appear to be reversed in the euglenoids examined to date. Therefore, at least three different spindle types and mechanisms of chromosomal separation are present in euglenoids and kinetoplastids. A spindle with multiple subspindles and reversed anaphase A/B sequence is characteristic of euglenoids. A central spindle with radial bundles of microtubules characterizes trypanosomes. The bodonids, represented only by *Trypanoplasma*, utilize three to four large bundles of microtubules and make use of the nuclear envelope in segregating chromosomes (Skarlato 1987). Furthermore, chromosomal separation in *Diplonema* appears to follow the anaphase A, B sequence characteristic of most eukaryotes (Triemer unpubl.) unlike that of euglenoids and has a spindle similar to that found in higher plants and animals. Fourth, nucleolar behavior varies during mitosis. In euglenoids the nucleolus remains intact, elongates, becomes dumbbell-shaped and eventually pinches in two. In kinetoplastids, the nucleolar material fragments and disperses over the spindle. Lastly, the chromosomes in kinetoplastids are not permanently condensed as they are in the euglenoids but undergo condensation and decondensation during the cell cycle and unwind into thin fibers at mitosis. Based on the previous discussion it appears that the fine structural features of mitosis cited previously (Kivic and Walne 1984) do not support the statement that "Nuclear architecture and division in euglenoids . . . are essentially the same as in bodonids, trypanosomatids, and *Isonema* ... " There are many features which are distinctly separate and only a single unifying feature, the closed nuclear envelope, which is present in a number of other protists as well. However, before the features of mitosis are eliminated from phylogenetic considerations it may be useful to examine a few more bodonids and euglenoids which are believed to have retained features ancestral to the entire phagotrophic line. For this reason we have studied mitosis in *Petalomonas cantuscygni*. This euglenoid has a rigid cell surface with few pellicle strips and has a feeding apparatus which is less structurally complex than that found in most euglenoid phagotrophs. These features lead us to propose that this organism is more closely related to the ancestral euglenoid than other phagotrophs. Preliminary studies of mitosis show that *P. cantuscygni* may develop a central spindle early in mitosis (Fig. 2), similar to the trypanosomes. Later it forms a number of sub-spindles as in other euglenoids (Fig. 3). The nucleolus also undergoes some degree of fragmentation as it separates on the central spindle into the two daughter nuclei (Fig. 4). These features temptingly suggest a common ancestry with the kinetoplastids but until such studies are completed and mitosis is examined in detail in more bodonids, one cannot conclude that the similarities that do exist between euglenoid and kinetoplastid mitoses are the result of synapomorphies.

Feeding apparatus

The ultrastructural features which are most likely to provide key information for determining the phylogeny of the euglenoids and their relationships to the kinetoplastids are to be derived from the feeding apparatus. Information is available on the feeding apparatus for a number of bodonid genera, including *Bodo* (Brooker 1971, Burzell 1975, Eyden 1977, Brugerolle et al. 1979), *Cryptobia* (Brugerolle et al. 1979, Nohynkova 1984), *Cephalothamnion* (Hitchen 1974), *Trypanoplasma* (Brugerolle et al. 1979), and *Rhynchomonas* (Burzell 1973, Swale 1973). In all cases the feeding apparatus consists of a pocket originating at the cell's anterior end adjacent to the flagellar opening. The pocket is supported by interconnected microtubules running along its length. Brugerolle et al. (1979) termed this complex of reinforcing microtubules the "MTR". The number and arrangement of microtubules varies between organisms and within genera. Most bodonids have one or more small bands of microtubules associated with the pocket. However, a few genera (e.g., *Cephalothamnion cyclopum*, Hitchen 1974; *B. curvifilus*, Burzell 1975; *Bodo designis*, Eyden 1977; *Phyllomitus apiculatus*, Mylnikov 1986) have supporting microtubules arranged in a small rod-like bundle which represents the limit of supporting rod complexity found in bodonids. The membrane adjacent to the MTR is usually denser and somewhat thicker than the rest of the cytostomal membrane.

The presence of a cytostomal complex is not restricted to the bodonids but also exists in some trypanosomes where it functions in pinocytosis rather than phagocytosis (e.g., *Trypanosoma mega*, Steinert and Novikoff 1960; *Trypanosoma raiae*, Preston 1969; *Crithidia fasciculata*, Brooker 1971). Furthermore, the cytostome may open into the flagellar pocket rather than directly to the cell surface (Brooker 1971). Feeding apparatuses with multiple supporting rods and/or vanes have not been reported in any kinetoplastid.

In comparison to the data available on feeding apparatus ultrastructure in the kinetoplastids there are published data on only two genera of euglenoids, in which sufficient details have been presented to warrant any useful comparisions (*Entosiphon*, Mignot 1966, Triemer and Fritz 1986; *Peranema*, Mignot 1966, Nisbet 1974). We have therefore engaged in an extensive study of euglenoid feeding apparatuses and have serially sectioned through eight phagotrophic euglenoid genera and the purportedly related genus *Diplonema*.

Scanning electron microscopy has proven useful in giving an overall view of the feeding apparatus and demonstrating that the feeding apparatus may be directly associated with and arise as a fold in the pellicle as in *Ploeotia* (Triemer 1986) or alternatively, that the feeding apparatus is separate from the pellicle as in *Entosiphon* (Triemer and Fritz 1986). Internally there are also a number of structural variations. The feeding apparatus may be supported by a few microtubules or by bundles of microtubules forming two or possibly three rods. Even within the rods there is variation in the number and organization of the microtubules. The cytostome may be a simple sac or may be surrounded by a diaphragm-like set of vanes. Striated fibers are associated with some of the feeding apparatuses (e.g., *Peranema*, Mignot 1966). The feeding apparatuses found in the euglenoids can be grouped into one of four types, the MTR/pocket type (Type I) as found in *Petalomonas cantuscygni*, the plicate type (Type II) as found in *Ploeotia costata*, the short extensible type (Type III) as found in *Peranema trichophorum*, and the siphon type (Type IV) as found in *Entosiphon sulcatum*.

The Type I feeding apparatus has the least structural complexity and is present in *Petalomonas cantuscygni* and *Calycimonas* sp. It consists of a cytoplasmic pocket which extends posteriorly for roughly one half to three quarters the length of the cell (Fig. 5). At its anterior end the feeding apparatus merges with the flagellar pocket. The feeding pocket is lined by microtubules which extend along its entire length (Figs. 6 and 7). These microtubules extend from the site where the feeding apparatuses and flagellar pocket merge and appear to be derived in part from one of the microtubular flagellar rootlets. A number of small vesicles border one side of the feeding apparatus. Type I feeding apparatuses are restricted to bacteriotrophic euglenoids and *Diplonema*.

The Type II feeding apparatus is found in *Ploeotia costata* (syn. *Serpenomonas costata*, Triemer 1986), *P. vitrea* (Farmer and Triemer 1988 b), and in the three species of *Diplonema* examined (Schuster et al. 1968, Porter 1973, Triemer and Ott 1990). In this type of apparatus the cytostome is supported by two rods which extend into the cytoplasm and a series of plicate folds, the vanes. The supporting rods contain few microtubules, but may possess a dense armorphous matrix. At the anterior end of the apparatus the vanes surround the membrane invagination which forms the cytostome. For most of the length of the apparatus a portion of the vane complex remains closely appressed to the supporting rods. At the base of the feeding apparatus the rods and vanes are intimately associated with one another forming a single complex. Individual microtubules may be associated with the vanes in some, if not all, Type II feeding apparatuses. As is the case for the Type I apparatus, all of the organisms known to have a Type II feeding apparatus are bacteriotrophs.

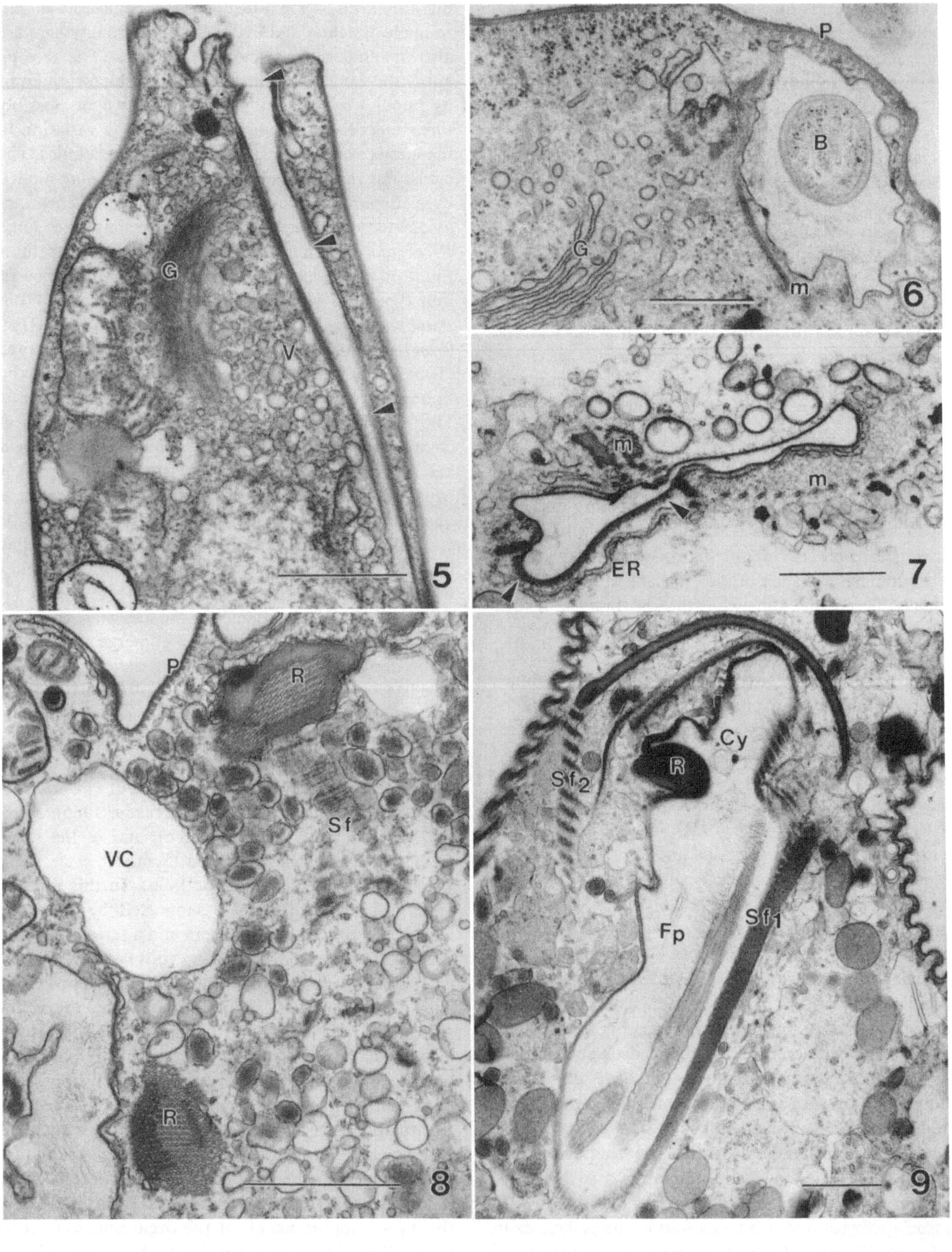
G
V
5
P
B
G
m
6
m
m
ER
7
P
R
Sf
VC
R
8
Cy
R
Sf2
Fp
Sf1
9

The type III feeding apparatus is found in *Peranema trichophorum* (Nisbet 1974), *Dinema sulcatum* (Farmer and Triemer 1988 a), *Urceolus cyclostomus* (Farmer 1988), and *Anisonema* sp. (Triemer 1985). All of these organisms are capable of engulfing eukaryotic prey. In the Type III apparatus the cytostome is surrounded by vanes and supported by two rods. In contrast to the Type II apparatus, the supporting rods are composed primarily of microtubules and the vanes do not remain appressed to the rods. The number and arrangement of microtubules in the rods varies by genus. In *Urceolus cyclostomus*, the rods are composed of a central cylinder of loosely packed microtubules surrounded by a matrix which itself is completely encircled by microtubules (Farmer 1988). A similar rod organization with more central microtubules and less matrix is present in *Peranema* (Nisbet 1974). Lastly, the rods found in *Anisonema* and *Dinema* are composed of a solid mass of microtubules with only a small amount of matrix material associated with them and located primarily at the anterior end (Fig. 8). Four centrally located vanes associate with the supporting rods for most of their length. The vanes arise from and are embedded in the rods at the base of the feeding apparatus and diverge from the rods towards the anterior of the apparatus. In addition, several fibrillar components can be associated with the Type III feeding apparatus. For example, in *Dinema* distinct striated fibers extend from the rods toward the vanes near the anterior end of the apparatus (Fig. 8) and in *Peranema* large striated fibers are associated with the cytostome (Fig. 9).

The last category of feeding apparatus, Type IV, is found only in the bacteriotroph, *Entosiphon sulcatum* (Mignot 1966, Triemer and Fritz 1986). It is similar to a Type III apparatus in that it has supporting rods composed of closely packed microtubules and four centrally located vanes surrounding the cytostome. However, near the base of the feeding apparatus one of the rods bifurcates giving rise to a total of three rods which extend nearly the length of the cell. In cross-section the feeding apparatus appears C-shaped, with the sides of the "C" being formed by the three roughly triangular bundles of microtubules. About one-third the distance down the apparatus (from the anterior end) the number of microtubules per rod increases dramatically and then decreases in number toward the base. This tapering gives the apparatus the overall appearance of a cone with one side open. Numerous vesicles are adjacent to the open side of the cone. Unlike the Type III, this feeding apparatus is usually in motion, extending toward the anterior of the cell and then with-drawing down into the cell for a distance of 3–5 μm. It is primarily the presence of the third supporting rod and the mechanism of movement which separates this feeding apparatus from the Type III apparatus.

In addition to these four types of feeding apparatuses, there exists another organelle which must be placed in this category. In 1985, Willey and Wibel (1985 a) described the existence of a pocket formed from the reservoir membrane in *Colacium*. A band of microtubules (MTR) lined the pocket and a dense fibrillar mesh was associated with the membrane. The similarity between this structure and the cytostome of colorless euglenoids (Type I) and bodonids raised the possibility that this pocket was a cytostomal homologue (Willey and Wibel 1985 b). Surek and Melkonian (1986) then demonstrated that similar pockets are present in three species of *Euglena*. By tracing the origin of the MTR in serial sections they discovered that it was continuous with, and in fact the same as, the ventral flagellar root. By comparing published information on bodonid cytostomes, colorless euglenoid cytostomes, and the newly discovered reservoir pockets, they proposed that the three structures were homologous, indicating that photosynthetic euglenoids arose from phagotrophic ancestors. Since that time the MTR/pocket complex has been reported in yet another green euglenoid, *Cryptoglena pigra* (Owens et al. 1988). In spite of the supporting data, the case for homology between reservoir pockets and cytostomes is not as clear as it may at first

---

Fig. 5. *Petalomonas cantuscygni*. Longitudinal section through MTR/pocket (▶). Vesicles (*V*) line the feeding apparatus. *G* Golgi. Bar: 1 μm

Fig. 6. *Petalomonas cantuscygni*. Cross-section through feeding apparatus adjacent to pellicle (*P*) ingested bacterium (*B*). *m* Microtubules. Bar: 0.5 μm

Fig. 7. *Calycimonas* sp. Cross-section through feeding apparatus showing thickened region of membrane (▶) and associated microtubules (*m*). Endoplasmic reticulum (*ER*) is adjacent to the apparatus. Bar: 0.5 μm

Fig. 8. *Dinema sulcatum*. Oblique section near cell anterior showing microtubular supporting rods (*R*) and associated striated fiber (*Sf*). The vestibular cavity (*VC*) which leads to the cytostome, is positioned between the supporting rods. *P* Pellicle. Bar: 1 μm

Fig. 9. *Peranema trichophorum*. Oblique section near cell anterior showing cytostome (*Cy*) and flagellar pocket (*Fp*). One of the two supporting rods (*R*) and two striated fibers (Sf_1, Sf_2) associated with the feeding apparatus are visible. Bar: 1 μm

seem. We have identified MTR/pockets in two phagotrophic euglenoids, *Ploeotia* and *Dinema*, and in *Diplonema* (Triemer and Ott 1990). *Ploeotia*, in addition to having a Type II feeding apparatus has an invagination of the reservoir lined with microtubules and surrounded by a dense fibrillar material. The microtubules can be traced back to the ventral basal body of the flagellar apparatus. A similar pocket exists in *Diplonema* which also has a Type II apparatus and in *Dinema*, which has a Type III apparatus. If these MTR/pockets are homologous with those of *Colacium* and *Euglena*, it would be difficult to explain why they are found in genera which already have a second, well-developed feeding apparatus composed of supporting rods and vanes. The implication is that some genera have developed a more elaborate feeding apparatus while still retaining a Type I apparatus derived from some ancestral form. Perhaps the MTR/pocket is retained for pinocytotic functions as in trypanosomes while the supporting rod/vane feeding apparatus is utilized primarily for phagocytosis. The need for retaining two feeding apparatuses remains enigmatic.

Flagellar apparatus

The arrangement of components in the flagellar apparatuses of both euglenoids and kinetoplastids is remarkably similar. The basic configuration of two basal bodies and three asymmetrically distributed microtubular roots is found in the majority of euglenoid and kinetoplastid species (Fig. 10). In addition to these structures, a striated connecting fiber is often associated with the two basal bodies (Fig. 11). This fiber is most prominent in those taxa whose flagella move heterodynamically. In some trypanosomes and euglenoids the second flagellum is so reduced that its basal body is little more than a barren stub or plate, thus making the cells functionally uniflagellate (Farmer and Triemer 1988a, Sherwin and Gull 1989). Despite these modifications the basic architecture of the flagellar apparatus is recognizable throughout the group. The specific configuration of the microtubular flagellar roots and the role each root plays in the construction of the cytoskeleton also seems to be nearly identical in the euglenoids and kinetoplastids (see below).

Other flagellar features that link the euglenoids and kinetoplastids include the presence of a paraxial rod (syn. paraflagellar rod) and flagellar hairs. Although paraxial rods are apparently absent from some members of the Kinetoplastida (Freymuller and Camargo 1981) they are present in the vast majority of euglenoids and kinetoplastids. Biochemical studies have shown that the paraxial rod of *Trypanosoma brucei* is composed of a single protein (Schlaeppi et al. 1989). That this protein can assume two different conformations explains why two distinct bands of approximately 70 kDa are observed in SDS-gel electrophoresis of paraxial rod proteins from euglenoids (Hyams 1982) and kinetoplastids (Russell et al. 1983, Cuhna et al. 1984). Despite the slight disparity in molecular weights and other differences in architecture of paraxial rods in the two groups (De Souza et al. 1980, Souto-Pardon et al. 1980, Hyams 1982, Farina et al. 1986), immunological studies show that the paraxial rod protein of *Euglena* is related to the paraxial rod protein of trypanosomes (Gallo and Schrevel 1985). Paraxial rods are present in other protists (e.g., dinoflagellates) but the lack of biochemical data prevent further speculation regarding the evolutionary relatedness of this structure in other groups (Cachon et al. 1988). Flagellar hairs when present consist of thin nontubular structures that are unilaterally arranged.

One feature that has been cited as an indicator of evolutionary relatedness of the euglenoids and the kinetoplastids is the flagellar transition zone (Kivic and Walne 1984). The hollow flagellar transition zone between the axoneme and basal body of *Euglena* was long thought to be characteristic of the euglenoids (Leedale 1967, Moestrup 1982). While the type of transition zone may prove to be useful in grouping certain taxa into families, a great deal of variability exists among different euglenoid species (Farmer and Triemer 1988a). This variability warrants caution in using the ultrastructure of the flagellar transition zone as proof that the euglenoids and kinetoplastids form a monophyletic group.

Cytoskeleton

Overall cell morphology is perhaps the oldest and most commonly used criterion by which the taxonomic relationships (phenetic) between protists are assessed. Superficially, the euglenoids and kinetoplastids resemble one another. Both are principally flagellate unicells that lack elaborate cell walls or plates and whose cell membranes are underlain by subpellicular microtubules. Unfortunately these features are shared by other protistans as well (e.g., Retortomonads, Brugerolle 1973, 1977; Proteromonads, Brugerolle and Joyon 1975; *Hemimastix*, Foissner et al. 1988). The one cytoskeletal feature that distinguishes the euglenoids from all other

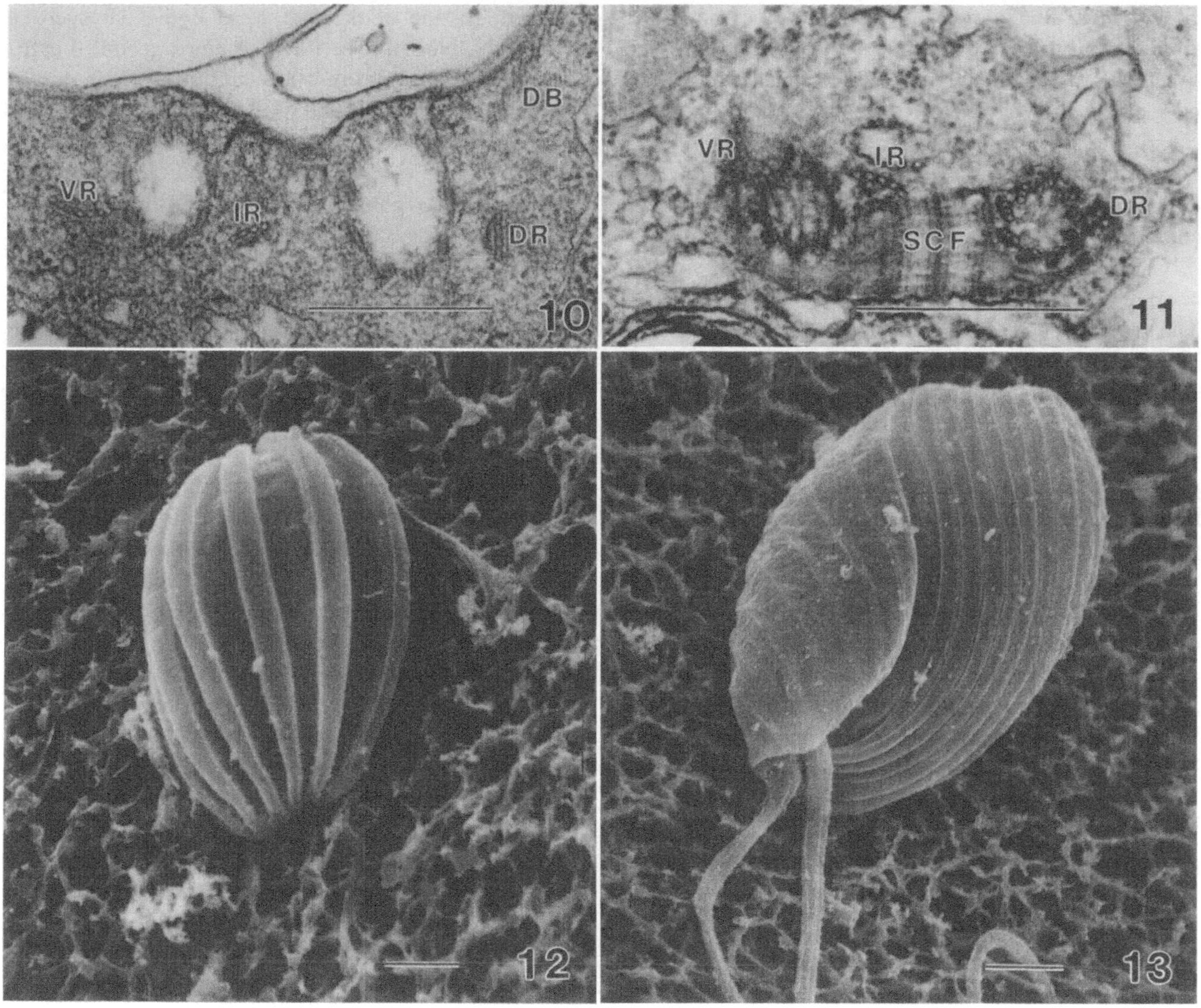

Fig. 10. Flagellar apparatus of *Euglena gracilis* showing the distribution of ventral (*VR*) and intermediate (*IR*) microtubular roots with the ventral basal body and the dorsal basal body with its associated dorsal root (*DR*) which gives rise to the dorsal band (*DB*) of microtubules. Bar: 0.5 μm

Fig. 11. Flagellar apparatus of *Ploeotia costata* showing large striated connecting fiber (*SCF*) between the two basal bodies and asymmetrical arrangement of microtubular roots. Bar: 0.5 μm

Fig. 12. SEM of *Entosiphon applanatum*, a species with an aplastic pellicle composed of ten longitudinally arranged pellicular strips. Bar: 5 μm

Fig. 13. SEM of *Dinema sulcatum*, a species with a plastic pellicle composed of multiple strips arranged in a helical fashion. Bar: 5 μm

protists is the presence of a pellicle composed of four or more strips that extend the length of the cell.

The euglenoid pellicle is believed to be responsible for allowing for euglenoid movement (i.e., "metaboly") in genera such as *Euglena*. Many euglenoids such as *Entosiphon, Petalomonas*, and *Ploeotia* have only a few pellicular strips that are arranged longitudinally (Fig. 12). Unlike *Euglena*, these euglenoids are completely rigid and are not capable of metaboly. In contrast, many species that have a pellicle composed of helically arranged strips are capable of altering their cell shape (Fig. 13). Based on these and other factors the euglenoids can be divided into two major groups; those possessing plastic pellicles (numerous helically arranged strips) and those with aplastic pellicles (fewer longitudinally arranged strips). Although aplastic species are incapable of altering their cell shape, metaboly alone can not be used as a distinguishing feature between the two pellicular types. Many euglenoids that have plastic pellicles are also essentially rigid (e.g., *Phacus, Cryptoglena*). Genera with plastic pellicles that do not undergo metaboly (e.g., *Euglena spirogyra, Phacus longicauda*) often have secondary elaborations or thickenings associated with their pellicles which may be responsible for their inability to change shape significantly.

The array of microtubules that underlies the cell membrane of the kinetoplastids has been cited as further evidence that the two groups are more closely related to one another than either is to other protistan groups (Kivic and Walne 1984, Willey et al. 1988). A comparison of the microtubular flagellar roots and the origin of the microtubules of the cytoskeleton suggests that the euglenoids and kinetoplastids do indeed have very similar architectures. The microtubules that lie beneath each of the pellicular ridges in euglenoids are continuous with those that line the flagellar canal and extend into the region of the flagellar reservoir (Willey and Wibel 1985 a, b). This group of microtubules has been termed the dorsal band (Willey and Wibel 1985 a, b, 1987) (Fig. 10) and appears to nucleate tangentially from one of the three microtubular flagellar roots (Willey and Wibel 1987, Owens et al. 1988). Development and replication of the pellicular microtubules in the euglenoid *Cyclidiopsis* is intimately related to the replication and separation of the microtubular flagellar roots (Mignot et al. 1987).

Likewise in some bodonids the majority of the cytoskeletal microtubules appear to be derived from a large band of microtubules that originate at the anterior end of the cell in the region of the flagellar opening. Like the dorsal band found in the euglenoids, this group of microtubules does not appear to be continuous with any of the microtubular flagellar roots. As in the euglenoids, the microtubules of this cytoskeletal band seem to nucleate adjacent to a small microtubular root that emanates from one of the two basal bodies (Brugerolle et al. 1979). Cell body deformations similar to those described for euglenoid metaboly have been observed in some bodonids (Swale 1973, Vickerman 1977) and may be a further indicator that the cytoskeletons of both groups are derived from a common ancestor.

The other major group of kinetoplastid cytoskeletal microtubules is the MTR (see section on feeding apparatus). As in the euglenoids (Willey and Wibel 1985 a, 1987; Surek and Melkonian 1986; Owens et al. 1988) the microtubules of the kinetoplastid MTR originate adjacent to one of the two basal bodies (Brugerolle et al. 1979, Nohynkova 1984). A third group of cytoskeletal microtubules that are continuous with one of the flagellar roots is found in the region of the flagellar opening in bodonids (Brugerolle et al. 1979, Nohynkova 1984).

The strongest argument in favor of all three microtubular flagellar roots and their derived cytoskeletal microtubules of bodonids being homologous to similar structures in the euglenoids is the fact that the distribution of these microtubules is identical in the two groups. The three microtubular flagellar roots are arranged asymmetrically. Of these three roots two are associated with one basal body while the third root emanates from the other basal body. In both the euglenoids and the bodonids it is this single microtubular root that gives rise to (but is not continuous with) the major group of cytoskeletal microtubules. Likewise the flagellar root that ultimately becomes the MTR of both euglenoids and bodonids is one of the two roots associated with the first basal body.

A number of studies have shown that many protists undergo a complex developmental process during cell division in which one parental basal body and its associated microtubular roots matures into a basal body that is identical to the alternate type of parental basal body (Melkonian et al. 1987; Moestrup and Hori 1989; Heimann et al. 1989 a, b). A similar process has been suggested for the euglenoids in which the basal body that has only one microtubular root associated with it develops into the type with two roots (Farmer and Triemer 1988 a) Although such developmental studies have not yet been done on any of the bodonids, it seems likely that a similar mechanism exists. If this proves to be the case the microtubular flagellar roots and the cytoskeletal microtubules that they ultimately give rise to would be structurally, positionally, functionally, and developmentally identical in both groups.

While similarities in cytoskeletal arrangement might suggest that euglenoids and bodonids are members of a monophyletic assemblage, the issue is less clear when one considers the trypanosomes. Unlike the bodonids and the majority of euglenoids, the trypanosomes are uniflagellate. Furthermore, discerning which group of microtubules gives rise to the cytoskeletal microtubules and which group is responsible for lining the cytostome remains difficult (Willey et al. 1988). Despite these discrepancies, the basic architecture of an encircling corset of cytoskeletal microtubules that are directly or indirectly derived from the flagellar microtubular roots is consistent with what is found in the euglenoids and bodonids. Immunocytochemical studies indicate that proteins that localize with isolated corset microtubules of trypanosomes may be responsible for the cross binding of the microtubules and determination of the cell architecture (Balaban et al. 1989, Bramblett et al. 1989). If the antibodies raised against these proteins prove to be cross reactive with the microtubular cytoskeleton of euglenoids and bodonids but not other protists, it would provide an additional link between these groups. The presence of a second barren basal body in all trypanosomes (Vickerman 1989) suggests that they are evolved from a biflagellate ancestor. Ultrastructural studies of trypanosome basal body replication (Paulin 1969) coupled with the fact that the single flagellum is homologous with the anterior flagellum of bodonids (Vickerman 1989) suggest that basal body development is identical to that observed in euglenoids. In both groups the basal body that has two microtubular roots represents the terminal or mature state. Together, the distribution of cytoskeletal microtubules and the unique arrangement of basal bodies and their microtubular roots strongly support the hypothesis that euglenoids and kinetoplastids are derived from a single common ancestor.

Conclusions

The data now available continue to provide support for the hypothesis that euglenoids and kinetoplastids are more closely related to each other than they are to other protists. However, the details of this relationship remain for the most part speculative. With only a single exception, mitosis in euglenoids is not similar to that of kinetoplastids in terms of spindle formation, chromosome organization and segregation, and nuclear division. Yet at least one euglenoid, *Petalomonas cantuscygni*, does undergo nucleolar dispersion and appears to form a central spindle at some point during mitosis.

The MTR/pocket (Type I) feeding apparatus can be found in both kinetoplastids and euglenoids. In the kinetoplastids this appears to be the limit of feeding apparatus development, whereas in the euglenoids the MTR/pocket is the least complex of the feeding apparatus types. Interestingly, this type of feeding apparatus is found in *Petalomonas cantuscygni*, the euglenoid with a mitosis most like that of kinetoplastids. The reduced MTR/pocket present in many photosynthetic euglenoids provides evidence for ancestry with a colorless phagotroph.

In both groups of organisms cytoskeletal microtubules underlie the plasma membrane forming a continuous or discontinuous supporting corset. In the euglenoids an additional glycoprotein layer is appressed to the inner surface of the plasma membrane generating the characteristic ridge and groove pattern of the pellicle. The rigid euglenoid phagotrophs have few pellicular ridges while the photosynthetic genera such as *Euglena* have many. Among those genera with the fewest ridges is *Petalomonas cantuscygni*.

The basic configuration of the flagellar apparatus with two basal bodies and three asymmetrically distributed microtubular roots is found in the majority of euglenoid and kinetoplastid species. In addition, both groups have paraxial rods which contain proteins of similar electrophoretic mobility and which exhibit some degree of immunological cross reactivity. Thin non-tubular flagellar hairs are also characteristic of both groups. The flagellar transition zones of some kinetoplastids are morphologically similar to those found in some phagotrophic euglenoids, yet the differences noted in the structure of the transition zones within the entire group of euglenoids is very diverse (Farmer and Triemer 1988 a) and provides evidence for a long and separate evolutionary history.

In summary, while we agree with Vickerman (1989) and others that the trypanosomes are descended from the bodonids, we do not believe that bodonids were ancestral to the euglenoids (Kivic and Walne 1984) but hypothesize that the euglenoids and bodonids both diverged from a common ancestor with yet a separate branch giving rise to *Diplonema*. A long and separate evolutionary history would explain why mitotic features and mechanisms differ substantially between the two groups. Furthermore, most bodonids have a Type I feeding apparatus supported by a few microtubules. This is also true for the euglenoid *Petalomonas cantuscygni*. However, some bodonids (e.g., *Bodo designis*, Eyden 1977) do form a small microtubular supporting rod and nearly all phagotrophic euglenoids have microtubular supporting rods. Therefore, we can trace the evolution of the feeding apparatus in bodonids from a Type I apparatus with few microtubules to one with a single supporting rod and no vanes. In the phagotrophic euglenoids we can present a similar, albeit more lengthy scenario moving from the Type I apparatus in *Petalomonas cantuscygni* to the Type IV apparatus. We interpret these patterns of supporting rod formation as representing homoplasies rather than evidence for direct descendence. We hypothesize that the ancestral form to both euglenoids and kinetoplastids was a heterodynamic phagotroph with a Type I feeding apparatus supported by a few microtubules, possessing two basal bodies with three asymmetrically distributed microtubular rootlets. Mitosis would occur within a closed nuclear envelope using a central spindle and the nucleolus would fragment either partially or completely. At present the kinetoplastid with features closest to this hypothetical ancestor would lie in the genus *Bodo* (Vickerman 1989) and within the euglenoids the organism which best represents the ancestral form is *Petalomonas cantuscygni*.

References

Balaban N, Waithaka HK, Njogu AR, Goldman R (1989) Isolation of a subpellicular microtubule protein from *Trypanosoma brucei* that mediates crosslinking of microtubules. Cell Motil Cytoskeleton 14: 393–400

Bianchi L, Rondanelli EG, Carosi G, Gerna G (1969) Endonuclear mitotic spindle in the leptomonad of *Leishmania tropica*. J Parasitol 55: 1091–1092

Bouck GB (1971) The structure, origin, isolation and composition of the tubular mastigonemes of the *Ochromonas* flagellum. J Cell Biol 50: 362–384

Bouck GB (1972) Architecture and assembly of mastigonemes. Adv Cell Mol Biol 2: 237–271

– Rogalski A, Valaitis A (1978) Surface organization and composition of *Euglena* II. Flagellar mastigonemes. J Cell Biol 77: 805–826

Bramblett GT Kambadur R, Flavin M (1989) Immunocytochemical studies with antibodies to three proteins prominent in the isolated microtubule cytoskeleton of a trypanosomatid. Cell Motil Cytoskeleton 13: 145–157

Brooker BE (1971) Fine structure of *Bodo saltans* and *Bodo caudautus*. (Zoomastigophora: Protozoa) and their affinities with the Trypanosomatidae. Bull Br Mus Nat Hist 22: 89–102

Brugerolle G (1973) Étude ultrastructurale du trophozoite et du kyste chez le genre *Chilomastix* Alexieff 1910 (Zoomastigophorea, Retortomonadida, Grassé 1952). J Protozool 20: 574–585

– (1977) Ultrastructure du genre *Retortomonas* Grassi 1879 (Zoomastigophorea, Retortomonadida, Wenrich 1932). Protistologica 8: 233–240

– Joyon L (1975) Étude cytologique ultrastructurale des genres *Proteromonas* et *Karotomorpha* (Zoomastigophorea, Proteromonadida, Grassé 1952). Protistologica 11: 531–546

– Lom J, Nohynkova E, Joyon L (1979) Comparison et des structures cellulaires chez plusiers especes de Bodonides et Cryptobiides appartenantaux genres *Bodo, Cryptobia* et *Trypanoplasma* (Kinetoplastida Mastigorphora). Protistologica 15: 197–221

Burzell LA (1973) Observations on the proboscis-cytopharynx complex and flagella of *Rhynchomonas metabolita* Pshenin, 1964 (Zoomastigophorea: Bodonidae). J Protozool 20: 385–393

– (1975) Fine structure of *Bodo curvifilus* Griessman (Kinetoplastida: Bodonidae). J Protozool 22: 35–39

Cachon J, Cachon M, Cosson M-P, Cosson J (1988) The paraflagellar rod: a structure in search of a function. Biol Cell 63: 169–181

Cavalier-Smith T (1981) Eukaryote kingdoms: seven or nine? Bio Systems 14: 461–481

Chaley N, Lord A, LaFontaine G (1977) A light and electron microscopic study of nuclear structure throughout the cell cycle in the euglenoid *Astasia longa*. J Cell Sci 27: 23–45

Corliss JO (1984) The kingdom Protista and its 45 phyla. BioSystems 17: 87–126

Croft SL (1979) Ultrastructural study of the nucleus of *Leishmania hertigi*. Protistologica 15: 103–110

Cuhna NL, De Souza W, Hasson-Voloch A (1984) Isolation of the flagellum and characterization of the paraxial structure of *Herpetomonas megaseliae*. J Submicrosc Cytol 16: 705–713

De Souza W, Meyer H (1974) On the fine structure of the nucleus in *Trypanosoma cruzi* in tissue culture forms. Spindle fibers in the dividing nucleus. J Protozool 21: 48–52

– Souto-Padron T (1980) The paraxial structure of the flagellum. Trypanosomatidae. J Parasitol 66: 229–235

– Rossi MA, Kitajima EW, Santos R, Roitman I (1976) An electron microscopic study of *Herpetomonas* sp. (*Leishmania pessoai*). Can J Microbiol 22: 197–203

Dodge JD (1975) The fine structure of *Trachelomonas* (Euglenophyceae). Arch Protistenk 117: 65–77

Dougherty EC (1955) Comparative evolution and the origin of sexuality. Syst Zool 4: 145–169

Eyden BP (1977) Morphology and ultrastructure of *Bodo designis* Skuja 1948. Protistologica 13: 169–179

Farina M, Attias M, Souto-Padron T, De Souza W (1986) Further studies on the organization of the paraxial rod of trypanosomatids. J Protozool 33: 552–557

Farmer MA (1988) A re-evaluation of the taxonomy of the Euglenophyceae based on ultrastructural characteristics. PhD thesis, Rutgers University, Piscataway, NJ

– Triemer RE (1988 a) Flagellar systems in the euglenoid flagellates. BioSystems 21: 283–291

– – (1988 b) A redescription of the genus *Ploeotia* Duj. (Euglenophyceae). Taxon 37: 319–325

Foissner W, Blatterer H, Foissner I (1988) The Hemimastigophora (*Hemimastix amphikineta* nov. gen., nov. spec.), a new protistan phylum from Gondwanian soils. Eur J Protistol 23: 361–383

Freymuller E, Camargo EP (1981) Ultrastructural differences between species of trypanosomatids with and without endosymbionts. J Protozool 28: 175–182

Fritsch FE (1935) Structure and reproduction of the algae, vol 1. Cambridge University Press, London

Gallo J-M, Schrevel J (1985) Homologies between paraflagellar rod proteins from trypanosomes and euglenoids revealed by a monoclonal antibody. Eur J Cell Biol 36: 163–168

Gillott MA, Triemer RE (1978) Ultrastructure of cell division in *Euglena*. J Cell Sci 31: 25–36

Heimann K, Reize IB, Melkonian M (1989a) The flagellar developmental cycle in algae: flagellar transformation in *Cyanophora paradoxa* (Glaucocystophyceae). Protoplasma 148: 106–110

– Benting J, Timmermann S, Melkonian M (1989 b) The flagellar developmental cycle in algae: two types of flagellar development in uniflagellated algae. Protoplasma 153: 14–23

Heywood P, Weinmann D (1978) Mitosis in the hemoflagellate *Trypanosoma cyclops*. J Protozool 25: 287–292

Hibberd DJ, Leedale GF (1970) Eustigmatophyceae – a new algal class with unique organization of the motile cell. Nature 225: 759–760

– – (1971) A new algal class – the Eustigmatophyceae. Taxon 20: 523–525

– – (1972) Observations on the cytology and ultrastructure of the new algal class, Eustigmatophyceae. Ann Bot 36: 49–71

Hitchen ET (1974) The fine structure of the colonial kinetoplastid flagellate *Cephalothamnion cyclopum* Stein. J Protozool 21: 221–231

Hollande A (1952) Classe des eugleniens (Euglenoidina Butschli, 1884). In: Grasse PP (ed) Traite de zoologie, vol 1, fasc 1. Mason, Paris, pp 238–284

Honigberg BM (1963) A contribution to systematics of the non-pigmented flagellates. In: Ludvik J, Lom J, Vavra J (eds) Progress in protozoology. Academic Press, New York, pp 68–69

– Balamuth M, Bovee EC, Corliss JO, Godjics M, Hall RP, Kudo RR, Levine ND, Loeblich AR Jr, Weiser J, Weinrich DH (1964) A revised classification of the phylum Protozoa. J Protozool 11: 7–20

Hyams JS (1982) The *Euglena* paraflagellar rod: structure, relationship to other flagellar components and preliminary biochemical characterization. J Cell Sci 55: 199–210

Inoki S, Ozeki Y (1969) Electron microscopic observations on division of kinetoplasts in *Trypanosoma gambiense*. Biken J 12: 31–41

Kivic PA, Walne PL (1984) An evaluation of a possible phylogenetic relationship between the Euglenophyta and Kinetoplastida. Orig Life 13: 269–288

Klebs G (1883) Über die Organisation einiger Flagellaten-Gruppen und ihre Beziehungen zu Algen und Infusorien. Unters Inst Tübingen 1: 233–362

Kugrens P, Rosowski JR (1972) Ultrastructure of mitosis in *Colacium*. J Phycol 8: 9 a

Leedale GF (1967) Euglenoid flagellates. Prentice Hall, Englewood Cliffs, NJ

– (1968) The nucleus in *Euglena*. In: Buetow DE (ed) The biology of *Euglena*. Academic Press, New York, pp 185–272

– (1970) Observations on the mitochondrial reticulum in living *Euglena gracilis*. Cytobiologie 1: 195–202

– (1978) Phylogenetic criteria in euglenoid flagellates. BioSystems 10: 183–187

– (1982) Ultrastructure. In: Buetow DE (ed) The biology of *Euglena*, vol 3. Academic Press, New York, pp 1–27

Margulis L, Corliss JO, Melkonian MM, Chapman DJ (eds) (1989) Handbook of Protoctista. Jones and Bartlett, Boston

Melkonian M, Reize IB, Preisig HR (1987) Maturation of a flagellum/basal body requires more than one cell cycle in algal flagellates: studies on *Nephroselmis olivacea* (Prasinophyceae). In: Wiessner W, Robinson DG, Starr RCJ (eds) Algal development. Molecular and cellular aspects. Springer Berlin Heidelberg New York Tokyo, pp 102–113

Mignot J-P (1964) Observations complementaires sur la structure des flagelles d'*Entosiphon sulcatum* (Duj.) Stein, flagelle euglenien. Acad Sci Paris 258: 3360–3363

– (1966) Structure et ultrastructure de quelques euglenomonadines. Protistologica 2: 51–117

– Brugerolle G, Bricheux G (1987) Intercalary strip development and dividing cell morphogenesis in the euglenid *Cyclidiopsis acus*. Protoplasma 139: 51–65

Moestrup Ø (1982) Flagellar structure in algae: a review, with new observations particularly on the Chrysophyceae, Phaeophyceae (Fucophyceae), Euglenophyceae, and *Reckertia*. Phycologia 21: 427–528

– Hori T (1989) Ultrastructure of the flagellar apparatus in *Pyramimonas octopus* (Prasinophyceae). II. Flagellar roots, connecting fibers, and numbering of individual flagella in green algae. Protoplasma 148: 41–56

Mylnikov AP (1986) Ultrastructure of a colorless flagellate *Phyllomitus apiculatus* Skuja 1948 (Kinetoplastida). Arch Protistenk 132: 1–10

Nisbet B (1974) An ultrastructural study of the feeding apparatus in *Peranema trichophorum*. J Protozool 21: 39–48

Nohynkova E (1984) A new pathogenic *Cryptobia* from freshwater fishes: a light and electron microscopic study. Protistologica 20: 181–195

Owens K, Farmer MA, Triemer RE (1988) The flagellar apparatus and reservoir/cytoskeleton of *Cryptoglena pigra* (Euglenophyceae). J Phycol 24: 520–528

Patterson WB, Woo PTK (1984) Ultrastructural studies on mitosis in *Trypanosoma danilewskyi* (Mastigophora : Zoomastigophorea). Can J Bot 62: 1167–1171

Paulin JJ (1969) Ultrastructural observations on the kinetosome cycle of *Crithidia fasciculata* and its significance to other members of the order (Kinetoplastida). Trans Am Microsc Soc 88: 400–410

Pickett-Heaps JD, Weik KL (1977) Cell division in *Euglena* and *Phacus* I. Mitosis. In: Rost TL, Gifford EM Jr (eds) Mechanisms and control of cell division. Academic Press, New York, pp 308–336

Porter D (1973) *Isonema papillatum* sp. n., a new colorless marine flagellate: a light and electron microscopic study. J Protozool 20: 351–356

Preston TM (1969) The form and function of the cytostome-cytopharynx of the culture forms of the elasmobranch haemoflagellate *Trypanosoma raiae* Laveran & Mesnil. J Protozool 16: 320–333

Raikov IB (1982) The protozoan nucleus. Springer, Wien New York [Alfert M et al (eds) Cell biology monographs, vol 9]

Russell DG, Newsam RJ, Palmer GCN, Gull K (1983) Structural and biochemical characterization of the paraflagellar rod of *Crithidia fasciculata*. Eur J Cell Biol 30: 137–143

Schlaeppi K, Deflorin J, Seebeck T (1989) The major component of the paraflagellar rod of *Trypanosoma brucei* is a helical protein that is encoded by two identical, tandemly linked genes. J Cell Biol 109: 1695–1709

Schuster FL, Goldstein S, Hershenov B (1968) Ultrastructure of a flagellate, *Isonema nigricans* nov. gen. nov. sp. from a polluted marine habitat. Protistologica 4: 141–149

Senn G (1900) Flagellata. In: Engler A, Prantl K (eds) Die natürlichen Planzenfamilien, vol 1. W Engelmann, Leipzig, pp 93–188

Sherwin T, Gull K (1989) The cell division cycle of *Trypanosoma brucei* brucei: timing of event markers and cytoskeletal modulations. Philos Trans R Soc Lond [Biol] 323: 573–588

Skarlato SO (1987) Fine structural pecularities of the nucleus of the parasitic flagellate *Trypanosoma borelli*. Rep Acad Sci USSR 293: 220–221

Sogin ML, Elwood HJ, Gunderson JH (1986) Evolutionary diversity of eukaryotic small-subunit rRNA genes. Proc Natl Acad Sci USA 83: 1383–1387

Solari AJ (1980 a) Function of the dense plaques during mitosis in *Trypanosoma cruzi*. Exp Cell Res 127: 457–460

– (1980 b) The 3-dimensional fine structure of the mitotic spindle in *Trypanosoma cruzi*. Chromosoma 78: 239–255

– (1983) The ultrastructure of mitotic nuclei of *Blastocrithidia triatomae*. Z Parasitenk 69: 3–15

Sommer JR, Blum JJ (1965) Cell division in *Astasia longa*. Exp Cell Res 39: 504–527

Souto-Padron T, Gon Calves de Lima VMQ, Roitman I, De Souza W (1980) An electron microscopic and cytochemical study of *Leptomonas samueli*. Z Parasitenk 62: 127–143

Steinert M, Novikoff AB (1960) The existence of a cytostome and the occurrence of pinocytosis in the trypanosome, *Trypanosoma mega*. J Biophys Biochem Cytol 8: 563–569

Surek B, Melkonian M (1986) A cryptic cytostome in *Euglena*. Protoplasma 133: 39–49

Swale EMF (1973) A study of the colourless flagellate *Rhynchonomonas nasuta* (Stokes) Klebs. Biol J Linn Soc 5: 255–264

Taylor FJR (1976) Flagellate phylogeny: a study in conflicts. J Protozool 23: 28–40

– (1980) The stimulation of cell research by endosymbiotic hypotheses for the origin of the eukaryotes. In: Schwemmer W, Schenk HEA (eds) Endocytobiology. W de Gruyter, Berlin, pp 917–947

Triemer RE (1985) Ultrastructural features of mitosis in *Anisonema* sp. (Euglenida). J Protozool 32: 683–690

– (1986) Light and electron microscopic description of a colorless euglenoid, *Serpenomonas costata* n. g., n. sp. J Protozool 33: 412–415

– (1988) Ultrastructure of mitosis in *Entosiphon sulcatum* (Euglenida). J Protozool 35: 53–59

– Farmer MA (1988) *Petalomonas cantuscygni*, a possible evolutionary link between the euglenoids and the kinetoplastids. J Phycol 24: 22 a

Triemer RE, Fritz L (1986) Structure and operation of the feeding apparatus in a colorless euglenoid, *Entosiphon sulcatum*. J Protozool 34: 154–162

– – (1988) Ultrastructural features of mitosis in *Ploeotia costata* (Heteronematales, Euglenophyta). J Phycol 24: 514–519

– Ott DW (1990) Ultrastructure of *Diplonema ambulator* Larsen & Patterson (Euglenozoa) and its relationship to *Isonema*. Eur J Protozool 25: 316–320

– Fritz L, Herman R (1986) Fine structure of mitosis in *Leishmania adlerii*. Protoplasma 134: 154–162

Vickerman K (1977) DNA throughout the single mitochondrion of a kinetoplastid flagellate: observations on the ultrastructure of *Cryptobia vaginalis* (Hesse 1910). J Protozool 24: 221–233

– (1989) Phylum Zoomastigina class Kinetoplastida. In: Margulis L, Corliss JO, Melkonian M, Chapman DJ (eds) Handbook of Protoctista. Jones and Barlett, Boston, pp 215–238

– Preston TM (1970) Spindle microtubules in the dividing nuclei of trypanosomes. J Cell Sci 6: 365–383

– – (1976) Comparative cell biology of the kinetoplastid flagellates. In: Lumsden HR, Evans DA (eds) Biology of the Kinetoplastida. Academic Press, London, pp 35–130

Willey RL, Wibel RG (1985 a) The reservoir cytoskeleton and a possible cytostomal homologue in *Colacium* (Euglenophyceae). J Phycol 21: 570–577

– – (1985 b) A cytostome/cytopharynx in green euglenoid flagellates (Euglenales) and its phylogenetic implications. BioSystems 18: 369–367

– – (1987) Flagellar roots and the reservoir cytoskeleton of *Colacium libellae* (Euglenophyceae). J Phycol 23: 283–288

– Walne PL, Kivic P (1988) Phagotrophy and the origins of the euglenoid flagellates. CRC Crit Rev Plant Sci 7: 303–340

Protoplasma (1991) 164: 105–122

PROTOPLASMA

The flagellar apparatus and cytoskeleton of the dinoflagellates

A comparative overview

K. R. Roberts* and **Julia E. Roberts**

Department of Biology, University of Southwestern Louisiana, Lafayette, Louisiana

Received June 30, 1990
Accepted December 14, 1990

Summary. Modern microscopical approaches have allowed more accurate investigations of the three-dimensional nature of the dinoflagellate flagellar apparatus (FA) and several other cytoskeletal protein complexes. Our presentation overviews the nature of the dinoflagellate FA and cytoskeleton in a number of taxa and compares them with those of other protists. As with other protists, the FA of the dinoflagellates can be characterized by the presence of fibrous and microtubular components. Our studies and others indicate that the dinoflagellate FA can be expected to possess a striated fibrous root on the basal body of the transverse flagellum and a multimembered microtubular root on the basal body of the longitudinal flagellum. Two other features that appear widespread in the group are the transverse striated root associated microtubule (tsrm) and the transverse microtubular root (tmr). The tsrm extends at least half the length of the transverse striated root while the tmr extends from the transverse basal body toward the exit aperture of the transverse flagellum. In most cases, the tmr gives rise to several cytoplasmic microtubules at a right angle. The apparent conserved nature of these roots leads us to the conclusion that the dinoflagellate FA can be compared to the FA of the cryptomonads, chrysophytes, and the ciliates for phylogenetic purposes. Of these groups, the chrysophytes possess an FA with the most structures in common with the dinoflagellates. Our immunomicroscopical investigations of the microtubular, actin and centrin components of the dinoflagellate cytoskeleton point to the comparative usefulness of these cytological features.

Keywords: Actin; Centrin; Confocal microscopy; Dinoflagellates; Dinophyceae; Flagellar apparatus; Immunomicroscopy; Microtubules; Tubulin.

Abbreviations: aptb apical transverse microtubular band; FA flagellar apparatus; lmr longitudinal microtubular root; mls multilayered structure; tmr transverse microtubular root; tmre transverse microtubular root extension; tsr transverse striated fibrous root; tsrm transverse striated root associated microtubule.

* Correspondence and reprints: Department of Biology, University of Southwestern Louisiana, Lafayette, LA 70504-2451, U.S.A.

Introduction

As with other protistan groups, our knowledge of the dinoflagellates has undergone a dramatic increase over the past thirty years. Their unusual external morphology, nuclear features, toxin production, nutritional peculiarities, and endosymbiotic characteristics are just some of the intriguing features of the group which have stimulated investigations. The phylogenetic treatment of the dinoflagellates has drawn considerable controversy due to their peculiar characteristics [see the comprehensive books by Spector (1984) and Taylor (1987)]. Two distinct phylogenetic ideas have emerged. Taylor (1987) continues to propose that the dinoflagellates are primitive flagellates being part of the chromophyte series of flagellates and closely aligned to the ciliates. Loeblich (1984), on the other hand, suggests that the dinoflagellates are a derived group arising from an *Oxyrrhis*-like organism with the subsequent loss of histone proteins. Despite these differences of interpretation and significant interest in the group, our knowledge of the fundamental and perhaps evolutionarily important dinoflagellate cell motility system has remained limited, primarily, because the taxa are notoriously difficult to preserve. Nevertheless, a number of important ultrastructural studies have provided much needed information concerning the nature of the FA and microtubular cytoskeleton in the group. The studies of Dodge and Crawford (1968, 1969 a, b, 1970 a, b, 1971 a, b, c) have given many insights into the complex nature of the dinoflagellate FA. Their work clearly illustrates the differences between the dinoflagellate FA and that of other groups. The last ten years have seen an advance in fixation protocol and in the number of additional taxa examined. The studies of Bradbury et al. (1983), Larsen (1988), Wedemayer and Wilcox (1984), Wedemayer et al. (1982), Wilcox (1989), Wilcox et al. (1982), and Gardiner et al. (1989), have contributed greatly to the quantity and quality of available comparative FA characters. Bullman and Roberts (1986), Farmer and Roberts (1990), Roberts (1985, 1986, 1989), and Roberts and Timpano (1989) have provided a number of investigations devoted to three-dimensional reconstructions of the FA in several taxa, while Farmer and Roberts (1989) have contributed to the accuracy of the data through the use of rapid-freeze and freeze-substitution analyses.

The structure of the dinoflagellate cytoskeleton has recently been investigated by Brown et al. (1988) and Roberts et al. (1987, 1988 a, b). The micro-architecture of the dinoflagellate microtubular cytoskeleton appears to be unique among the flagellate groups. Lemoine and Roberts (1989 a, b) reported on the three-dimensional nature of the actin- and centrin-based cytoskeleton in some dinoflagellates while Höhfeld et al. (1988) showed that the transverse flagellum of *Peridinium inconspicuum, Gyrodinium resplendens*, and *Scrippsiella trochoidea* possesses a centrin-based contractile strand adjacent to the microtubular axoneme.

Rapid advances have been made in the study of the cell motility system of other flagellate groups, as well. In our opinion, one of the greatest conceptual leaps for protistologists was the discovery that basal bodies transform from one generation to the next and finally mature in the third generation (Melkonian et al. 1987). Since this discovery, a number of other investigators have detailed flagellar development in additional groups (e.g., uniflagellate green algae, Heimann et al. 1989 a; a multi-flagellate green alga, Moestrup and Hori 1989; a glaucocystophyte, Heimann et al. 1989 b; a euglenoid, Farmer and Triemer 1988; a prymnesiophyte, Beech et al. 1988; a synurophyte, Beech and Wetherbee 1990 a,b; a chrysophyte, Wetherbee et al. 1988; ciliates, Allen 1969; Bohatier 1979; Bernard and Bohatier 1981; Eisler, 1988, 1989; Kamra and Sapra 1990). Unfortunately, developmental studies of the FA have not been reported for the dinoflagellates.

As a result of these structural and developmental studies, the comparative morphologist is much better equipped to make phylogenetic appraisals of dinoflagellate affinities to other protists. As such, our presentation will attempt to summarize the comparative aspects of the dinoflagellate FA and cytoskeleton with respect to similar components in other groups. We will focus only on comparing those FA and cytoskeletal components which we feel are conservative in the dinoflagellates. A different presentation (Roberts 1991) utilizes these and other cytological features to develop a cladistic analysis of the dinoflagellate taxa.

The dinoflagellate flagellar apparatus

Basal body incidence and connectives

The angle between basal bodies of this group vary from nearly parallel in *Prorocentrum* to anti-parallel (160–180°) in the *Amphidinium* spp. (Farmer and Roberts 1989). To our knowledge, basal body variation of this magnitude has been reported only in the green algae (see Moestrup 1982) and chrysophytes (Andersen 1987, 1989 a, b). The possession of parallel basal bodies and, therefore, anterior flagellation in the dinoflagellates is considered primitive by Taylor (1976, 1978, 1980, 1987), but this form of flagellation may be the result of anterior/posterior compression (Loeblich 1976, Taylor 1980, Honsell and Talarico 1985, Roberts and Ma-

Abbreviations used in the figures: alb anterior longitudinal microtubular bundle; aptb apical transverse microtubular band; at anterior transverse microtubular ribbon (after Lynn 1988); atb anterior transverse microtubular band; clb cingular longitudinal microtubular bundle; cp collared pit; cr two stranded microtubular root (after Mignot et al. 1968); db dorsal basal body; dc dorsal connective; fa flagellar apparatus region; kd kinetodesmal fibril; lb longitudinal basal body; lf longitudinal flagellum; lfo longitudinal flagellar opening; lmr longitudinal microtubular root; mls multilayered structure; plb posterior longitudinal microtubular bundle; ppc posterior postciliary microtubular ribbon (after Lynn 1988); pt posterior transverse microtubular ribbon (after Lynn 1988); ptb posterior transverse microtubular band; r 1, r 2, r 3, r 4 microtubular roots of the chrysophytes (after Andersen 1987, 1989 a); rhiz rhizoplast; rhs rhizostyle; sc striated collar; src striated root connective; tb transverse basal body; tf transverse flagellum; tfo transverse flagellar opening; tmr transverse microtubular root; tmre transverse microtubular root extension; tsr transverse striated root; tsrm transverse striated root microtubule; vb ventral basal body; vc ventral connective; vr ventral ridge; vroot ventral root; 1 developmentally mature basal body; 2 developmentally immature basal body

Fig. 1. Diagrammatic illustrations of the FA of several flagellate taxa. **A** Chrysophyceae, **B** Xanthophyceae, **C** *Hydrurus/Chrysonebula* (Chrysophyceae), **D** cryptomonads, **E** *Oxyrrhis* (Dinophyceae), **F** generalized dinoflagellate showing cytoskeletal elements, **G** basal body pair (view from cell's right), **H** basal body pair rotated 180° on basal body # 1 axis to compare with **I** "stylized" *Oxyrrhis* type of flagellar apparatus. Drawings not to scale. A–C after Andersen (1989 a), D after Roberts (1984), G and H after Peck (1977)

tese 1990, Roberts 1991). The predominant basal body condition in the flagellates appears to be the open posture with an angle between 45° and anti-parallel. It is noted that in these flagellar apparatuses the greatest number of FA components are found. Despite this separation of the basal bodies, the dinoflagellates rarely possess prominent fibrous connectives between them, unlike the other protists where one or more fibrous bands are frequently encountered (Moestrup 1982, Sleigh 1989). In the dinoflagellates, linkage of basal

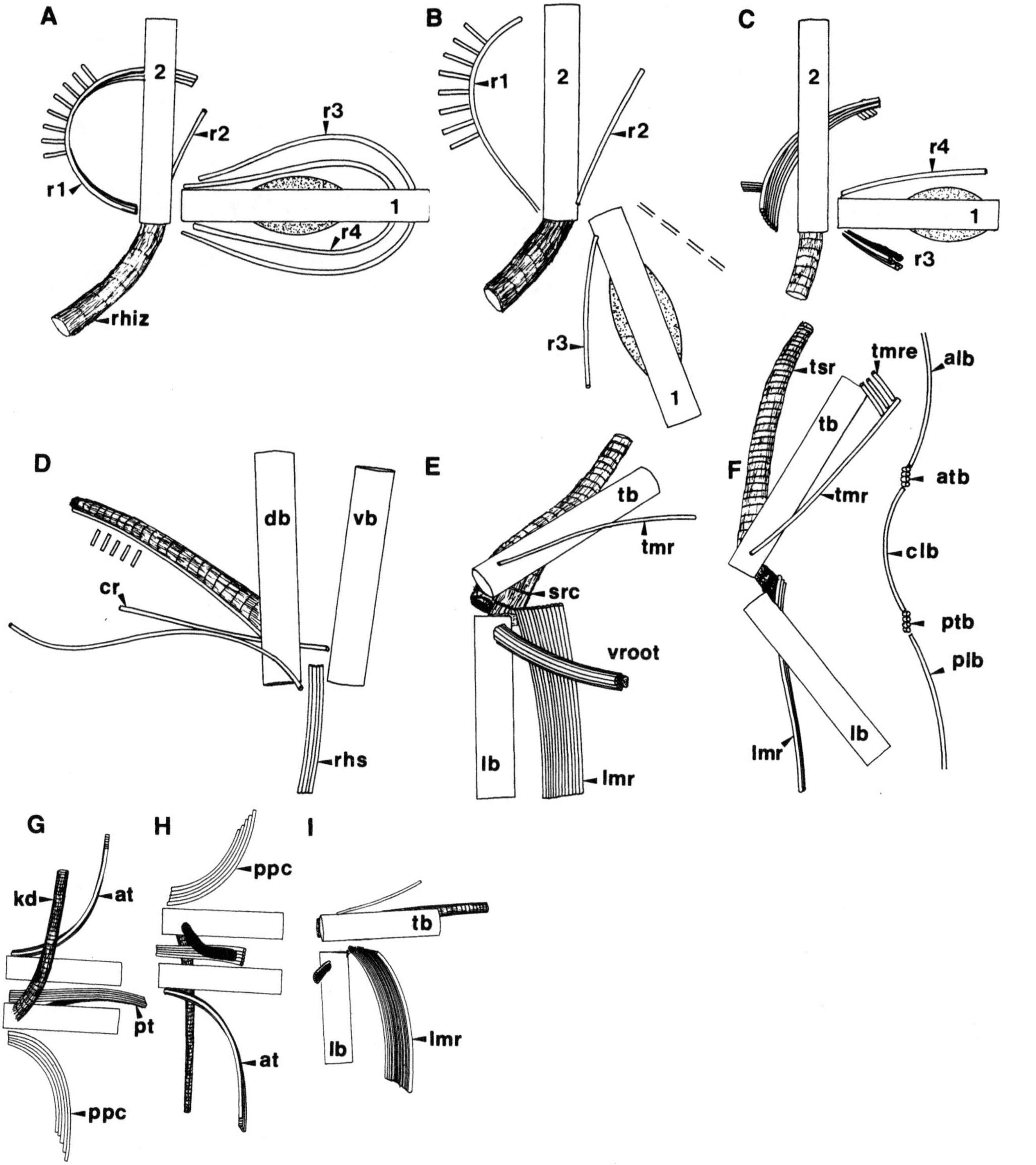

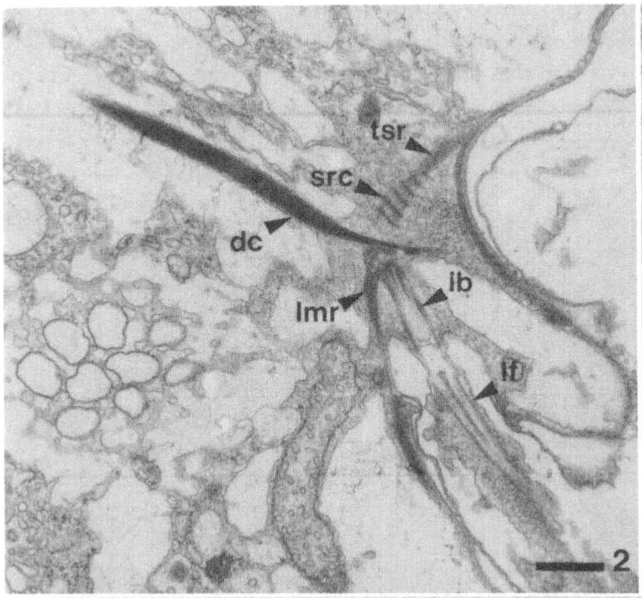

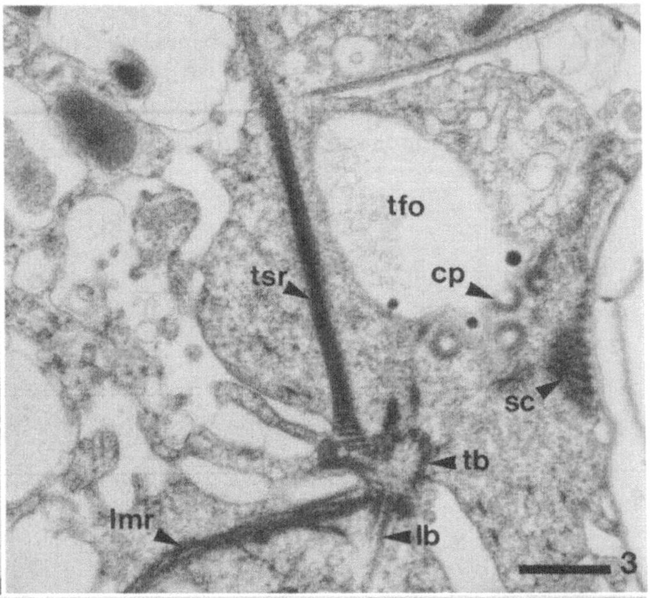

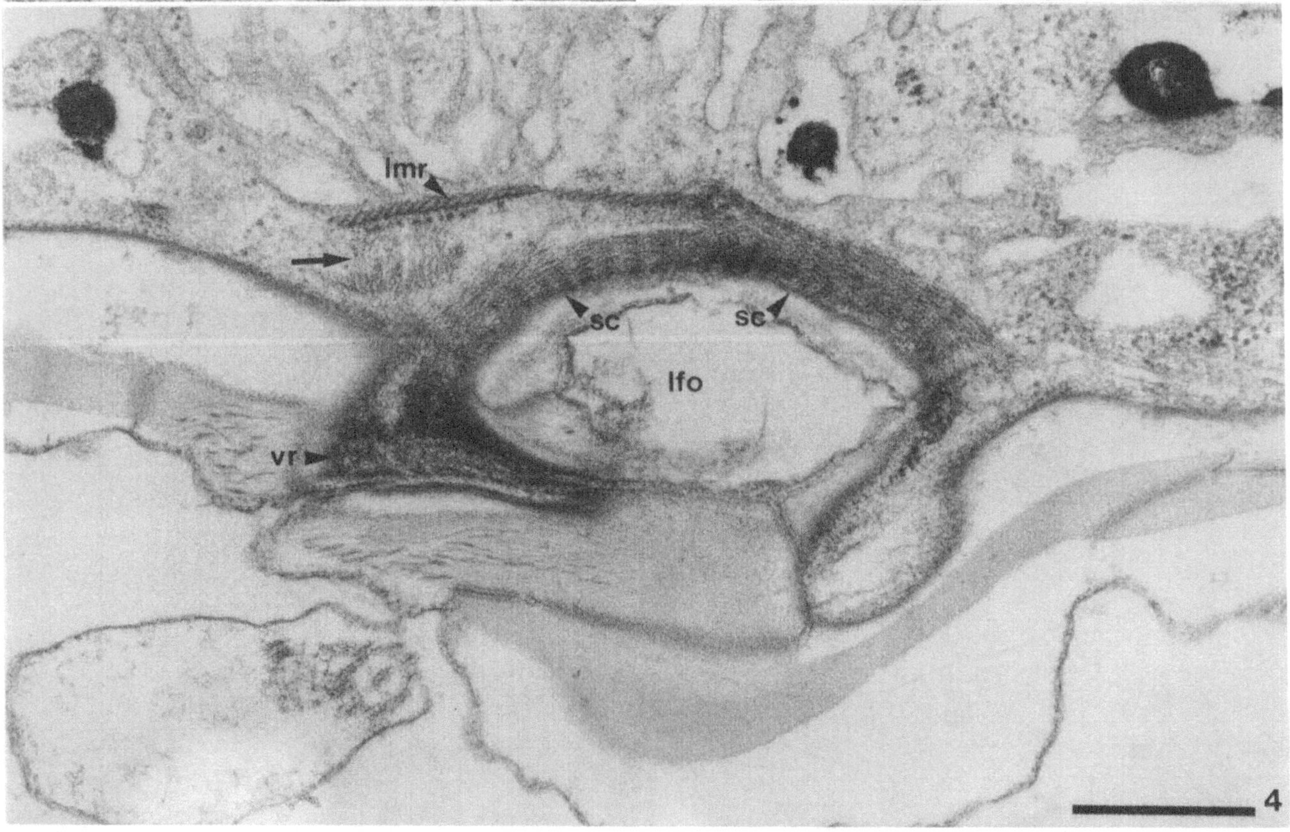

Fig. 2. TEM of *Gymnodinium acidotum* showing many of the flagellar components. Anterior view. The distal connective attaches to the longitudinal microtubular root (*lmr*) and extends to the cell's dorsal cingular groove. The transverse striated root (*tsr*) extends to the cell's left from the transverse basal body (not shown). The striated root connective (*scr*) links the tsr to the lmr. The lmr extends posteriorly adjacent to the longitudinal basal body. (From Farmer and Roberts 1990, with permission of J. Phycol.) Bar: 0.5 μm

Fig. 3. TEM of *Pheopolykrikos beauchampi*. Dorsal view. Bar: 0.5 μm

Fig. 4. TEM of *Woloszynskia limnetica*. Posterior view. Note the connection of the striated collar (*sc*) to the ventral ridge (*vr*). At this level only half of the intact ring is visible. Also note the small connective between the lmr and the sc (arrow). Bar: 0.5 μm

bodies occurs indirectly via components attaching to one another and then to the respective basal body (Fig. 1 E and F). To our knowledge, this intriguing variation has no parallel in other flagellate groups.

The longitudinal basal body and associated components

This basal body gives rise to the longitudinal flagellum which usually lies in the ventral sulcal groove of the cell (Figs. 1 E and F, 2, 3, 5, 7, and 8). The disposition of the flagellum may vary between taxa, but the longitudinal basal body can be readily denoted by its consistent association with a multi-membered longitudinal microtubular root (lmr). The lmr is always located on the same side of the longitudinal basal body as the transverse basal body (Fig. 1 E and F). This constancy of lmr and basal body placement permits the assignment of direction within the cell despite its external morphological aberrations. Even with the wide variation in basal body angle within the group, the position of the lmr is always on the left side of the longitudinal basal body (as viewed from the ventral surface) and, therefore, to the cell's left as are the overwhelming majority of other FA components (Fig. 1 E and F). This simple concept becomes very important when comparing the FA of other groups (see below).

The lmr of the dinoflagellates varies in the number of microtubules from as few as 8–12 in *Heterocapsa* sp. (Bullman and Roberts 1986) and *Amphidinium rhynchocephalum* (Farmer and Roberts 1989) to several dozen in *Ceratium* (Roberts 1989), *Oxyrrhis* (Roberts 1985), and *Gymnodinium sanguineum* (Roberts and Melkonian 1986). A correlation appears to exist between the numbers of microtubules in the root and the size of the cell. Immuno-localization studies of microtubules (Roberts et al. 1988 a, b) and the serial thin section analyses indicate that the lmr is separate from the cytoskeletal microtubules (Fig. 17). The lmr extends posteriorly along the cell's left side of the sulcal depression terminating at or near the cell's posterior (Fig. 4). At its origin (Fig. 1 E and F), the lmr is often associated with a number of fibrous components that link it to either the longitudinal basal body, transverse basal body, striated collars (Fig. 4), and/or transverse striated fibrous root (Figs. 2 and 3). In *Amphidinium cryophilum, Gymnodinium* sp., *G. sanguineum, G. acidotum* (Fig. 2) the lmr is usually associated with a fibrous component that extends dorsally. It gives rise to striated fibrous ventral connectives in *Polykrikos kofoidii* and *G. sanguineum.* This variability in lmr connectives restricts the comparative usefulness at the phylum level but may be important at lower taxonomic levels (Roberts 1991). The recent discovery of a multilayered structure (mls) on and near the origin of the lmr adds a new comparative feature to this complex root. Wilcox (1989) found a mls on the lmr's of *Katodinium campylops* and *Woloszynskia pascheri.* In taxa we have examined, the mls appears to constitute the basal portion of the fibrous extensions that connect the lmr to other FA components as occurs in *W. pascheri* (Figs. 9 and 10 of Wilcox 1989). In *Amphidinium rhynchocephalum* (Figs. 5 and 6) and other dinoflagellates, the base of the striated root connective appears nearly identical to the mls. The base of the large ventral striated connective of *G. sanguineum* (Figs. 7 and 8) is also comparable. Hence, we believe that the mls described by Wilcox (1989) is homologous to the fibrillar bases of all fibrous components associated with the longitudinal microtubular root of the dinoflagellates. Although multilayered structures have been reported in a number of other flagellate groups, it is important to note that the multilayered structures of these groups have never been found to be a direct part of a fibrous component as occurs in the dinoflagellates. This suggests that the mls of the dinoflagellates is unique to the group, but it is not clear if it is a conserved feature or will be useful in comparisons with other natural groups.

The transverse basal body and associated components

The transverse basal body extends to the cell's left in all known cases. This basal body gives rise to the transverse striated fibrous root (tsr) that extends to the left and anterior (Figs. 1 E and F, 2, and 3). The tsr exhibits little variation in direction or periodicity and occurs in all taxa examined. In general, the tsr extends beyond the collared pit zone of the flagellar opening (Fig. 3) and terminates beyond the striated collar that surrounds the exit aperture of the flagellum (Fig. 2). The tsr terminates at the sub-thecal membrane without connecting to any other cell component. Our studies indicate that a lone microtubule often parallels the tsr either embedded in it or lying to its side. This transverse striated root microtubule (tsrm) has not been observed in all taxa (Roberts 1991), but this may be due to fixation variations or fortuitous sectioning. We have re-investigated a number of taxa in which the tsrm was not observed and have found it (Roberts 1991). We feel strongly that this microtubule/striated fibrous root combination is a conservative feature of the dinoflagellate FA. Although striated fibrous roots have been

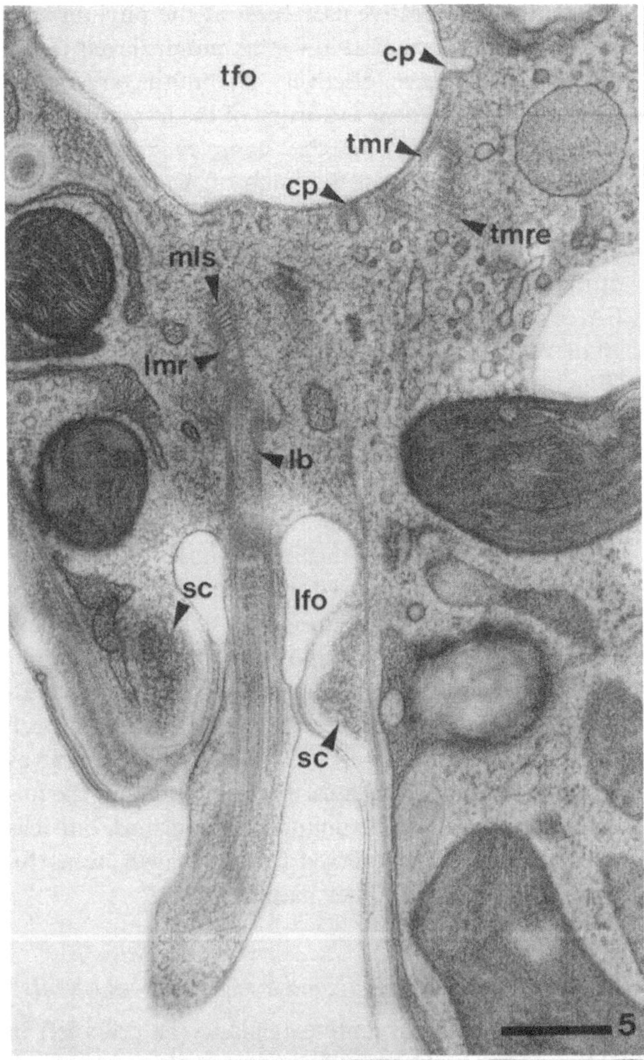
tfo
cp
tmr
cp
tmre
mls
lmr
lb
sc
lfo
sc
5

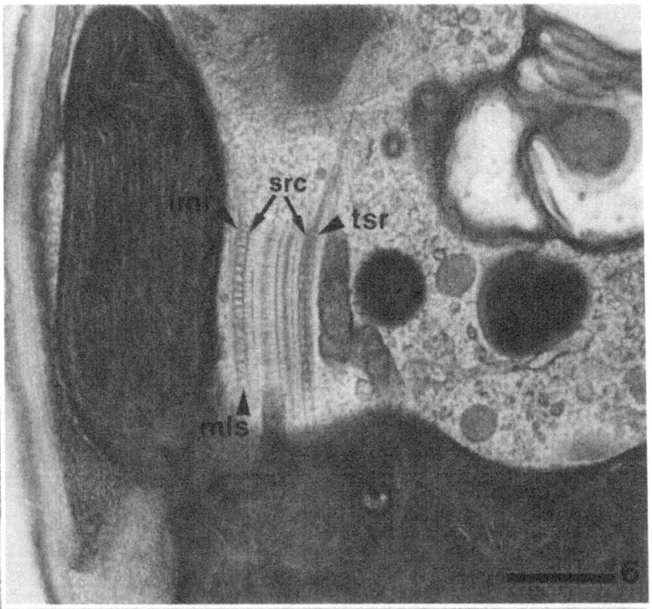
src
tsr
mls
6

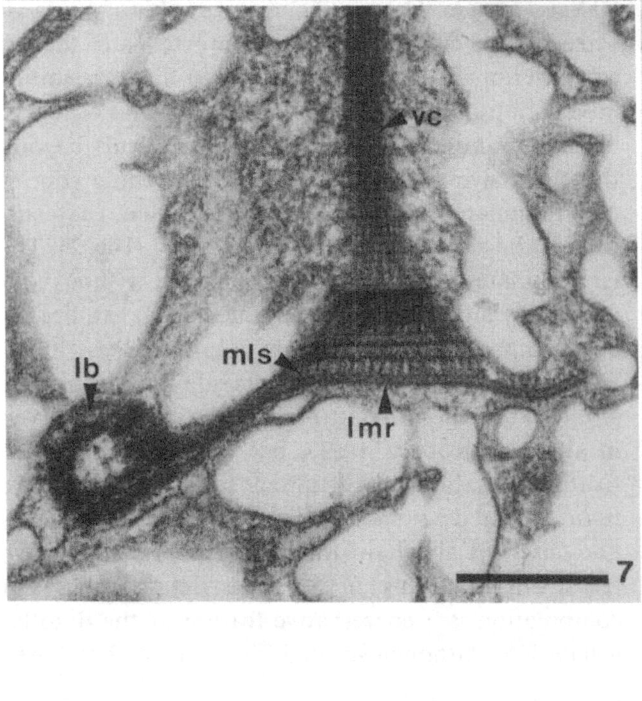
vc
lb
mls
lmr
7

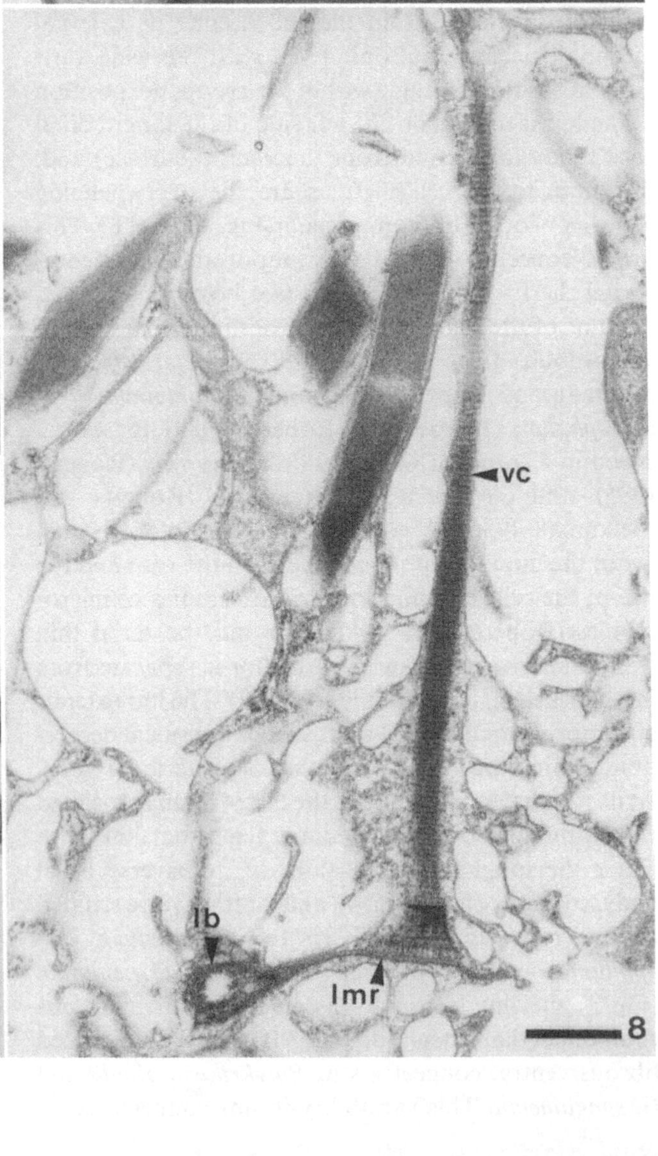
vc
lb
lmr
8

reported in a number of phytoflagellates (see Moestrup 1982), only the striated root of the cryptomonads is accompanied by a microtubular root (Gillot and Gibbs 1983, Roberts et al. 1981, Roberts 1984). It is important to note that the banding pattern of the cryptomonad striated fiber (Roberts 1984 Fig. 12) appears quite different from that of the dinoflagellates (Figs. 2 and 3) thus signaling, perhaps, an analogous rather than a homologous root type in the two groups. It would be very interesting to investigate these fibrous structures with various immunological probes to determine whether their protein content is similar.

Farmer and Roberts (1989) discovered, in their freeze-substitution studies of *Amphidinium rhynchocephalum*, that a single microtubule emanates from nearly the opposite side of the transverse basal body from that of the tsr. This transverse microtubular root (tmr; Fig. 1 F) extends to the zone of the collared pits of the flagellar canal at a slight angle to the transverse basal body. At this point the tmr gives rise to microtubular extensions (tmre; Figs. 1 F and 5) that extend into the cytoplasm at right angles. The point of termination of the tmre microtubules is not known. Farmer and Roberts (1989) pointed out that the tmr and tmre complex is highly reminiscent of the right angle emanation of the cytoplasmic microtubules from the R 1 microtubular root of chrysophytes (including synurophytes, see Andersen 1987, 1989 a, b). The newly discovered flagellate *Aulacomonas submarina* (Brugerolle and Patterson 1990) also possesses a microtubular root similar in appearance to the R 1 root and it too gives rise to microtubules that extend into the cytoplasm. We find resemblance of basal body 1 of the synurophytes (now referred to as #2, Beech and Wetherbee 1990 a, b, following their developmental studies of *Mallomonas*) to the transverse basal body of the dinoflagellates to be intriguing (Fig. 1 A–C, E, and F). In both the dinoflagellates and synurophytes this basal body gives rise to both a striated fibrous root and a microtubular root that is believed to nucleate cytoplasmic microtubules. It should also be noted that the microtubules beneath the striated root in the cryptomonad *Cryptomonas* sp. φ (Gillot and Gibbs 1983) give rise to microtubules in dividing cells. The significance of this observation is unclear, since microtubules are also nucleated from the striated fibers in dividing cells of the synurophyte *Mallomonas* (Beech and Wetherbee 1990).

The ciliates also possess a striated fiber similar in appearance (see Peck 1977; Lynn 1988, 1990) to the tsr of dinoflagellates (Fig. 1 G and H). The basal body that gives rise to the striated fibrous root also gives rise to a microtubular root. However, this root has not been observed to nucleate microtubules. It may also be significant that this striated fibrous root bearing basal body may be the developmentally *mature* basal body in the ciliates. This basal body would, therefore, be developmentally homologous to the short basal body #2 of the synurophytes that lacks a striated fiber at maturity. The striated root bearing basal body of the ciliates, by definition, issues its striated fibrous root to the cell's right rather than to the left as in the dinoflagellates. This basal body of the ciliates also produces, in some taxa, an electron dense extension that meets a transverse set of microtubules between the basal bodies imparting this microtubular root with an electron dense core. A similar microtubular root occurs in the enigmatic dinoflagellate *Oxyrrhis marina* (Dodge and Crawford 1971 b, Roberts 1985) in which a curved microtubular root arises from the longitudinal basal body in conjunction with an electron dense core (Fig. 1 E and F). The differences between the two phyla are apparent in Fig. 1 G–I with the electron dense core arising from the longitudinal basal body in *Oxyrrhis* and from the striated fibrous root bearing basal body in the ciliates. A similar microtubular root with an electron dense core is found in several chrysophytes and is quite prominent

---

Fig. 5. TEM of *Amphidinium rhynchocephalum*. Oblique longitudinal view from near ventral. Near the proximal portion of the longitudinal microtubular root (*lmr*) a multilayered structure (*mls*) occurs in the vicinity of the striated root connective between the longitudinal microtubular root (*lmr*) and the transverse striated root (not visible in this plane). Note the transverse microtubular root (*tmr*) and its associated microtubular extension (*tmre*). (From Farmer and Roberts 1990, with permission of J. Phycol.) Bar: 0.5 μm

Fig. 6. TEM of *A. rhynchocephalum*. This view shows the nature of the striated root connective (*src*) between the longitudinal microtubular root (*lmr*) and the transverse striated root (*tsr*). Note the multilayered structure (*mls*) region adjacent to the microtubules of the lmr. Bar: 0.5 μm

Fig. 7. TEM of *Gymnodinium sanguineum*. This view shows the anterior region of the broad longitudinal microtubular root (*lmr*) and the proximal region of the ventral connective (*vc*). Note that the base of the vc is reminiscent of a multilayered structure (*mls*). Bar: 0.5 μm

Fig. 8. TEM of *G. sanguineum*. This view shows the extent of the ventral connective (*vc*) as it extends to the sub-thecal plate region of the cell's ventral surface. Bar: 0.5 μm

in the *Hydrurus*- (Hoffman et al. 1986) and *Chrysonebula*- (Hibberd 1977) type FA illustrated here (Fig. 1 C).

The dinoflagellate cytoskeleton

Microtubular cytoskeleton

The dinoflagellate microtubular cytoskeleton is unique among the flagellates. For years it has been known that microtubules lie beneath the theca of dinoflagellates (see Dodge 1983) but only recently has their microarchitecture been elucidated. Brown et al. (1988) were the first to illustrate the microtubular cytoskeleton of the dinoflagellates with immunomicroscopic techniques, while Roberts et al. (1988 a, b, 1991) detailed the three distinct microtubular regions that comprise the system in *Amphidinium rhynchocephalum*. The longitudinal microtubules appear to be of three types in the dinoflagellates (Figs. 9–11, 15, and 16), anteriorly directed, posteriorly directed and those that are confined to and lie beneath the cingular depression. The anteriorly directed microtubules arise from the anterior transverse band of microtubules in the region of the sulcus and extend around the cell in a clock-wise fashion (apical view). The precise origin of the transverse band is unclear and the clock-wise direction is implied (Roberts et al. 1988 a) by the emergence of the anterior longitudinal microtubules from the band (Fig. 18). The transverse microtubular band extends around the cell and back to the cell's left edge. The anteriorly directed longitudinal microtubules extend to the apex of the cell in the majority of taxa (Figs. 15 and 16). An interesting exception occurs in *Gymnodinium sanguineum* (Figs. 9–11) where the anterior longitudinal microtubules abut an apical transverse microtubular band (aptb). This apical band of microtubules is associated with the contractile protein centrin (Salisbury et al. 1987) or its homologue (see below). It is peculiar that the two ends of the helical aptb are at different heights on the cell's ventral surface (Figs. 9 and 11) as are the ends of the associated centrin component (Fig. 12). The posterior longitudinal microtubules arise from the posterior transverse microtubular band and abut one another at the posterior end of the cell (Figs. 9 and 10). The longitudinal microtubules that are restricted to the cingulum abut the posterior transverse band and the anterior transverse band, repectively. In all taxa examined, the components of the FA are *independent* of the cytoskeletal microtubules.

With the exception of the apical transverse band found in *G. sanguineum*, the three-dimensional nature of the microtubular cytoskeleton appears conserved in the taxa we have studied; *Amphidinium rhynchocephalum, Gymnodinium splendens, G.* sp., *Woloszynskia limnetica, Heterocapsa* sp., *Pheopolykrikos beauchampi*, and *Scrippsiella sweeneyae*. It is interesting that the microtubular cytoskeleton of *Oxyrrhis marina* has a different appearance and only possesses one distinct transverse band of microtubules (Fig. 15). This may reflect the peculiar external morphology of this organism, since it lacks a well defined cingulum. While it is not surprising to find an unusual microtubular cytoskeleton in *Oxyrrhis*, we are continuing to investigate the location of a transverse band of microtubules homologous to that of the posterior transverse microtubular band. The studies of Brown et al. (1988) suggest that the tentacle region may be a strongly reduced hypocone. If this is the case and *Oxyrrhis* is related to the dinoflagellates, one would expect to find a transverse band of microtubules giving rise to the microtubules of the tentacle. This band would be located at the base of the tentacle and the microtubules located in the slightly concave ventral posterior region (Fig. 15) would then be homologous to the cingular longitudinal microtubular bundles. The transverse band of *Oxyrrhis* that has already been identified would be homologous to the anterior transverse microtubular band if the previous assumptions are proven.

Actin cytoskeleton

We have successfully determined the presence and disposition of F-actin in the dinoflagellates *Gymnodinium*

Figs. 9–11. Confocal laser scanning micrographs of indirect immuno-labelled microtubules in *Gymnodinium sanguineum*. See text for details. The helically wound apical transverse band of microtubules (*aptb*) extends from the viewer's left, around the dorsal surface of the cell, and returns to the ventral surface at a level more anterior than at its origin (asterisks; Fig. 9). Note posterior longitudinal microtubular bundles (*plb*) appear to attach to the brightly stained posterior edge of the cell between its antapical lobes (arrowheads; Fig. 10). The proximal and distal ends (*p* and *d*, respectively; Fig. 11) of the helical apical transverse microtubular band is shown as are the connections between it and the alb's (arrows). Bars: 5.0 µm

Fig. 12. Scanning confocal laser micrograph of indirect immuno-labelled cell for the protein centrin or its homologue in *G. sanguineum*. Note the association of the centrin staining with the apical transverse microtubular band (arrow) and the large unidentified bands of centrin containing structures in the FA region (arrowheads). Bar: 0.5 µm

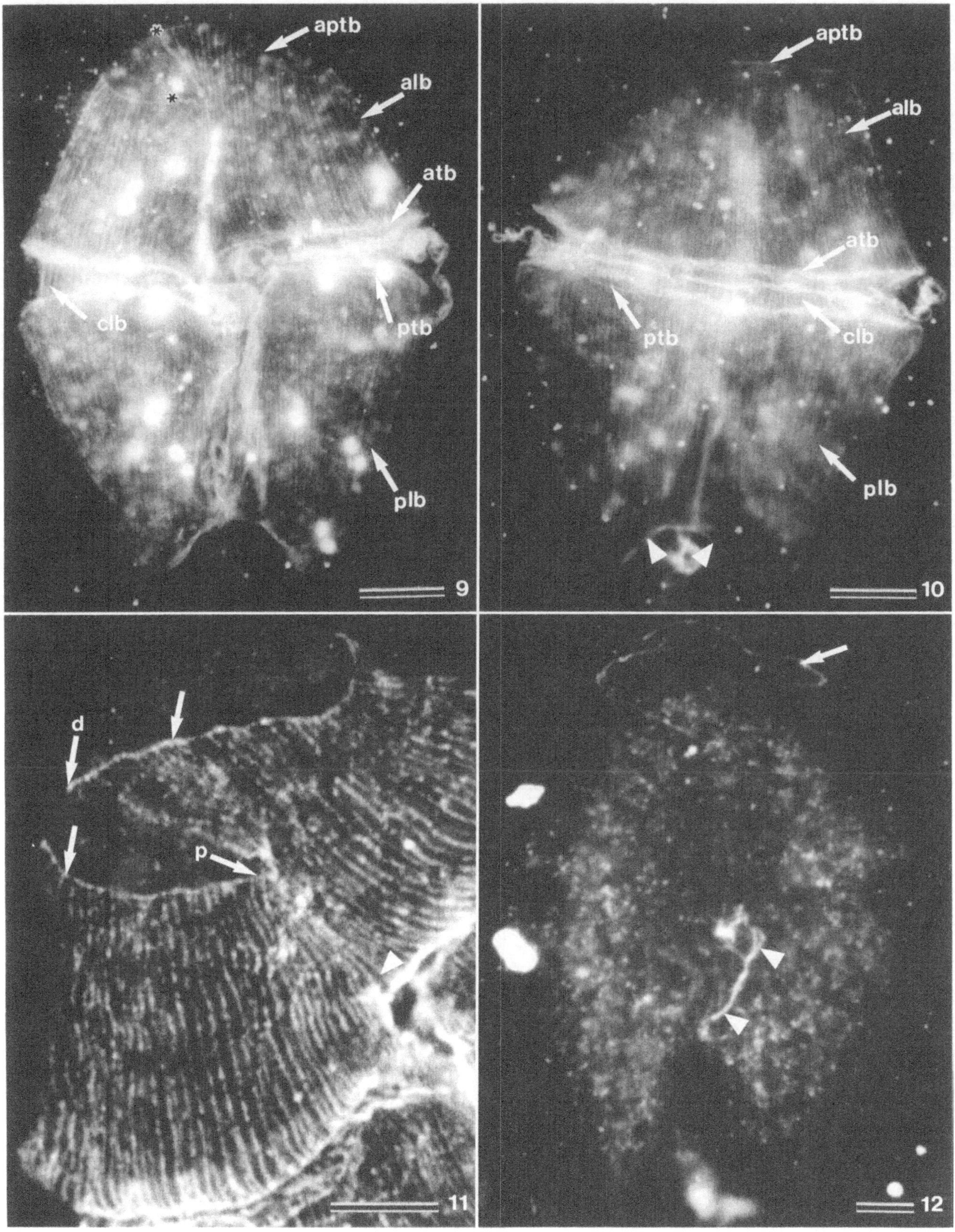
aptb
alb
atb
clb
ptb
plb
9
aptb
alb
atb
ptb
clb
plb
10
d
p
11
12

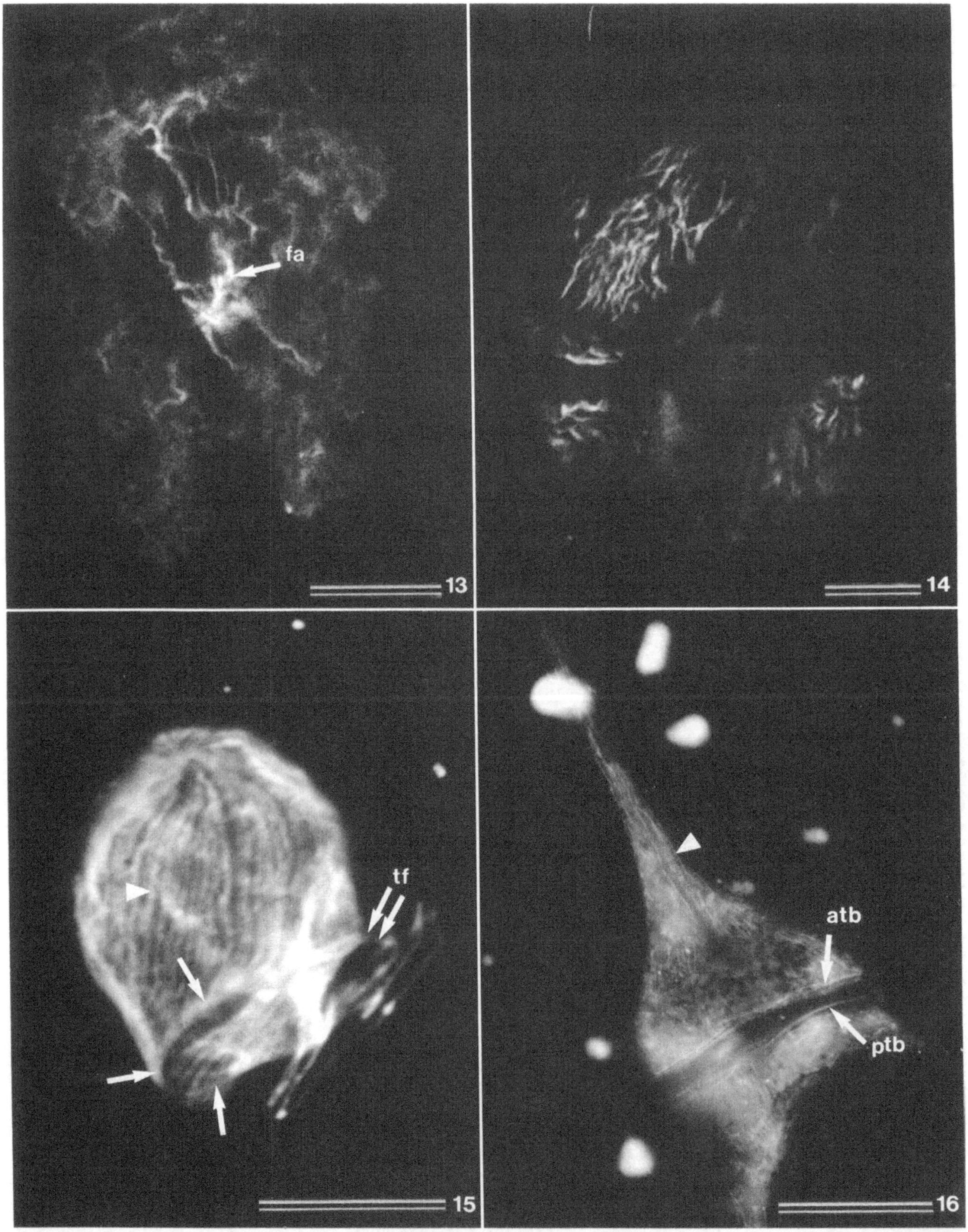
fa
13
14
tf
15
atb
ptb
16

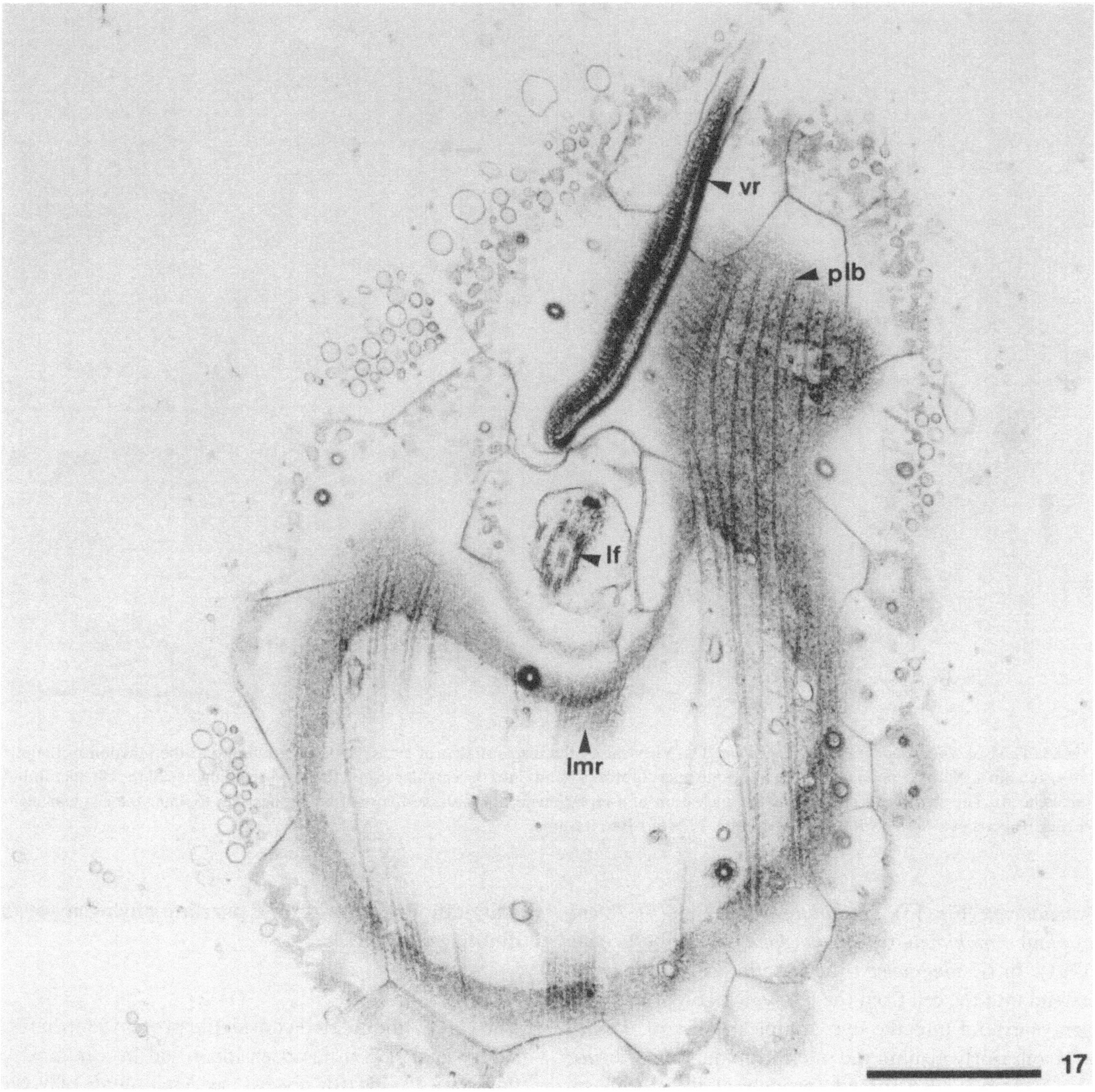

Fig. 17. TEM of *Amphidinium rhynchocephalum*. This view shows the displacement of cytoskeletal microtubules, microtubules of the lmr and the ventral ridge between the transverse and longitudinal flagella. Note (1) that microtubules of the lmr are distinct from the plb microtubules and (2) the fibrillar connections between the electron dense ventral ridge and the extreme plb's. (From Roberts et al. 1988, with permission of the J. Phycol.) Bar: 1.0 µm

---

Fig. 13. Epifluorescence confocal laser scanning micrographs of direct phalloidin/FITC labeling of F-actin in *Gymnodinium sanguineum*. Ventral internal optical section of the cell showing dispersion of F-actin from FA region (*fa*). Bar: 10.0 µm

Fig. 14. Epifluorescence confocal laser scanning micrographs of direct phalloidin/FITC labeling of F-actin in *Peridinium willei*. Ventral surface optical section showing F-actin localization under the thecal plates of the cell and near the anterior edge of the cingulum. Bar: 10.0 µm

Fig. 15. Epifluorescence micrograph of indirectly immuno-labelled microtubules in *Oxyrrhis marina*. View from cell's right ventral side showing a band of microtubules (arrows) that may be homologous to the anterior transverse microtubular band of other dinoflagellates. Anteriorly directed longitudinal microtubules (arrowhead) meet at the cell apex. Note the replicated transverse flagella (*tf*). Bar: 5.0 µm

Fig. 16. Epifluorescence micrograph of indirectly immuno-labelled microtubules in *Ceratium hirundinella*. This view shows the anterior (arrowhead) and posterior longitudinal microtubular bundles emanating from the atb and ptb, respectively. Bar: 10.0 µm

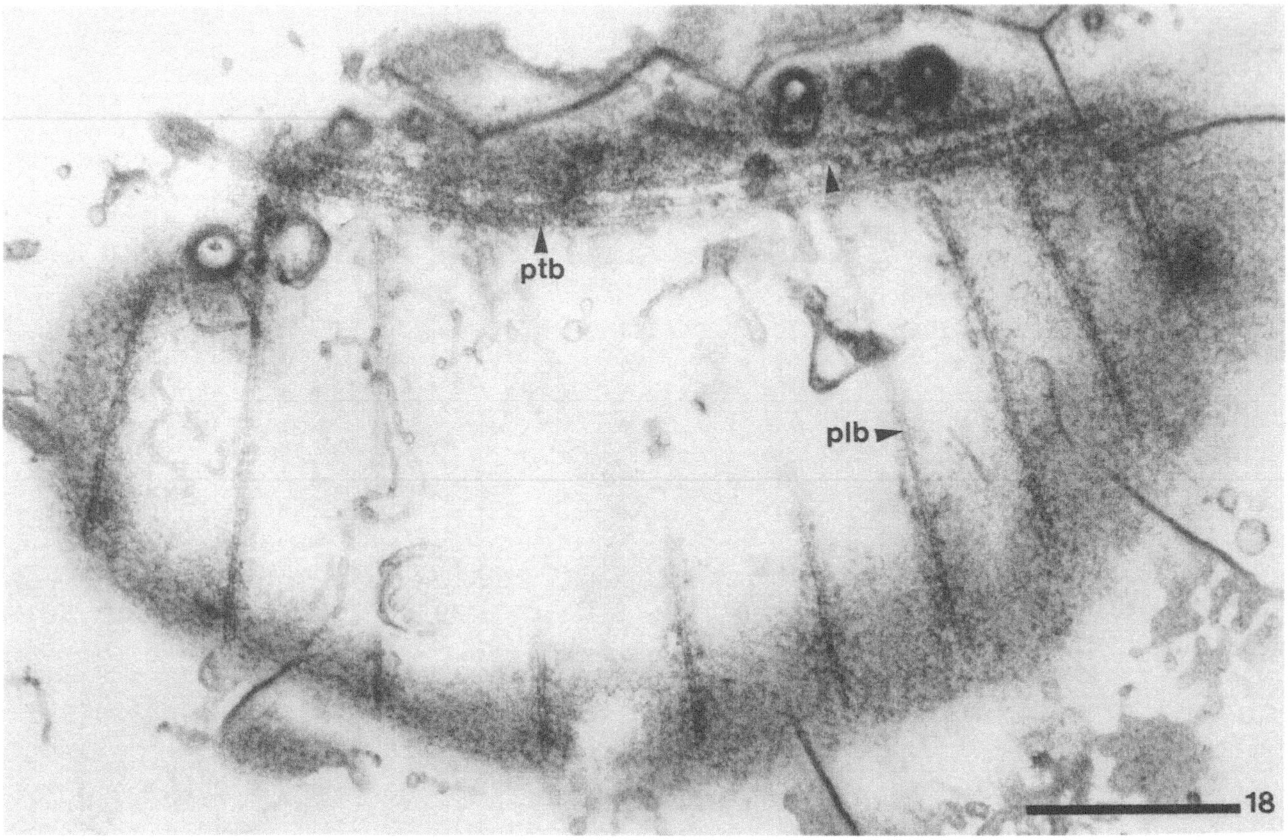

Fig. 18. TEM of *Amphidinium rhynchocephalum*. This view shows the displacement of cytoskeletal microtubules at the junction of the ptb's, clb's, and plb's. Note the point where one of the ptb microtubules extends into the cingular region thereby becoming one of the clb microtubules (arrowhead). The arrowhead also marks the nucleation of a replacement microtubule that will be the next clb to enter the cingular region. (From Roberts et al. 1988, with permission of J. Phycol.) Bar: 0.5 µm

sanguineum (Fig. 13), *Peridinium willei* (Fig. 14), *P. volzii*, and *Amphidinium rhynchocephalum* (Roberts et al. 1991). In *G. sanguineum* the F-actin arrays appear to extend into the cell from the FA region (Fig. 13). These arrays extend into the two antapical lobes of the cell and anteriorly around the nuclear domain. In *P. willei* the F-actin is also spread throughout the cytoplasm but is heavily localized near the edge of the cingular groove (Fig. 14). We have also noted that the F-actin concentration is high near the FA in *A. rhynchocephalum*. The presence of F-actin bundles adjacent to the theca in *P. willei* may indicate that actin is involved in the disposition of thecal plate vesicles during the cytokinetic process. Schnepf et al. (1990) have provided clear evidence of the important role of actin in the mitotic process of *Prorocentrum micans*. While the evolutionary significance of their findings must await further analyses, the prospect of finding a similar involvement of the actin cytoskeleton with the mitotic system of other protists is quite exciting and will most likely add additional clues to the puzzling phylogeny of the dinoflagellates.

Centrin cytoskeleton

The Ca^{++}-modulated contractile protein centrin (Salisbury et al. 1987) has been identified in a number of flagellate groups (for reviews, see Melkonian 1989, Salisbury 1989). Centrin has been reported to occur primarily in components of the FA. In the dinoflagellates, Höhfeld et al. (1988) discovered centrin in the transverse flagellum of three dinoflagellates. We have been fortunate to discover it not only in the FA region but also in the apical transverse ring of *G. sanguineum* (Fig. 12). It is not clear at present what structures, if any, of the FA contain centrin or its homologue. Figure 12 shows a heavily stained band that lies between the flagellar apertures at an angle of approximately 45°. In the majority of dinoflagellates this would be the region of the ventral ridge or connective between the striated collars. In *G. sanguineum* these structures have not been identified at the electron microscopical level, so it is not clear to which, if any, of these structures the immunofluorescent structures correspond. It is likely, however, that the large ventral connective that links the longitudinal microtubular root to the ventral region of the cell would be contractile (Figs. 7 and 8). We have shown that the apical transverse band of microtubules in *G. sanguineum* consists of microtubules and non-striated fibrous band (Lemoine and Roberts 1989 b). Immunofluorescent staining of this ring in our preparations (Fig. 12) indicates that centrin or its homologue is a component of the fibrous ring. One of our goals at this point is to investigate the presence of centrin in the apical pore fibrous complex (Roberts et al. 1987) of *Scrippsiella sweeneyae* and other dinoflagellates that are known to possess an apical pore complex. The presence of centrin in this complex may have significant phylogenetic importance, if the apical band of *G. sanguineum* can be shown to consist of the same proteins as the apical pore complex of some peridinioid dinoflagellates. Our preliminary data and that of Melkonian (pers. comm.) suggest that the striated roots of the dinoflagellates may not contain the protein centrin as do the striated roots of other algae (see Salisbury 1989, Melkonian 1989). Similarly, centrin may not be a part of the large striated collars (unpubl. obs.; Melkonian pers. comm.), but, it may be an important part of the various fibrous connectives found in the dinoflagellate flagellar apparatus. Farmer and Roberts (1990) noted that the striated root connective between the longitudinal microtubular root and the transverse striated fibrous root may be contractile because of its different periodicities observed in chemical versus rapid-freeze fixations. The striated root connective may, therefore, possess centrin or its homologue. In order to fully understand the evolutionary affinities of the group, though, it will be crucial to isolate the striated fibrous proteins and determine their homologue in other flagellate groups.

Phylogenetic considerations

One of the most prominent similarities between dinoflagellates and any other flagellate group occurs on the basal body that possesses the striated fibrous root of dinoflagellates and heterokont flagellates. In heterokonts (Andersen 1989 a, b) (Fig. 1A–D) the R 1 microtubular root nucleates cytoplasmic microtubules and is similar in position to the tmr and tmre of the dinoflagellates. Both flagellate groups possess striated fibrous roots on this basal body. A number of features are similar between the components of the other basal body in both flagellate assemblages as well. The R 4 root of the heterokonts is spatially similar to the longitudinal root of the dinoflagellates. Finally, the R 3 root of many heterokonts possesses an electron dense attribute adjacent to the microtubules of the root and spatially and morphologically resembles the ventral root of the dinoflagellate *Oxyrrhis*. Structural similarities also occur between FA components in the raphidiophytes *Gonyostomum semen* (Heywood 1980) and *Heterosigma akashiwo* (Vesk and Moestrup 1978) and some dinoflagellates. In *Gonyostomum*, the non-striated connective between the microtubular root and the nucleus is highly reminiscent of the similarly positioned connective in the dinoflagellates *Polykrikos* (Bradbury et al. 1983) and *Gymnodinium* sp. (Roberts 1986). In *Heterosigma*, the major microtubular root has a striated connective to the nucleus that imitates the ventral connective of *Gymnodinium sanguineum* (Fig. 7).

Although a number of similarities exist between the FA of heterokonts and dinoflagellates, several discrepancies occur as well. One of the most obvious is the variable morphology of the striated fibrous roots between the groups. Furthermore, it is unexplained why the striated fiber of the dinoflagellates is so similar in appearance to that found in the cryptophytes and ciliates and quite unlike that found in the other groups. It is also noted that the striated fibrous root in the majority of heterokonts arises from the proximal end of the presumed developing basal body. In the dinoflagellates, cryptophytes, and ciliates, the striated root arises from the side of the basal body. Further, no microtubule has been found in association with the striated fiber of the heterokonts, but one is present along the striated fiber of the cryptophytes and the dinoflagellates. The ciliates also possess a microtubular root with an electron dense core, but the electron dense material and the microtubules arise from the presumed mature (striated fiber possessing) basal body. Again this is opposite from what occurs in the heterokonts.

Another perplexing problem is that the cytoskeletal microtubules of the dinoflagellates do not appear to arise from FA born microtubules as they do in the heterokonts, cryptophytes, and ciliates. In all flagellate groups except the dinoflagellates, the microtubular cytoskeleton is involved in some way with one or more components of the FA. This fundamental difference in the dinoflagellate microtubular cytoskeleton signals the need for greater scrutiny of the dinoflagellate FA during developmental phases to insure that it is indeed distinct from the cytoskeleton. If the microtubular cytoskeleton is distinct from the FA in the dinoflagellates, as it appears to be, then this would indicate a highly derived condition from that of other flagellate groups. It would be very difficult to envision a dinoflagellate type ancestor giving rise to other flagellate groups which require such a drastic cytoskeletal shift. Thus, our data concerning the cytoskeleton indicates that the dinoflagellates are a derived flagellate group that no longer relies on the FA components for development of the cytoskeleton or never did!

Cluster analyses using FA components, cytological, and external morphological features indicate that the synurophytes share more characters with dinoflagellates than do the cryptomonads or ciliates (Schneider and Roberts 1989, Roberts 1991). Cladistic analyses of the same data indicate that, of the dinoflagellates, *Oxyrrhis marina* always diverges near the base of the most parsimonious trees, separate from the other dinoflagellate taxa. This is the case whether or not the synurophytes, cryptophytes or ciliates are designated as the out group. This supports the contentions of Loeblich (1984) that an *Oxyrrhis*-like organism may have given rise to the dinoflagellates. The inclusion of nuclear features in the data matrix is partially responsible for the outcome of the cladistic analyses, since *Oxyrrhis* possesses a different mitotic spindle (Triemer 1982) and histone proteins unlike the other dinoflagellates. Nevertheless, the FA of *Oxyrrhis* is different in many respects from that of other dinoflagellates and these differences have likely contributed to the segregation of *Oxyrrhis* in the cladistic analysis. It should be noted that when *Oxyrrhis* was used as the out group the synurophytes, cryptophytes, and ciliates formed a distinct clade from the dinoflagellates. For this presentation, we have included the majority of known FA characters for the dinoflagellates in Table 1. Although the data generated several hundred most-parsimonious trees using the branch and bound algorithm in the 2.4 version of PAUP (D. L. Swofford, see Theriot 1989), some consistent associations persisted and are apparent in the three consensus trees (Fig. 19 a–c) generated from this version of PAUP. Unlike our analyses of flagellar apparatus, cytological and external morphological data

Table 1. Flagellar apparatus characters of the dinoflagellates and selected groups

	A	B*	C	D	E	F	G	H	I	J	K	L
Amphidinium rhynchocephalum	1	0	0	0	1	1	1	0	0	0	0	1
A. cryophilum[a]	1	0	1	0	1	1	9	1	1	1	0	1
Ceratium hirundinella	1	1	0	0	0	1	1	0	0	1	0	1
Gymnodinium acidotum	1	0	1	0	0	1	1	0	1	1	0	1
Gymnodinium sp.	1	0	1	0	0	1	1	0	1	0	0	1
G. sanguineum	0	1	0	1	0	1	9	0	0	1	0	1
Heterocapsa pygmeae	1	0	0	0	1	1	1	1	0	0	0	1
Oxyrrhis marina	0	1	0	0	0	1	9	0	0	0	1	9
Pheopolykrikos beauchampi	1	1	0	0	0	1	1	0	0	1	0	1
Polykrikos kofoidii[b]	0	1	0	1	0	1	9	0	1	0	0	1
Prorocentrum micans	1	0	0	0	1	1	9	0	0	1	0	1
Woloszynskia limnetica	1	0	0	0	1	1	1	1	0	0	0	1
Woloszynskia sp.	1	0	0	0	0	1	1	1	0	0	0	1
Cryptophytes	0	0	0	0	0	1	0	0	0	0	0	0
Synurophytes	0	0	0	0	0	1	1	0	0	0	1	0
Ciliates	0	0	0	0	0	1	0	0	0	0	0	0

A Striated collars
B lmr < or > 15 microtubules. **0* < 15 microtubules, *1* > 15 microtubules
C Anti-parallel basal body overlap
D Ventral connective on lmr
E Transverse striated root associated microtubule
F Transverse microtubular root
G Transverse microtubular root extensions
H Microfibrils between lmr and theca
I Dorsal connective on lmr
J Peduncle
K Ventral root
L Striated root connective

0 Absence of character, *1* presence of character, *9* missing data

[a] Inferred from micrographs in Wilcox et al. (1982)
[b] Inferred from micrographs in Bradbury et al. (1983)

(Schneider and Roberts 1989, Roberts 1991), cladistic analysis of the FA data alone always groups *Oxyrrhis* within the dinoflagellate clade. Furthermore, when the cryptophytes are used as the out group (Fig. 19 B), the synurophytes appear more closely related to the dinoflagellate clade than do the ciliates. It is interesting that *Ceratium* and *Polykrikos* appear together in one clade, while *Amphidinium cryophilum* and *Gymnodinium acidotum* are united in a separate clade. These associations do not support the current concepts of dinoflagellate systematics (Taylor 1987). Unfortunately, it is clear that the FA data matrix is inadequate to resolve the phylogenetic relations of the Dinophyceae at this time. Nonetheless, our morphological and cytological based conclusions do not support the close association between the dinoflagellates and the ciliates inferred by the molecular studies of Maroteaux et al. (1985), Lenaers et al. (1988, 1989), Lynn and Sogin

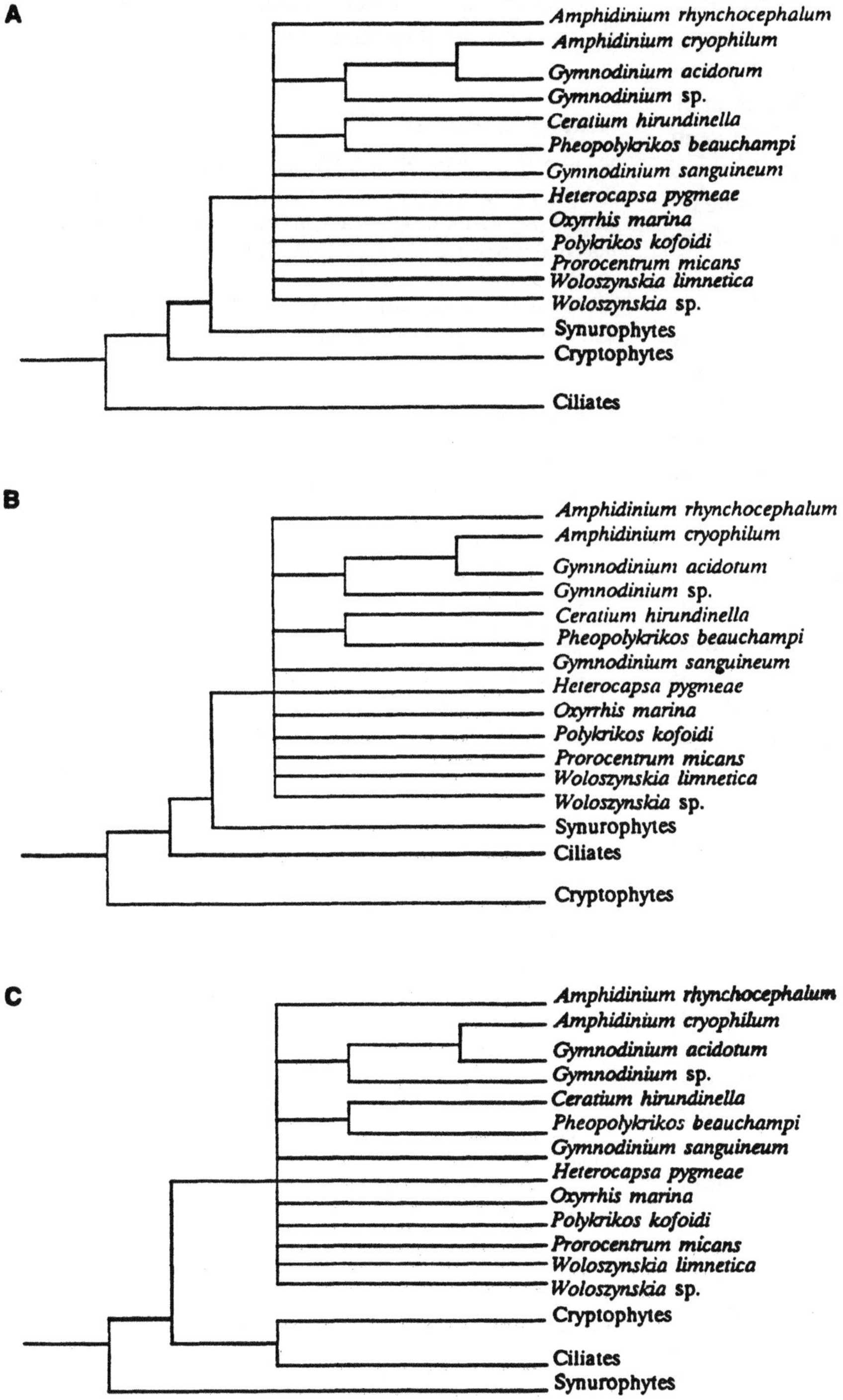

Fig. 19. Strict consensus trees generated from PAUP version 2.4 and using the ciliates as the designated out group (**A**), the cryptophytes as the designated out group (**B**), and the synurophytes as the designated out group (**C**)

(1988) but do agree with their conclusion that the dinoflagellates are a derived eukaryote group (Lenaers et al. 1989). Of interest here are the recent findings of Gajadhar et al. (1991) which further support the phylogenetic relations between the dinoflagellates and the ciliates but also indicate that the dinoflagellates and Apicomplexa share common ancestors. This is intriguing, since structural details of the apical complex of the Apicomplexa are strictly similar to the apical pore fibrous complex in some dinoflagellates (Roberts et al. 1987).

The significant studies into the development of prymnesiophyte and synurophyte basal bodies (Beech et al. 1988; Beech and Wetherbee 1990 a, b; respectively) and ciliate basal bodies (Allen 1969, Bohatier 1979, Bernard and Bohatier 1981, Eisler 1989) must be weighed heavily when considering the evolutionary affinities of the FA of different natural flagellate assemblages. In these heterokont flagellates, the developing basal body has the striated fibrous root and the mature basal body bears only microtubular roots. Although basal body development and transformation is not well documented for the ciliates, the available information (see Bohatier 1979, Eisler 1989) suggests that the mature basal body of ciliates possesses the striated fiber. This would represent an opposite development pattern from that which occurs in the heterokont flagellates. This prominent potential difference in developmental pattern of the FA between these groups is interesting. Future developmental studies of the basal bodies as well as elucidation of the protein constitution of presumably similar FA components in dinoflagellates and ciliates will doubtless help support or reject current molecular-based phylogenetic hypotheses.

It now seems clear that the dinoflagellates represent a derived group of protists. The FA data suggest a link between the heterokonts and dinoflagellates, and the cytoskeleton data indicates the highly derived status of the group. It is also apparent that neither the FA nor the cytoskeleton of any dinoflagellate corresponds sufficiently to that of a heterokont to indicate a clear connection between the groups at this time. This suggests that the cytoskeleton of *Oxyrrhis* may be crucial to further our understanding of the evolution of the group. Several questions immediately come to mind. First, does the cytoskeleton of *Oxyrrhis* represent an intermediate type in the evolution of the dinoflagellates, and if so, does this support the variation noted in the FA and mitotic apparatus of the group? Second, are any components of the FA in *Oxyrrhis* connected to its cytoskeleton, and if so, can these components be correlated with components from other flagellate groups? Third, does the possibility exist that the microtubular cytoskeleton evolved separately from the FA in the dinoflagellates, and if so, is it possible that the evolution of either microtubular system is linked to one or more endosymbiotic events? The answers to these and other questions will be sought through concentrated study of FA and cytoskeleton ontogeny as well as a thorough compilation and analysis of key cytological components considered to be homologous throughout the flagellate and ciliate groups.

Acknowledgements

We wish to thank the National Science Foundation (Grant BSR 85-06413) and the Louisiana Educational Quality Trust Fund (Grant LEQSF RD-A-16, LEQSF/Epscor Grant USL[1]-126-07) for support of this research. We also acknowledge the University of Southwestern Lousiana Electron Microscopy Center and the gracious assistance of Dr. Dennis Lynn, Dr. Michael Zavada, Ms. Robin Schneider, and Ms. Stephania Cormier.

References

Allen RD (1969) The morphogenesis of basal bodies and accessory structures of the cortex of the ciliated protozoan *Tetrahymena pyriformis*. J Cell Biol 40: 716–733

Andersen RA (1987) Synurophyceae classis nov., a new class of algae. Amer J Bot 74: 337–353

– (1989 a) Absolute orientation of the flagellar apparatus of *Hibberdia magna* comb. nov. (Chrysophyceae). Nord J Bot 8: 653–665

– (1989 b) The Synurophyceae and their relationship to other golden algae. In: Kristiansen J, Cronberg G, Geissler U (eds) Chrysophytes: developments and perspectives. Proceedings of the Second International Chrysophyte Symposium. Nova Hedwigia 95: 1–26

Beech PL, Wetherbee R (1990 a) Direct observations on flagellar transformation in *Mallomonas splendens* (Synurophyceae). J Phycol 26: 90–95

– – (1990 b) The flagellar apparatus of *Mallomonas splendens* (Synurophyceae) at interphase and its development during the cell cycle. J Phycol 26: 95–111

– – Pickett-Heaps JD (1988) Transformation of the flagella and associated flagellar components during cell division in the coccolithophorid *Pleurochrysis carterae*. Protoplasma 145: 37–46

Bernard F, Bohatier J (1981) Ultrastructure et mise en place des organelles buccaux au cours de la régénération orale chez *Stentor coeruleus* (Cilié Hétérotriche). Can J Zool 59: 2306–2318

Bohatier J (1979) Morphogenèse de régération chez le cilié *Condylostoma magnum* (Spiegel): etude ultrastructurale. J Protozool 26: 404–414

Bradbury PC, Westfall JA, Townshend JW (1983) Ultrastructure of the dinoflagellate *Polykrikos* II. The nucleus and its connections to the flagellar apparatus. J Ultrastruct Res 85: 24–32

Brown DL, Cachon J, Cachon M, Boillot A (1988) The cytoskeletal microtubular system of some naked dinoflagellates. Cell Motil Cytoskeleton 9: 361–474

Brugerolle G, Patterson DJ (1990) A cytological study of *Aulacomonas submarina* Skuja 1939, a heterotrophic flagellate with a novel ultrastructural identity. Eur J Protistol 25: 191–199

Bullman V, Roberts KR (1986) Structure of the flagellar apparatus in *Heterocapsa pygmaea* (Pyrrophyta). Phycologia 25: 558–571

Dodge JD (1983) Dinoflagellates: investigation and phylogenetic speculation. Br Phycol J 18: 335–356

– Crawford RM (1968) Fine structure of the dinoflagellate *Amphidinium carteri* Hulbert. Protistologica 4: 231–242

– – (1969a) The fine structure of *Gymnodinium fuscum* (Dinophyceae). New Phytol 68: 613–618

– – (1969b) Observations on the fine structure of the eyespot and associated organelles in the dinoflagellate *Glenodinium foliaceum*. J Cell Sci 5: 479–493

– – (1970a) The morphology and fine structure of *Ceratium hirundinella* (Dinophyceae). J Phycol 6: 137–149

– – (1970b) A survey of thecal fine structure in the diophyceae. Bot J Linn Soc 63: 53–67

– – (1971a) Fine structure of the dinoflagellate *Oxyrrhis marina* I. The general structure of the cell. Protistologica 7: 295–304

– – (1971b) Fine structure of the dinoflagellate *Oxyrrhis marina* II. The flagellar system. Protistologica 7: 399–409

– – (1971c) A fine-structural survey of dinoflagellate pyrenoids and food-reserves. Bot J Linn Soc 64: 105–115

Eisler K (1988) Electron microscopical observations on the ciliate *Furgasonia blochmanni* Fauré-Fremiet, 1967. Part I: an uptate on morphology. Eur J Protistol 24: 75–93

– (1989) Electron microscopical observations on the ciliate *Furgasonia blochmanni* Fauré-Fremiet, 1967. Part II: morphogenesis and phylogenetic conclusions. Eur J Protistol 24: 181–199

Farmer MA, Roberts KR (1989) Comparative analyses of the dinoflagellate flagellar apparatus. III. *Amphidinium rhynchocephalum.* J Phycol 25: 280–292

– – (1990) Comparative analysis of the dinoflagellate flagellar apparatus IV. *Gymnodinium acidotum.* J Phycol 26: 122–131

– Triemer RE (1988) Flagellar systems in the euglenoid flagellates. BioSystems 21: 283–291

Gajadhar AA, Marquardt WC, Hall R, Gunderson J, Ariztia-Carmona EV, Sogin ML (1991) Evolutionary relationships among apicomplexans, dinoflagellates, and ciliates: ribosomal RNA sequences of *Sarcocystis muris, Theileria annulata*, and *Crypthecodinium cohnii.* Mol Biochem Parasitol (in press)

Gardiner WE, Rushing AE, Dawes CJ (1989) Ultrastructural observations of *Gyrodinium estuariale* (Dinophyceae). J Phycol 25: 178–183

Gillott MA, Gibbs SP (1983) Comparison of the flagellar rootlets and periplast in two marine cryptomonads. Can J Bot 67: 1964–1978

Heimann K, Benting J, Timmermann S, Melkonian M (1989a) The flagellar developmental cycle in algae. Two types of flagellar development in uniflagellated algae. Protoplasma 153: 14–23

– Reize IB, Melkonian M (1989b) The flagellar developmental cycle in algae: flagellar transformation in *Cyanophora paradoxa* (Glaucocytstophyceae). Protoplasma 148: 106–110

Heywood P (1980) Chloromonads. In: Cox E (ed) Developments in marine biology, vol 2, phythoflagellates. Elsevier North-Holland, New York, pp 351–380

Hibberd DJ (1977) The cytology and ultrastructure of *Chrysonebula holmsii* Lund (Chrysophyceae), with special reference to the flagellar apparatus. Br Phycol J 12: 369–383

Höhfeld I, Otten J, Melkonian M (1988) Contractile eukaryotic flagella: centrin is involved. Protoplasma 147: 16–24

Hoffman LR, Vesk M, Pickett-Heaps JD (1986) The cytology and ultrastructure of zoospores of *Hydrurus foetidus* (Chrysophyceae). Nord J Bot 6: 105–122

Honsell G, Talarico L (1985) The importance of flagellar arrangement and insertion in the interpretation of the theca of *Prorocentrum* (Dinophyceae). Bot Mar 28: 15–21

Kamra K, Sapra GR (1990) Partial retention of parental ciliature during morphogenesis of the ciliate *Coniculostomum monilata* (Dragesco & Njiné, 1971) Nijné, 1978 (Oxytrichidae, Hypotrichida). Eur J Protistol 25: 264–278

Larsen J (1988) An ultrastructural study of *Amphidinium poecilochroum* (Dinophyceae), a small phagotrophic dinoflagellate feeding on small species of cryptophytes. Phycologia 27: 366–377

Lemoine JE, Roberts KR (1989a) The microtubular and centrin cytoskeleton of *Gymnodinium sanguineum.* J Phycol 25 [suppl]: 9a

– – (1989b) The centrin and actin cytoskeleton of the dinoflagellate *Gymnodinium sanguineum.* J Cell Biol 109: 337a

Lenears G, Nielsen H, Engberg J, Herzog M (1988) The secondary structure of large-subunit rRNA divergent domains, a marker for protist evolution. BioSystems 21: 215–222

– Maroteaux L, Michot B, Herzog M (1989) Dinoflagellates in evolution. A molecular phylogenetic analysis of large subunit ribosomal RNA. J Mol Evol 29: 40–51

Loeblich AR III (1976) Dinoflagellate evolution: speculation and evidence. J Protozool 23: 13–28

– (1984) Dinoflagellate evolution. In: Spector DL (ed) Dinoflagellates. Academic Press, Orlando, pp 481–522

Lynn DH (1988) Cytoterminology of cortical components of ciliates: somatic and oral kinetids. BioSystems 21: 299–307

– (1991) The implications of recent descriptions of kinetid structure to the systematics of ciliated protists. Protoplasma 164: 123–142

– Sogin ML (1988) Assessment of phylogenetic relationships among ciliated protists using partial ribosomal RNA sequences derived from reverse transcripts. BioSystems 21: 249–254

Maroteaux L, Herzog M, Soyer-Gobillard M-O (1985) Molecular organization of dinoflagellate ribosomal DNA: evolutionary implications of the deduced 5.8 S rRNA secondary structure. Bio Systems 18: 307–319

Melkonian M (1989) Centrin-mediated motility: a novel cell motility mechanism in eukaryotic cells. Bot Acta 102: 3–4

– Reize IB, Preisig HR (1987) Maturation of a flagellum/basal body requires more than one cell cycle in algal flagellates: studies on *Nephroselmis olivacea* (Prasinophyceae). In: Wiessner W, Robinson DG, Starr RC (eds) Algal development. Molecular and cellular aspects. Springer, Berlin Heidelberg New York Tokyo, pp 102–113

Mignot J-P, Joyon L, Pringsheim EG (1968) Compléments à l'étude cytologique des Cryptomonadines. Protistologica 4: 493–506

Moestrup Ø (1982) Flagellar structure in algae: a review, with new observations particularly on the Chrysophyceae, Phaeophyceae (Fucophyceae), Euglenophyceae, and *Reckertia.* Phycologia 21: 427–528

– Hori T (1989) Ultrastructure of the flagellar apparatus in *Pyramimonas octopus* (Prasinophyceae) II. Flagellar roots, connecting fibres, and numbering of individual flagella in green algae. Protoplasma 148: 41–56

Peck RK (1977) Cortical ultrastructure of the scuticociliates *Dexiotricha media* and *Dexiotricha copidiopsis* (Hymenostomata). J Protozool 24: 122–134

Roberts KR (1984) Structure and significance of the cryptomonad flagellar apparatus. I. *Cryptomonas ovata* (Cryptophyta). J Phycol 20: 590–599

– (1985) The flagellar apparatus of *Oxyrrhis marina* (Pyrrophyta). J Phycol 21: 641–655

– (1986) The flagellar apparatus of *Gymnodinium* sp. (Dinophyceae). J Phycol 22: 456–466

– (1989) Comparative analyses of the dinoflagellate flagellar apparatus. II. *Ceratium hirundinella.* J Phycol 25: 270–280

– (1991) The flagellar apparatus and cytoskeleton of dinoflagellates: organization and use in systematics. In: Patterson DJ, Larsen J (eds) The biology of free-living heterotrophic flagellates. Clarendon Press, Oxford, pp 285–302 (Systematics Association special volume 43)

– Matese JC (1990) Ultrastructure of the flagellar apparatus and allied components in *Prorocentrum micans* (Dinophyceae). J Phycol 26 [Suppl]: 9

– Melkonian M (1984) Comparative fine structure of the flagellar apparatus in the large gymnodinoid *Gymnodinium splendens.* J Phycol 20 [suppl]: 28

– Timpano P (1989) Comparative analyses of the dinoflagellate flagellar apparatus. I. *Woloszynskia* sp. J Phycol 25: 26–36

– Roberts JE, Cormier SA (1991) The dinoflagellate cytoskeleton. In: Menzel D (ed) The cytoskeleton of the algae. CRC Press, Boca Raton (in press)

– Stewart KD, Mattox KR (1981) The flagellar apparatus of *Chilomonas paramecium* (Cryptophyceae) and its comparison with certain zooflagellates. J Phycol 17: 159–167

– Timpano P, Montegut AE (1987) The apical pore fibrous complex: a new cytological feature of some dinoflagellates. Protoplasma 137: 65–69

– Farmer MA, Schneider RM, Lemoine JE (1988 a) The microtubular cytoskeleton of *Amphidinium rhynchocephalum* (Dinophyceae). J Phycol 24: 544–553

– Lemoine JE, Schneider RM, Farmer MA (1988 b) The microtubular cytoskeleton of three dinoflagellates: an immunofluorescence study. Protoplasma 144: 68–71

Salisbury JL (1989) Centrin and the algal flagellar apparatus. J Phycol 25: 201–206

– Sanders MA, Harpst L (1987) Flagellar root contraction and nuclear movement during flagellar regeneration in *Chlamydomonas reinhardtii.* J Cell Biol 105: 1799–1805

Schneider RM, Roberts KR (1989) A preliminary phylogenetic analysis of selected dinoflagellates using Wagner parsimony. J Phycol 25 [Suppl]: 19

Schnepf E, Winter W, Storck I, Quader H (1990) A complementary experimental study of cell division in the dinoflagellate *Prorocentrum micans.* Eur J Protistol 25: 234–242

Sleigh MA (1989) Protozoa and other protists. Edward Arnold, London

Spector DL (ed) (1984) Dinoflagellates. Academic Press, Orlando

Taylor FJR (1976) Flagellate phylogeny: a study in conflicts. J Protozool 23: 28–40

– (1978) Problems in the development of an explicit hypothetical phylogeny of the lower eukaryotes. BioSystems 10: 67–89

– (1980) On dinoflagellate evolution. BioSystems 13: 65–108

– (ed) (1987) The biology of the dinoflagellates. Blackwell, Oxford (Botanical monographs, vol 21)

Theriot E (1989) Phylogenetic systematics for phycology. J Phycol 25: 407–411

Triemer RE (1982) A unique mitotic variation in the marine dinoflagellate *Oxyrrhis marina* (Pyrrophyta). J Phycol 18: 399–411

Vesk M, Moestrup Ø (1987) The flaggellar root system in *Heterosigma akashiwo* (Raphiodphyceae). Protoplasma 137: 15–28

Wedemayer GJ, Wilcox LW (1984) The ultrastructure of the freshwater, colorless dinoflagellate *Peridiniopsis berolinense* (Lemm.) Bourrelly (Mastigophora, Dinoflagellida). J Protozool 3: 444–453

– – Graham LE (1982) *Amphidinium cryophilum* sp nov (Dinophyceae), a new freshwater dinoflagellate. I. Species description using light and scanning electron microscopy. J Phycol 18: 13–17

Wetherbee R, Platt SJ, Beech PL, Pickett-Heaps JD (1988) Flagellar transformation in the heterokont *Epipyxis pulchra* (Chrysophyceae): direct observations using image enhanced light microscopy. Protoplasma 145: 47–54

Wilcox LW (1989) Multilayered structures (MLSs) in two dinoflagellates, *Katodinium campylops* and *Woloszynskia pascheri.* J Phycol 25: 785–789

– Wedemayer GJ, Graham LE (1982) *Amphidinium cryophilum* sp nov (Dinophyceae), a new freshwater dinoflagellate. II. Ultrastructure. J Phycol 18: 18–30

Protoplasma (1991) 164: 123–142

PROTOPLASMA

The implications of recent descriptions of kinetid structure to the systematics of the ciliated protists

D. H. Lynn*

Department of Zoology, University of Guelph, Guelph, Ontario

Received June 30, 1990
Accepted December 14, 1990

Summary. Review of the last decade has generally confirmed the presence of nine major kinetid patterns of ciliates. These patterns form the basis of several recent classification schemes that are not able, however, to recognize several variant patterns: *Phacodinium, Transitella, Plagiopyla, Lechriopyla, Mesodinium*, and *Schizocaryum* are examples of taxa whose position remains unclear even with ultrastructural evidence. The development of cell structures has provided some striking insights into the phylogenetic relationship of some taxa, notably the stomatogenesis of *Coleps* and kinetid replication of *Eudiplodinium*. Systematic implications of these data are also discussed.

Keywords: Ciliates; Kinetodermal fibril; Microtubular ribbon; Microtubular root; *Paramecium*; striated root; *Tetrahymena*.

Introduction

The past two decades of research on the comparative ultrastructure of protists has clearly established the utility of this approach in resolving some questions of relatedness among contemporary taxa. This has been particularly fruitful for the ciliates, beginning with early work by Grain (1969) and Pitelka (1969), which stimulated reevaluations of our ideas of phylogeny. Several years later, de Puytorac et al. (1974) and Corliss (1974, 1979) presented revisions to the classification, which borrowed from the ideas of Jankowski (1967) and accomodated the results of ultrastructural research.

These revisions essentially began with the assumption that the oral features were most indicative of common ancestry. Lynn (1976, 1981) made the case that somatic features might be more conserved, and Small and Lynn (1981, 1985) followed with substantially different notions of the major ciliate lineages. De Puytorac et al. (1984, 1987) have presented revised systems that clearly converge in many respects on that proposed by Small and Lynn (1985).

There has been substantial ultrastructural research published since my 1981 review (Lynn 1981). It is the purpose of this paper to summarize the work since 1981, provide some schematic figures of the somatic kinetids of these additional taxa, and discuss the implications of this data on our ideas of the phylogeny of the ciliates.

The approach

Lynn (1981) provided definitions and detailed descriptions of the variations of the microtubular and fibrillar components of the ciliate kinetid. The terminology introduced in that study will be followed here. Since Small and Lynn (1981, 1985) have provided formal taxonomic categories for the kinetid patterns described by Lynn (1981), these categories will be used to organize the present review.

Somatic kinetid pattern and variation

Class Karyorelictea

Grolière et al. (1980) have provided the only additional description of a karyorelictean cortex. Their analysis of the ultrastructure of *Protocruzia* indicated that it possesses dikinetids characterized as follows: a ciliated anterior kinetosome bears a tangential transverse ribbon at triplets 4 and 5; a ciliated posterior kinetosome bears a smaller transverse microtubular ribbon of 1–2 microtubules near the triplet 4, a set of 10–15 divergent

* Correspondence and reprints: Department of Zoology, University of Guelph, Guelph, Ont. N1G 2W1, Canada.

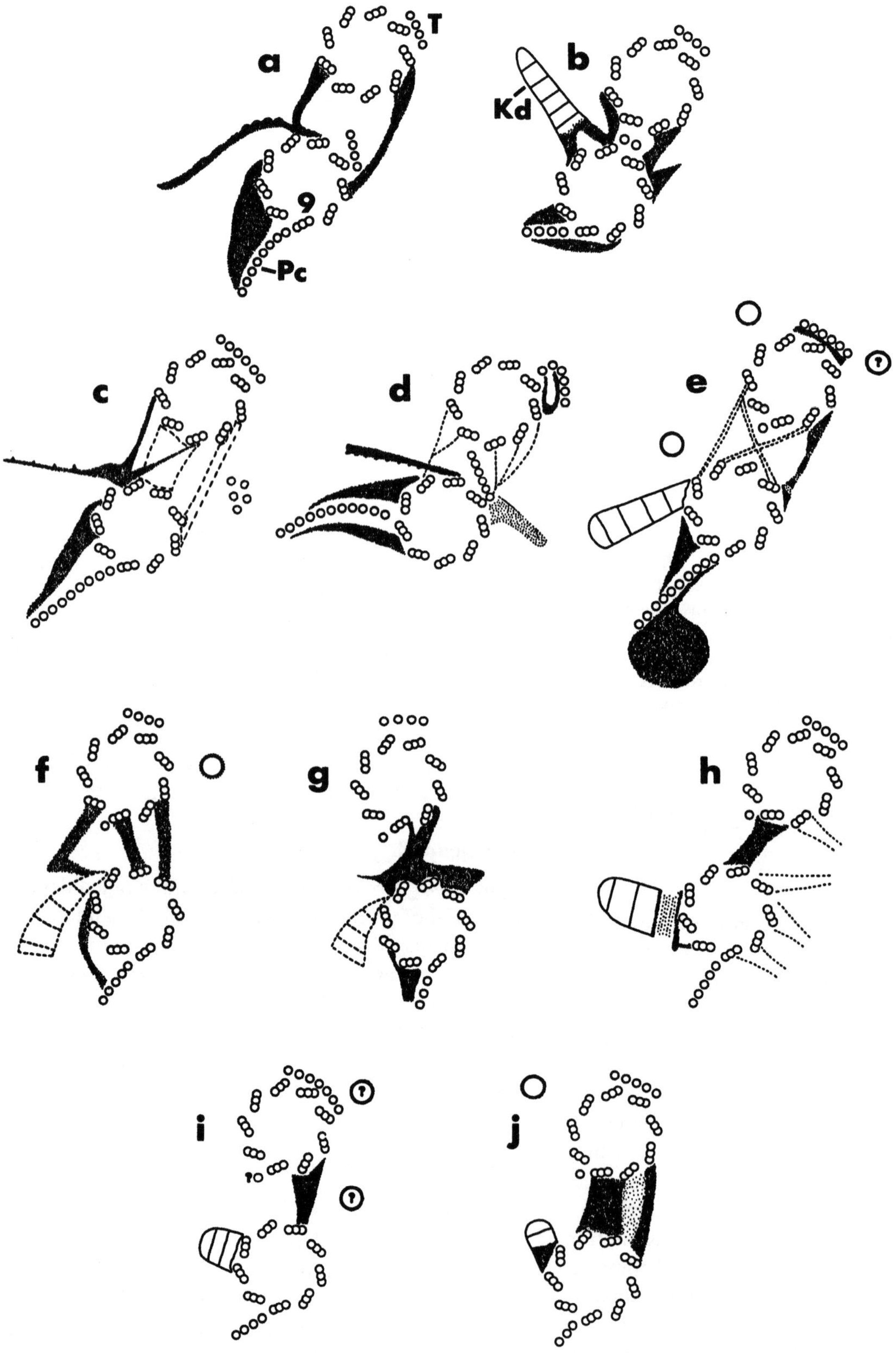
a
T
9
Pc
b
Kd
c
d
e
f
g
h
i
j

postciliary microtubules that extend to overlap adjacent sets, and a striated kinetodesmal fibril at triplets 5 and 6 that extends to the right and anterior but does not overlap the fibril of the next kinetid (Fig. 1 b).
Comparison of this pattern with the karyorelicteans *Geleia* and *Remanella* (see Lynn 1981) demonstrates significant variability in the group. The kinetodesmal fibrils are different: *Geleia* has a fibril similar to the heterotrich spirotrichs (cf. Fig. 1 a, c, and d); *Remanella* and *Loxodes* have two striated fibrils; and *Protocruzia* has a single striated fibril (Fib. 1 b). Moreover, although the postciliary ribbons overlap, only those in the family Geleiidae (e.g., *Avelia, Geleia*) unambiguously demonstrate a postcilidesma like the heterotrich spirotrichs (see below). Based on these variations in kinetid diversity, this class is very likely paraphyletic.

Class Spirotrichea

The "heterotrichs"

The only heterotrich described recently has been *Eufolliculina* (Mulisch et al. 1981). Its cortex exhibits a dikinetid, and like other heterotrichs sensu stricto is characterized as follows: a ciliated anterior kinetosome with a curved transverse ribbon; a posterior kinetosome, ciliated in the rear of the cell, with transverse ribbon of 5–6 microtubules at triplets 4 and 5, a divergent postciliary ribbon of about 15 microtubules that extends posteriorly to form a postciliodesma sensu stricto, a laterally-directed non-striated kinetodesmal fibril at triplets 5 and 6, and a dense fibre near triplets 3 and 4 that extends laterally left (Fig. 1 d).
Transitella, described by Iftode et al. (1983), has been considered a relative of the heterotrichs (Corliss 1979, Small and Lynn 1985). The features of its kinetid, however, argue strongly against a close relationship. The dikinetids are characterized as follows: a ciliated anterior kinetosome with a tangential transverse ribbon of about 6 microtubules near triplets 3, 4, and 5, a single postciliary microtubule, and possibly two microtubules near triplet 1, which could be posterior transverse microtubules; and a ciliated posterior kinetosome with a large divergent postciliary ribbon of about 10 microtubules from the base of which arises a large dense structure as the origin of a nematodesma and a posteriorly-directed striated kinetodesmal fibril at triplet 6 (Fig. 1 e).
Lynn (1981) illustrated an initial diversity in "heterotrich" kinetids, demonstrating that *Sicuophora* differed from *Condylostoma, Stentor*, and *Climacostomum* in having a striated fibril, possibly an extension of the kinetodesmal fibril, extending to the left. With some slight variation in the shape of the anterior transverse ribbon and the interkinetosomal linkages, *Eufolliculina* can be placed with the latter three genera while, as pointed out by Iftode et al. (1983), *Transitella* clearly differs again, forming a new category.

The "hypotrichs"

Based on features of the dorsal somatic kinetids, Small and Lynn (1985) suggested that this group be divided between their two subphyla Postciliodesmatophora and Cyrtophora. This conclusion differed from the ones presented in other classifications, which maintained the classical, close affinity between all hypotrich taxa (Corliss 1974, 1979; de Puytorac et al. 1974, 1984, 1987). Lynn and Sogin (1988) demonstrated that ribosomal RNA sequences confirmed the relatedness of the hypotrichs while demonstrating that there were at least two major groups, more genetically diverse than previously realized, represented by *Euplotes* on one hand and *Oxytricha* and *Stylonychia* on the other. Thus, the distinction, which Small and Lynn (1985) had originally suggested based on marked differences in dorsal somatic kinetid pattern, was partially corroborated, although the subphylum separation was undoubtedly incorrect as present sequence data suggest.
Since the pioneering work on *Euplotes* (Ruffolo 1976), the cortex has been described for the euplotids *Euplotes, Certesia, Kiitricha, Uronychia*, and *Aspidisca*, the discocephalid *Discocephalus*, the stichotrichines *Kahliella*,

Fig. 1. Kinetid patterns of the somatic cortex of ciliates. In this and following figures the kinetodesmal fibril (*Kd*) is usually striated; transverse (*T*) and postciliary (*Pc*) microtubular ribbons are illustrated in solid circles if sectioned with kinetosome and by dotted circles if sectioned out of the plane of the kinetosome transection; parasomal sacs are larger circles associated with the kinetid; and dense material and connectives are stippled. Triplets are numbered downward clockwise from triplet 9 (**a**) with which the postciliary microtubular ribbon (*Pc*) is associated: thus, the kinetodesmal fibril is usually associated with triplets 5, 6, and 7 while the transverse microtubular ribbon is associated with triplets 3, 4, and 5. **a, b** Karyorelictean patterns: **a** *Geleia*; **b** *Protocruzia* (after Grolière et al. 1980). **c–e** Heterotrichian pattern: **c** *Climacostomum* (from Lynn 1981); **d** *Eufolliculina* (after Mulisch et al. 1981); **e** *Transitella* (after Iftode et al. 1983). **f–j** Hypotrichian pattern: **f** dorsal bristle dikinetid of *Paraurostyla* (after Jerka-Dziadosz 1982) and **g** *Stylonychia* (from Lynn 1981) in which the kinetodesmal fibrils (dashed lines) are only transient; **h** dikinetid unit of ventral polykinetids of *Engelmanniella* (after Wirnsberger-Aescht et al. 1989); **i, j** dorsal bristle dikinetids of **i** *Aspidisca* (after Rosati et al. 1987) and **j** *Discocephalus* (after Wicklow 1982)

Engelmanniella, and *Parastrongylidium*, the urostyline *Thigmokeronopsis*, and the sporadotrichines *Stylonychia, Histriculus*, and *Laurentiella*. The dorsal dikinetids are not described for all these genera.

Görtz (1982) confirmed the presence of lasiosomes surrounding the base of the axoneme of the anterior ciliated kinetosome of *Euplotes*. Wicklow (1983) observed these in *Certesia*, concluding this to be a significant feature for euplotids. *Aspidisca* (Rosati et al. 1987) and *Discocephalus* (Wicklow 1982) apparently lack lasiosomes. However, all these genera exhibit kinetids characterized as follows: a ciliated anterior kinetosome with a tangential transverse ribbon, probably near triplets 3, 4, and 5 and sometimes a single postciliary microtubule; a posterior kinetosome with a short condylocilium, a divergent postciliary ribbon at triplet 9, and a laterally-directed and probably striated kinetodesmal fibril at triplets 6 and 7 (Fig. 1 i and j). Elongate vesicles surround the dikinetids of *Euplotes* and *Certesia*, in which they are called ampules, and smaller spherical ones are observed in *Aspidisca* and *Discocephalus*. *Kiitricha* and *Uronychia* (Fleury et al. 1985 a, 1986) apparently show euplotine features, although the micrographic evidence is not yet definitive.

Görtz (1982) noted that the dorsal dikinetid of *Stylonychia mytilus* differed from that of *S. pustulata* (Grimes and Adler 1976) in that the posterior kinetosome was resorbed after fission. *Kahliella* and *Parastrongylidium* (Fleury et al. 1985 b) have a dorsal dikinetid characterized as follows: the ciliated anterior kinetosome bears a tangential transverse ribbon; the posterior kinetosome bears a short axoneme and has a small divergent postciliary ribbon but no kinetodesmal fibril. The dorsal dikinetids of *Laurentiella* (Torres et al. 1986) and *Engelmanniella* (Wirnsberger-Aescht 1989) are characterized as follows: a ciliated anterior kinetosome with a tangential transverse ribbon; a posterior kinetosome with a radial postciliary ribbon but no kinetodesmal fibril. No lasiosomes or ampule-like vesicles are observed around the kinetids of any of these stichotrichs sensu lato.

The ventral kinetids of *Engelmanniella* can vary from dikinetids to small polykinetids. The polykinetids are apparently composed of dikinetids linked laterally. Each dikinetid is characterized as follows: an anterior ciliated kinetosome with a single postciliary microtubule and a tangential transverse ribbon at triplets 4 and 5; and a posterior ciliated kinetosome with a divergent postciliary ribbon and a kinetodesmal fibril indirectly linked by fine filaments near triplets 6, 7, and 8 (Wirnsberger-Aescht et al. 1989) (Fig. 1 h). Although smaller, this bears great similarity to the detailed model of the somatic polykinetid or cirrus of *Histriculus* derived by Matsusaka et al. (1984), and it further corroborates the view of Jerka-Dziadosz (1980) on the development of these structures from dikinetid units. Thus, the division of hypotrichs into two major groups, called respectively the stichotrichs and hypotrichs by Small and Lynn (1985) and the euhypotrichs and pseudohypotrichs by Fleury et al. (1986), is supported by features of the dorsal dikinetids. Fleury et al. (1986) concluded that the pseudohypotrich or hypotrich group is probably not monophyletic: a number of the features of the cirri and cirral development and the apparent diversity in dorsal dikinetid structure indicate this. For example, lasiosomes and ampules are apparently not universal features of hypotrichs while the kinetodesmal fibril of the dorsal dikinetid is clearly not a feature that is unique to this group. It will be exciting to see some additional molecular sequence data to test further these conclusions based primarily on morphology.

Class Colpodea

The recent descriptions of colpodean genera continue to substantiate the conservative nature of the somatic cortex of ciliate taxa, and this group in particular. Aspects of the cortex of *Sorogena, Bryophrya, Bresslaua, Pseudoplatyophrya, Bursaria*, and *Bryometopus* have been described.

The somatic dikinetids of *Bryophrya* (Grain et al. 1979) and *Pseudoplatyophrya* (de Puytorac et al. 1983) share features as follows: the anterior ciliated kinetosome bears a tangential transverse ribbon at triplets 3, 4, and 5 and a single to several postciliary microtubules; and the posterior ciliated kinetosome bears a divergent postciliary ribbon, a laterally-directed, usually fan-shaped kinetodesmal fibril at triplets 5 and 6, and a long posteriorly-directed transverse ribbon that arises obliquely in the region of triplets 4 and 5 (Fig. 2 a and b). Bradbury and Olive (1980) present micrographic evidence that *Sorogena* has a somatic dikinetid that is essentially colpodean, although they identified it as a haptorid (Fig. 2 c). Schematic figures of the kinetid of *Bryometopus* show similar features (Wirnsberger et al. 1985). Although the published micrographs do not permit an independent confirmation of the authors' interpretation, *Bryometopus* is undoubtedly a colpodean. Garcia-Rodriguez et al. (1981) confirmed the description of the somatic dikinetids of *Bresslaua* (Lynn 1979). The colpodean somatic dikinetid remains as characterized by Lynn (1981).

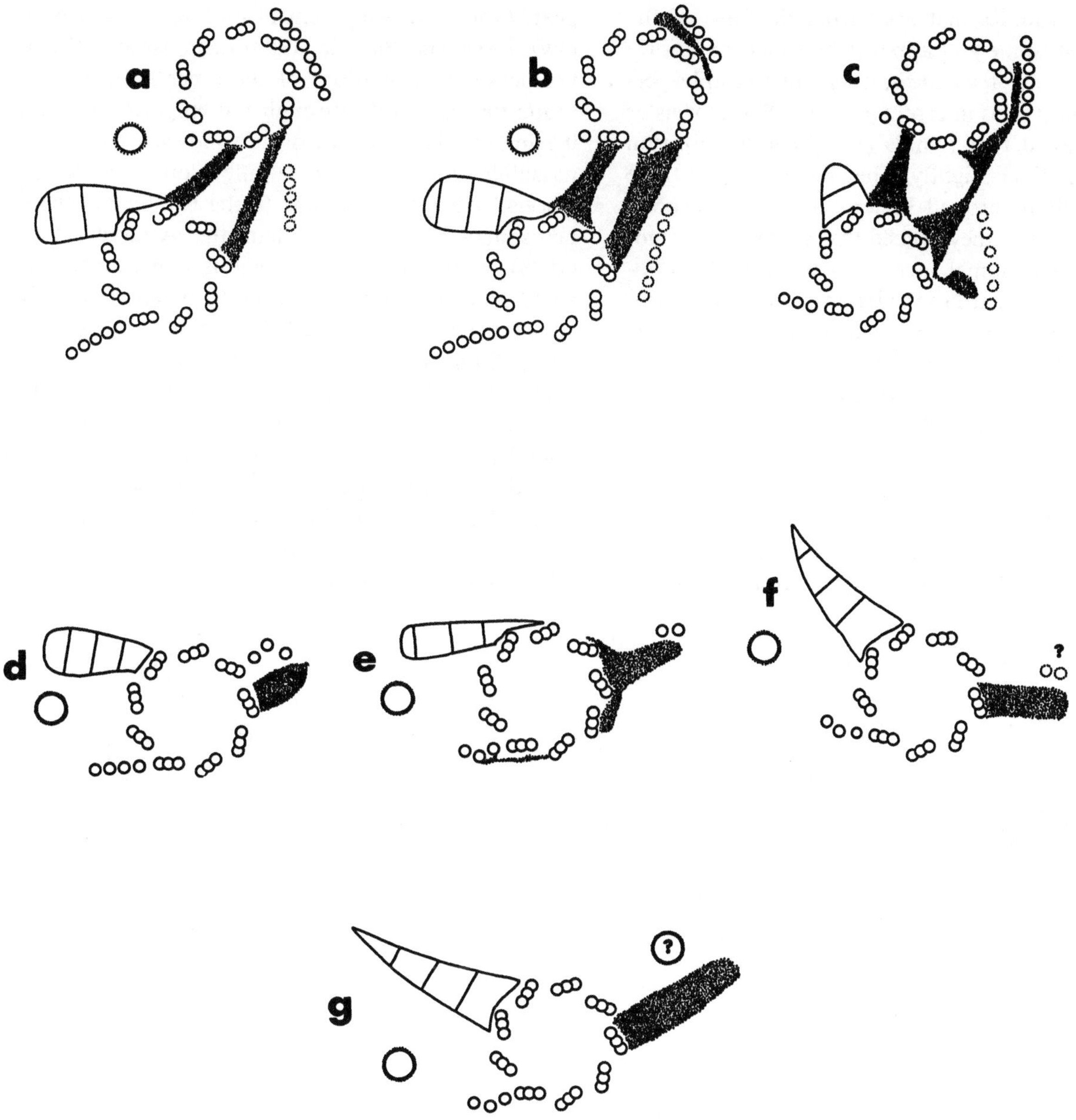

Fig. 2. Kinetid patterns of the somatic cortex of ciliates. **a–c** Colpodean patterns: **a** *Bryophrya* (after Grain et al. 1979); **b** *Pseudoplatyophrya* (after de Puytorac et al. 1983); **c** *Sorogena* (after Bradbury and Olive 1980). **d–g** Phyllopharyngean patterns: **d** *Spirochona* (after Fahrni 1982); **e** *Hypocoma* (after Grell and Meister 1983); **f** *Acinetopsis* (after Grell and Meister 1982 b); **g** *Trichophrya* (after Mogensen and Butler 1984)

Class Phyllopharyngea

Descriptions of the cortex of the cryptophorids *Chilodonella* and *Trithigmostoma*, the hypocomid *Hypocoma*, rhynchodid *Sphenophrya*, the chonotrich *Spirochona*, and the suctorians *Acinetopsis* and *Trichophrya* have generally confirmed the characterization of the kinetid in Lynn (1981). The somatic kinetid of *Chilodonella* (Hofmann 1987), *Trithigmostoma* (Hofmann and Bardele 1987), *Hypocoma* (Grell and Meister 1983), *Sphenophrya* (de Puytorac et al. 1980), *Spirochona* (Fahrni 1982), *Acinetopsis* (Grell and Meister 1982 b), and *Trichophrya* (Mogensen and Butler 1984) can be characterized as follows: a monokinetid with a slightly convergent postciliary ribbon at triplet 9, a short, rapidly tapering kinetodesmal fibril at triplets 6 and 7, and a transverse fibre at triplet 3. *Spirochona* and *Hypocoma* have transverse microtubules associated with triplets 4 and 5 (Fig. 2 d and e). All phyllopharyngeans have subkinetal microtubules that arise from the base of the somatic kinetosome and apparently extend anteriorly in *Hypocoma* and *Acinetopsis*, where micrographs permit an unambiguous interpretation. Grell and Meister (1983) suggested that Lynn's (1981) characterization should be modified slightly. They indicated that there is some variability around 3 microtubules for the postciliary ribbon, and they argued that suctorians do not always have parasomal sacs on both sides of the kinetosome. With regard to the latter point, there is usually a parasomal sac to the right of the kinetosome, and apparently sometimes a parasomal sac to its left (e.g., Fig. 5c of Grell and Meister 1982a, Fig. 39 of Mogensen and Butler 1984).

Class Nassophorea

There have been no studies on new genera of nassophoreans, although several detailed accounts have appeared on the development of oral structures (see below). Eisler and Bardele (1983) demonstrated that alveolocysts, invaginations of the inner alveolar membrane that penetrate through the epiplasm for about 0.5 µm, are characteristic of nassulids sensu lato while nassulids sensu stricto have kinetosomes whose cartwheels have spokes running to the B-tubule of the kinetosomal triplet rather than the A-tubule. Eisler (1988), confirming these previous conclusions, provided a detailed ultrastructural documentation of the somatic monokinetids and dikinetids of *Furgasonia*, which differs from the interpretation provided by Grain et al. (1978). The kinetids are now characterized as follows: a divergent postciliary at triplet 9; an anterior-laterally directed kinetodesmal fibril at triplets 5 and 6; and a small tangential transverse ribbon at triplets 3 and 4 (Fig. 3a and b). This is very similar to the mono- and dikinetid pattern characterized for *Nassula* (Eisler 1986, 1988) (Fig. 3c and d).

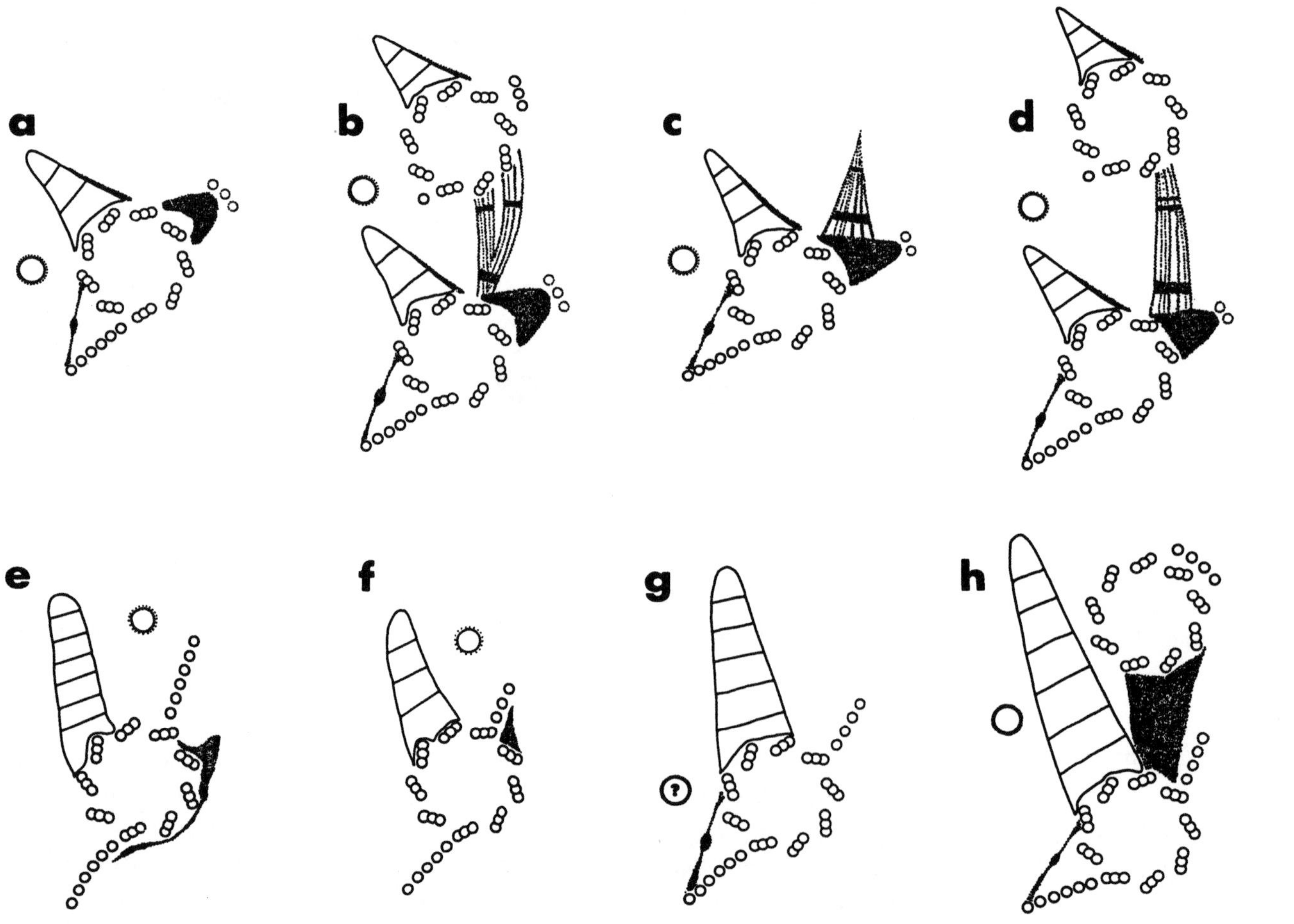

Fig. 3. Kinetid patterns of the somatic cortex of ciliates. **a–d** Nassophorean patterns: **a** monokinetid and **b** dikinetids of *Furgasonia* (after Eisler 1988); **c** monokinetid and **d** dikinetids of *Nassula* (after Eisler 1988). **e, f** Oligohymenophorean patterns: **e** *Colpidium* (from Lynn 1981); **f** *Ichthyophthirius* (after Chapman and Kern 1983). **g, h** Prostomatean pattern: **g** monokinetid and **h** dikinetid of *Coleps* (after Huttenlauch 1987, Lynn 1985)

Class Oligohymenophorea

New data have been provided on the cortex of the hymenostomes *Ichthyophthirius*, *Colpidium*, and *Turaniella*, the peritrichs *Thuricola*, *Trichodina*, and *Tripartiella*, and the apostomes *Hyalophysa* and *Conidophrys*. The first published micrographs of the kinetid of *Ichthyophthirius* (Chapman and Kern 1983) corroborated the schematic figure of Lom and Corliss (1971), clearly demonstrating that this ciliate is a hymenostome. Its kinetid is characterized as follows: a ciliated monokinetid with a divergent postciliary ribbon at triplet 9, an anteriorly-extending and tapering kinetodesmal fibril at triplets 5, 6, and 7, and a radial transverse ribbon of several microtubules near triplet 4 (Chapman and Kern 1983) (Fig. 3 f). Iftode et al. (1984) argued that similarities in the stomatogenesis, which interrupts kineties on the left of the oral opening, and the non-ciliated paroral kinetosomes merit placing *Colpidium* and *Turaniella* in the family Turaniellidae.

Bradbury and Tyson (1982) presented only one micrograph of the somatic kinetids of *Conidophrys*. However, detailed presentation of the features of the rosette together with the obvious presence of overlapping kinetodesmal fibrils of somatic kinetids clearly demonstrates its affinities with other apostomes.

The mobiline peritrichs that have been described have highly modified somatic kinetids, lacking the standard fibrillar associates in whose place are highly modified striated rootlets and filaments (Hausmann and Hausmann 1981, Maslin-Leny and Bohatier 1984).

Class Prostomatea

Although there have been several studies on these ciliates, except for those on *Coleps*, the published micrographs have not been detailed enough to provide unambiguous interpretation. The somatic kinetid of *Coleps* has an obviously divergent postciliary ribbon and a conspicuous anteriorly-directed and tapering kinetodesmal fibril that may begin as a bifurcation at triplets 5, 6, and 7 (Huttenlauch 1987, Lynn 1985). Lynn (1985) was unable to demonstrate a radial transverse ribbon, but Huttenlauch (1987) argued from features of the dikinetid that the transverse ribbon is radial (Fig. 3 g). Lynn (1985) further suggested that the prostome dikinetid possessed an anterior kinetosome with a tangential transverse ribbon while Huttenlauch (1987) suggested it is more radial but of a different orientation than that of the posterior kinetosome (Fig. 3 h).

Class Litostomatea

There has been a significant amount of new information on genera in this class: the haptorians *Homalozoon*, *Helicoprorodon*, *Spathidium*, *Lepidotrachelophyllum*, *Enchelydium*, and *Fuscheria*; the vestibuliferidans *Isotricha*, *Dasytricha*, and *Vestibulongum*; the entodiniomorphids *Epidinium*, *Eudiplodinium*, and *Troglodytella*; and *Mesodinium*.

Williams et al. (1981) were the first to clearly demonstrate that the somatic monokinetids of *Spathidium* had two transverse ribbons: T 1, a tangential set at triplets 3 and 4, which extended anteriorly into the cortical ridge, and T 2, a radial set near triplet 5, which extended laterally across to the adjacent kinety (Fig. 4 a). Williams et al. (1981) identified these two ribbons in previously published micrographs of *Bryophyllum* and *Spathidium*. Subsequent descriptions of the kinetids of the haptorians *Lepidotrachelophyllum* (Lynn and Nicholls 1985), *Enchelydium* (Foissner and Foissner 1985), *Fuscheria* (Foissner and Foissner 1988), and *Homalozoon* (Leipe and Hausmann 1989) have also demonstrated these two transverse ribbons. Moreover, the somatic kinetids of the vestibuliferidans *Isotricha* (Gerassimova 1981, Grain 1966), *Dasytricha* (Paul et al. 1989), and *Vestibulongum* (Grim 1988) demonstrate these in the morphostatic state while, during kinetid replication of the entodiniomorphid *Eudiplodinium*, Furness and Butler (1986) suggested that a single T 2 microtubule appears transiently (Fig. 17 of Furness and Butler 1986). The morphostatic somatic kinetids of *Epidinium* (Furness and Butler 1983), *Eudiplodinium* (Furness and Butler 1985 a), *Entodinium* (Furness and Butler 1985 b), and *Troglodytella* (Bonhomme et al. 1989) exhibit only the tangential T 1 ribbon.

Small and Lynn (1981, 1985) included all these ciliates in their class Litostomatea. The litostome kinetid is now characterized as follows: a monokinetid with a ciliated kinetosome bearing a slightly convergent postciliary ribbon at triplet 9, a laterally directed kinetodesmal fibril at triplets 6 and 7, and two transverse ribbons of which a tangential one, T 1, is associated with triplets 3 and 4 and a somewhat radial one, T 2, is associated with triplet 5 at some time during the kinetid development (Fig. 4 a–k). The postciliary microtubules of each kinetid characteristically change into a two-layered structure of "n + 1-over-n" microtubules as they reach the cortex: normally this is a "4-over-3" assemblage. Nematodesmal microtubules may extend from the kinetosome into the cytoplasm, and, in the trichostome lineage the postciliary microtubules can be

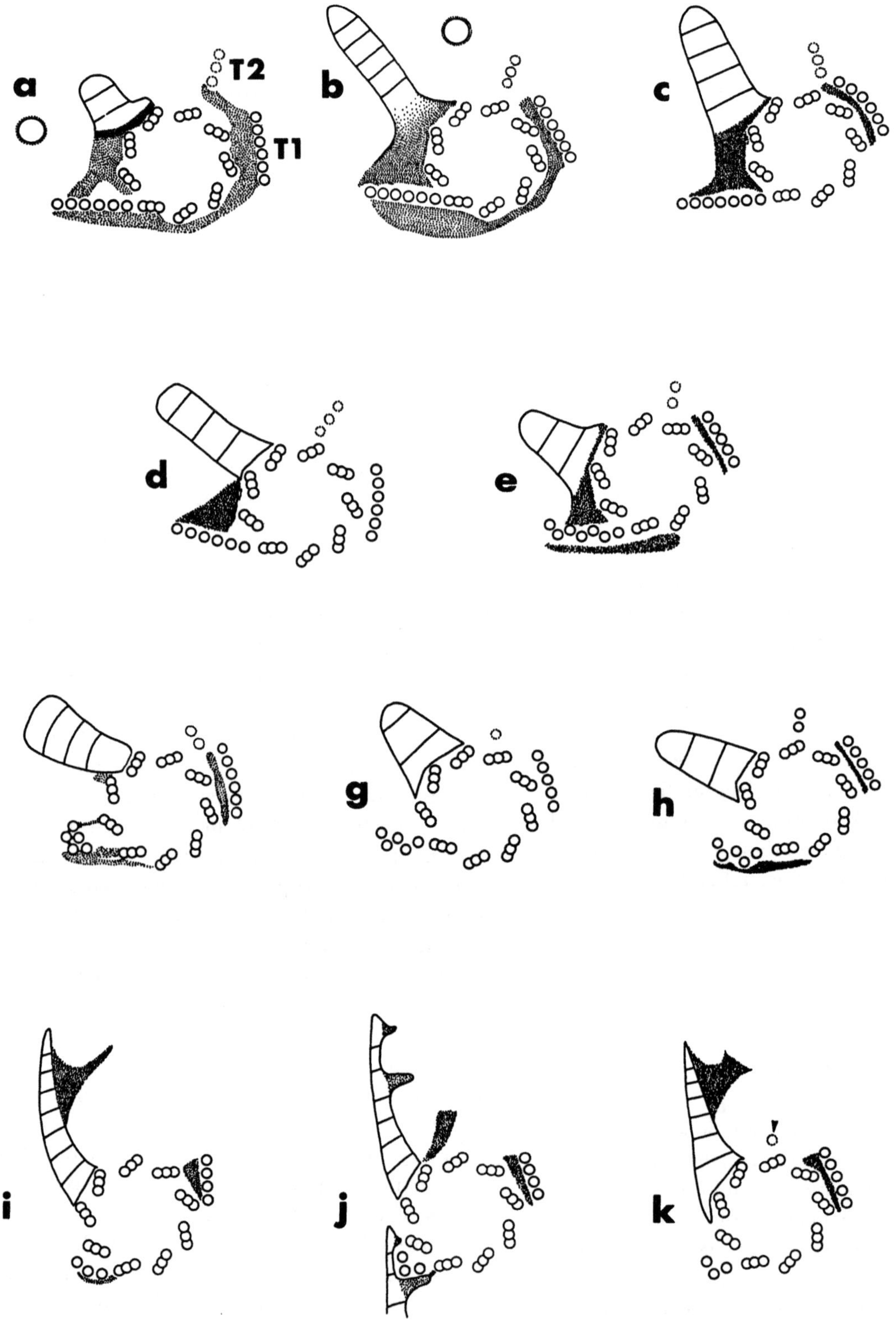

Fig. 4. Kinetid patterns of the somatic cortex of litostomate ciliates. **a–e** Haptorian patterns: **a** *Spathidium*, *T1* transverse microtubular ribbon 1, *T2* transverse microtubular ribbon 2 (after Williams et al. 1981); **b** *Lepidotrachelophyllum* (after Lynn and Nicholls 1985); **c** *Enchelydium* (after Foissner and Foissner 1985); **d** *Fuscheria* (after Foissner and Foissner 1988); **e** *Homalozoon* (after Leipe and Hausmann 1989). **f–k** Trichostomatian patterns: **f** *Isotricha* (after Gerassimova 1981); **g** *Dasytricha* (after Paul et al. 1989); **h** *Vestibulongum* (after Grim 1988); **i** *Epidinium* (after Furness and Butler 1983); **j** *Entodinium* (after Furness and Butler 1985 b); **k** *Euplodinium* where the arrowhead indicates the transient appearance of the second transverse microtubular ribbon (after Furness and Butler 1985 a, 1986)

clustered in a small group. The entodiniomorphid kinetid, in addition to lacking the T 2 in the morphostatic state, shows a characteristic linkage of kinetodesmal fibrils and postciliary ribbons between adjacent kinetids in a somatic kinety (see Bonhomme et al. 1989; Furness and Butler 1983, 1985 a, b).

The somatic kinetids of *Mesodinium* are arranged in two girdles. The anterior girdle is composed of polykinetids whose kinetosomes are arrayed in two or three uneven rows (Grain 1982, Lynn unpubl. obs.). The posterior girdle is composed of two files of over a dozen kinetosomes, the rightmost kinetosomes may bear postciliary ribbons (Grain et al. 1982). The somatic structures are so dramatically different from any other litostomes that *Mesodinium*, if it is indeed a litostome, has diverged significantly from the ancestral stock.

Taxa of uncertain affinities

There are several genera whose affinities based on morphological features at the light microscopic level have not been confirmed by ultrastructural examination. For example, *Phacodinium* has been considered a heterotrich (Corliss 1979) or related to the heterotrichs (de Puytorac et al. 1987, Small and Lynn 1985) although its somatic kinetids are dramatically different (Lynn 1981). The same might be said for *Transitella*, described above as a heterotrich.

Plagiopyla, Lechriopyla, and *Schizocaryum* are three genera whose cortical ultrastructure does not reveal their affinities. The kinetid of the plagiopylids *Plagiopyla* and *Lechriopyla* (Berger and Lynn 1984, de Puytorac et al. 1985) is characterized as follows: a ciliated monokinetid with a divergent postciliary ribbon at triplet 9; and an anteriorly-directed and relatively long kinetodesmal fibril originating near triplets 5, 6, and 7 and extending to overlap the fibrils of more anterior kinetids. Berger and Lynn (1984) described an unusual transverse ribbon that originates near triplets 1, 2, and 3 and extends anteriorly apparently parallel to the cell surface (Fig. 5 a), while de Puytorac et al. (1985) doubt this interpretation. If true, no other major group exhibits such a kinetid pattern.

Schizocaryum has been placed as a trichostome, incertae sedis in the Plagiopylidae (Corliss 1979). Again, ultrastructural features of its kinetid do not provide significant clues to its relationships. The somatic kinetid is characterized as follows: a loosely organized polykinetid usually with 12–15 ciliated kinetosomes; the posterior-right kinetosome only bears the fibrillar associates, which include a divergent postciliary ribbon at triplet 9, an anteriorly-directed kinetodesmal fibril at triplets 6 and 7, and a presumed transverse ribbon at triplets 4 and 5 that extends laterally behind the other kinetosomes of the polykinetid (Lynn and Frombach 1987) (Fig. 5 b). Again, no other such kinetid has been described.

Oral kinetid pattern and variation

De Puytorac and Grain (1976) reviewed the oral structures of ciliates and speculated on their evolution. They redefined a number of terms, such as polykinety, membranoid, membranelle, and peniculus, used to describe

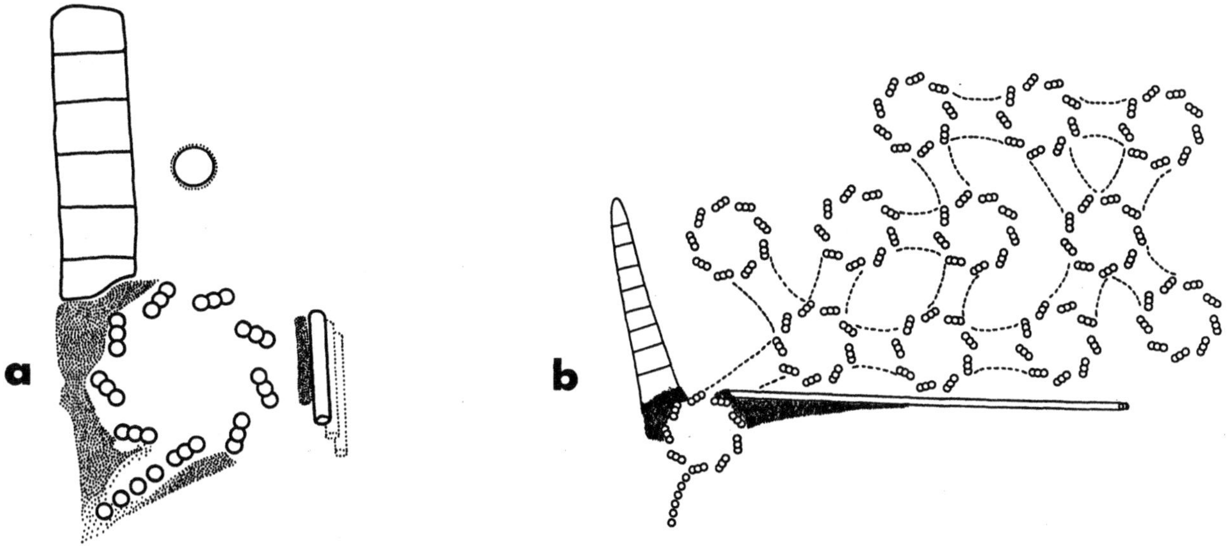

Fig. 5. Kinetid patterns of the somatic cortex of ciliates of uncertain taxonomic affinities. **a** *Plagiopyla* and *Lechriopyla* (after Berger and Lynn 1984). **b** *Schizocaryum* (after Lynn and Frombach 1987)

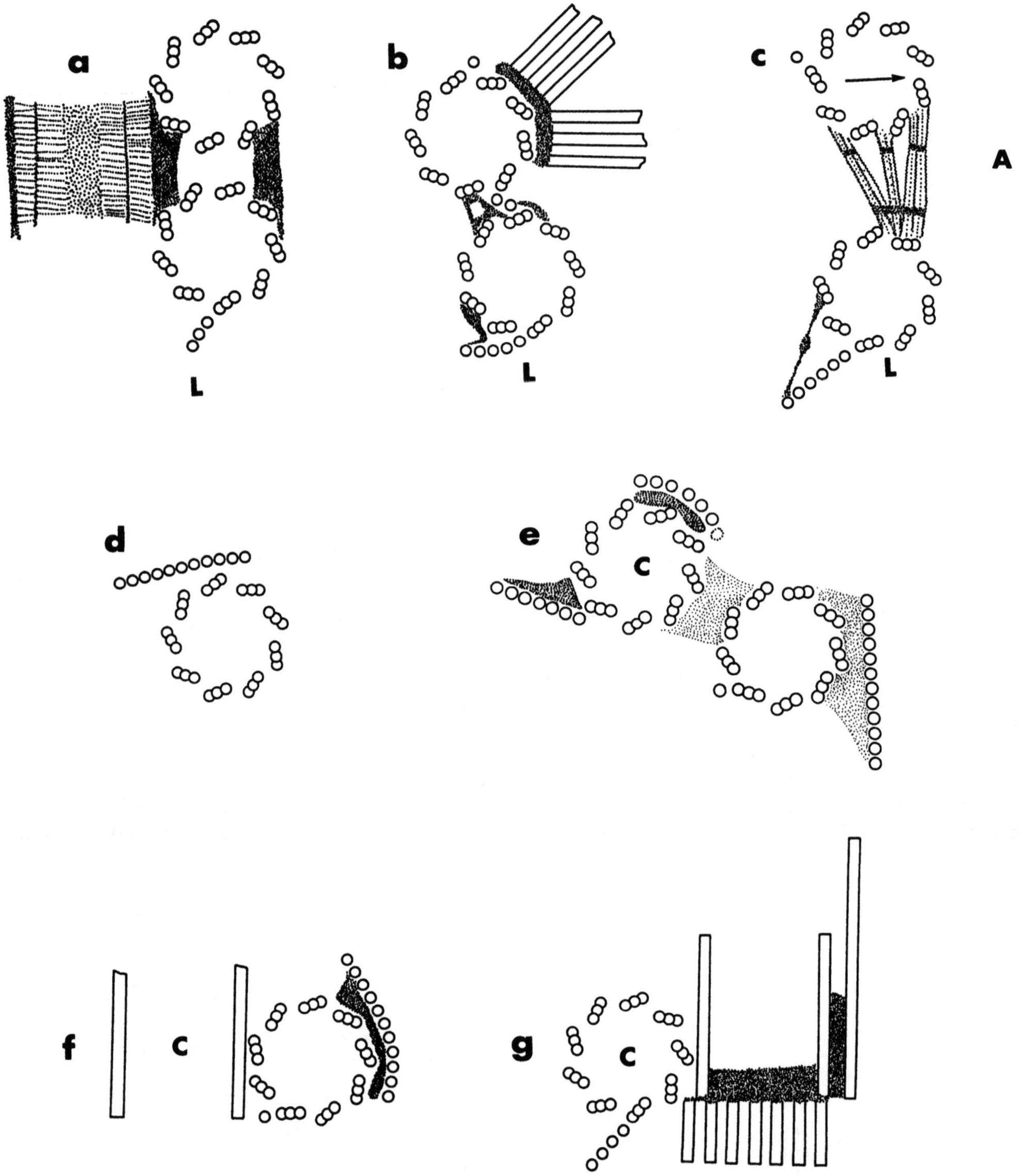

Fig. 6. Kinetid patterns of the oral cortex of ciliates. **a–c** Paroral structures; *A* anterior of cell, *L* left: **a** *Protocruzia* (after Grolière et al. 1980); **b** *Eufolliculina* (after Mulisch and Hausmann 1984); **c** *Furgasonia* (after Eisler 1988). **d–g** Litostomate patterns: **d** monokinetid of *Enchelydium* (after Foissner and Foissner 1985); **e** *Lepidotrachelophyllum*, *C* ciliated kinetosome (after Lynn and Nicholls 1985); longitudinal sections of **f** the ciliated posterior kinetosome and **g** non-ciliated anterior kinetosome of the oral dikinetid of *Homalozoon* (after Kuhlmann et al. 1980)

left oral or adoral structures in the light of new ultrastructural data and defined two new terms based only on ultrastructural observations, paramembranelles, and heteromembranelles. They also proposed a number of new terms to describe the right oral or paroral structures, including stichodyad, stichomonad, and diplostichomonad. These structures differ from each other on the basis of what microtubular groups are associated with the kinetosomes, how the kinetosomes are linked, how the organellar complexes are joined to each other, and how the oral structures are oriented with respect to the axes of the oral region and the cell. Lynn (1981) extended the perspective of viewing somatic structures as monokinetids, dikinetids, and polykinetids to the oral structures. As will be demonstrated below, the new ultrastructural information since 1980 clearly shows that few of the terms as defined by de Puytorac and Grain (1976) will escape emendation, and it is perhaps best to abandon this kind of definition and merely qualify the kind of kinetid with the taxon name from which it derives (Lynn 1988).

Class Karyorelictea

Grolière et al. (1980) provided a brief documentation of the paroral and oral polykinetids of *Protocruzia*. Both kinetosomes of the paroral are ciliated with the inner one bearing a postciliary ribbon. The kinetosomes of each dikinetid are linked to each other by dense desmoses while a less dense, almost fibrous linkage unites each dikinetid to its neighbour (Fig. 6 a). Observations on other karyorelicteans do not permit any detailed comparison.

The oral polykinetids of *Protocruzia* are composed of 4 rows of kinetosomes: the spacing of the rows, one from another, is variable; kinetosomes of the posterior row bear postciliary ribbons; and kinetosomes of the anterior row probably bear transverse ribbons (Grolière et al. 1980). These oral polykinetids do not precisely fit the definition of paramembranelles (de Puytorac and Grain 1976), since the anterior two rows are closer together than the posterior two rows while the anterior row is not often noticeably shorter than the rest.

Class Spirotrichea

The "heterotrichs"

The oral structures of the heterotrich *Eufolliculina* include a paroral and a series of oral polykinetids. The oral polykinetids precisely fit the definition of paramembranelles (see Mulisch and Hausmann 1984), with transverse ribbons associated with the anterior row of kinetosomes and postciliary ribbons associated with the posterior row of kinetosomes. Mulisch and Hausmann (1984) point out that within the heterotrichs there is some variability, both in the association of transverse ribbons with the kinetosomes and in the number of rows per polykinetid, although all have some kind of nematodesmata that extend interiorly from the kinetosomal bases.

The paroral shows great variability among the heterotrichs, both within one species depending upon position in the paroral and between species (Mulisch and Hausmann 1984). The peristomial or adoral portion of the paroral structure of *Eufolliculina* (Fig. 6 b) most closely resembles that of *Stentor* differing in having only one postciliary ribbon but two transverse ribbons (Mulisch and Hausmann 1984). The wide variation in paroral structure among "heterotrichs" may be related to the great variation in oral cavity structure, probably as a result of adaptive evolution for feeding (Mulisch and Hausmann 1984), and demonstrates again the observation that oral structures are less conservative than somatic ones (Lynn 1981).

The "hypotrichs"

The oral structures of a number of hypotrich genera have been examined recently: *Laurentiella* (Torres et al. 1986), *Kahliella, Parastrongylidium, Spiretella* (Fleury et al. 1985 a, b), *Uronychia* (Fleury et al. 1986), *Thigmokeronopsis* (Wicklow 1981), *Discocephalus* (Wicklow 1982), *Certesia* (Wicklow 1983), *Engelmanniella* (Wirnsberger-Aescht et al. 1989), and *Paraurostyla* (Jerka-Dziadosz 1981 a). However, the micrographic evidence to demonstrate components is often not substantial.

Generally, the left oral polykinetids can be characterized as follows: postciliary ribbons on the posterior row; in cases where the micrographs permit it, single postciliary microtubules on the internal kinetosomes; transverse ribbons on the anterior kinetosomes of the ultimate and penultimate rows; a square packing of the kinetosomes, which are linked by dense connectives at proximal and mid-heights; and nematodesmata that extend internally. These fit the definition of paramembranelle except that internal postciliary microtubules are not mentioned in the definition (de Puytorac and Grain 1976, de Puytorac et al. 1987).

The right oral structures of "hypotrichs" show more variability than the left, a situation comparable to the "heterotrichs". The endoral structure is a single file of kinetosomes or stichomonad (de Puytorac and Grain 1976) usually with a postciliary ribbon and a transverse ribbon associated with each kinetosome in *Thigmokeronopsis, Discocephalus, Laurentiella*, and *Kiitricha*. The first three genera also have as a paroral structure, a polykinetid of from 2 to 6 rows with which postciliary ribbons are attached to the cytostomal kintosomes while transverse ribbons are associated with kinetosomes away from the cytostome. The right oral structures in the hypotrichs *Kahliella, Parastrongylidium*, and *Spiretella* unites the paroral and endoral structures as a diplostichomonad (de Puytorac and Grain 1976, de Puytorac et al. 1976), which is composed of two files of kinetosomes whose bases oppose each other; the kinetosomes of one file bear postciliary ribbons while those of the other file bear transverse ribbons.

Class Colpodea

The oral structures of colpodeans have been described for genera as follows: *Bryophrya* (Grain et al. 1979), *Sorogena* (Bradbury and Olive 1980), *Bresslaua* (Garcia-Rodriguez et al. 1981), *Pseudoplatyophrya* (de Puytorac et al. 1983), and *Bryometopus* (Wirnsberger et al. 1985). These further descriptions of colpodean genera serve to emphasize the tremendous diversity of oral features in this class: *Bryophrya* showing features akin to some nassophoreans, even feeding on filamentous cyanobacteria (Grain et al. 1979); *Pseudoplatyophrya* bearing a suctorian-like oral structure (de Puytorac et al. 1983); *Bryometopus* being like a heterotrich (Wirnsberger et al. 1985); and *Sorogena* having a prostomous oral region with simplified oral kinetids (Bradbury and Olive 1980). Garcia-Rodriguez et al. (1981) demonstrated clearly that, contrary to Lynn (1979), the oral features of *Bresslaua* were differentiated into clearly distinct left and right oral structures, homologous to those described for other colpodeans.

As with the spirotrichs, there is much more variability in right oral structures than in left oral structures. However, the colpodeans show relatively more variability between genera even in the left oral structures although all seem to be built on a common plan.

The left oral polykinetids, sometimes called cirromembranelles, are characterized in *Bryophrya, Bresslaua, Pseudoplatyophrya*, and *Bryometopus* as follows: kinetosomes of the posterior or right or cytostomal kinetosomal row bear postciliary ribbons; there are often single internal postciliary microtubules; the kinetosomes are square-packed; and nematodesmata extend internally from the kinetosome bases. *Sorogena* may also have some reduced polykinetids exhibiting this general structure (Figs. 12 and 14 of Bradbury and Olive 1980).

As noted above, the right oral structures show much more variability, and a consistent pattern does not emerge: *Pseudoplatyophrya* has a single file of kinetosomes; *Bryophrya* has many files of kinetosomes with internal postciliary microtubules; *Bresslaua* has an irregularly-arrayed field of kinetosomes bordered on the cytostomal side with what may be dikinetids; and *Bryometopus* has a file of dikinetids whose posterior or cytostomal kinetosome bears a postciliary ribbon and whose anterior kinetosome bears a transverse ribbon.

Class Phyllopharyngea

Among the genera whose somatic kinetids have been described, the oral structures of *Spirochona* (Fahrni 1982), *Hypocoma* (Grell and Meister 1982 c, 1983), *Acinetopsis* (Grell and Meister 1982 a), and *Trichophrya* (Mogensen and Butler 1984) have been examined. Fahrni (1982) suggests that the "X field" of *Spirochona* may be the vestiges of an oral structure although it is composed of irregularly-arrayed and non-ciliated kinetosomes. The "tentacle" arrays of *Hypocoma, Acinetopsis*, and *Trichophrya* are all composed of radially arrayed ribbons of microtubules. These phyllae are also found surrounding the pseudatrium of *Spirochona*. These features have been observed in many previous studies (Corliss 1979, Lynn and Corliss 1991).

Class Nassophorea

Detailed descriptions of the polykinetids of *Nassula* (de Puytorac and Njiné 1980, Eisler 1986) and *Furgasonia* (Eisler 1988) have appeared. The left oral polykinetids are characterized as follows: kinetosomes of the posterior row bear a postciliary ribbon; single transverse microtubules may be associated with all kinetosomes (Eisler 1988); parasomal sacs are arranged diagonally between kinetosomes; and there are three rows of square-packed kinetosomes. *Nassula* lacks a paroral structure in the morphostatic stage while the stichodyad or paroral structure of *Furgasonia* is composed of dikinetids whose longitudinal axis is perpendicular to the long axis of the paroral. Eisler (1988) provided some significant details about kinetosomal orientation of its paroral: the posterior or cytostomal non-ciliated kinetosome is turned 90° to the somatic kinetids so that triplet 7 is most posterior and the postciliary ribbon faces the cytostome; the anterior ciliated kinetosome has a single postciliary microtubule demonstrating that the orientation of this kinetosome is identical to that of the somatic kinetids (Fig. 6 c).

Class Oligohymenophorea

From a systematic perspective, there are relevant ultrastructural studies on the apostomes *Conidophrys* (Bradbury and Tyson 1982) and *Hyalophysa* (Bradbury 1989), the hymenostomes *Colpidium* and *Turaniella* (Iftode et al. 1984), and the peritrichs *Trichodina* (Hausmann and Hausmann 1981, Maslin-Leny and Bohatier 1984), *Tripartiella* (Maslin-Leny and Bohatier 1984), and *Thuricola* (Eperon and Grain 1983). Similarities both in the paroral structures of *Colpidium* and *Turaniella* and in their stomatogenesis were used to justify placing *Colpidium* in the family Turaniellidae.

The ultrastructure of the tomite of the pilisuctorian *Conidophrys* revealed a rosette and canaliculi surrounded by dense inclusion bodies and opening to the cell surface. These features together with stacked kinetodesmal fibrils of the somatic kinetid place this genus among the apostomes (Bradbury and Tyson 1982), although the structure of the trophont is very divergent from other apostomes.

Bradbury (1989) described the ultrastructure of the ciliature and infraciliature surrounding the subapical lateral canal, the homologue of the cytostome in apostomes. To its right, a paroral structure is composed of two files of staggered non-ciliated kinetosomes. The falciform and ogival kinetosomal fields are considered homologues of the left oral ciliature of other oligohymenophoreans, although they show few similarities apart from position and the presence of postciliary ribbons on the right-most kinetosomes (Bradbury 1989).

The peritrichs have paroral dikinetids of the stichodyad type (Maslin-Leny and Bohatier 1984) while the oral polykinetids are only accompanied by rightmost postciliary ribbons either during stomatogenesis (Eperon and Grain 1983) or at certain levels along the long length of these oral polykinetids (Maslin-Leny and Bohatier 1984).

Class Prostomatea

The studies of Lynn (1985), Huttenlauch (1987), and Huttenlauch and Bardele (1987) on *Coleps* species provided further details on the oral structures of prostomes. Lynn (1985) continued the historical misinterpretation of the cytostomally-directed microtubular ribbons as transverse ribbons while Huttenlauch and Bardele (1987) presented clear evidence from an electron microscopic study of stomatogenesis in *Coleps* that these were postciliary ribbons. Huttenlauch (1987) and Huttenlauch and Bardele (1987) provided details of the oral polykinetids, which are characterized as follows: convergent postciliary ribbons on the right row of kinetosomes, which are non-ciliated; kinetosomes of the left row are ciliated with dilated ciliary membranes but bear no fibrillar associates.

Class Litostomatea

There has been active research on a variety of litostome genera: the haptorians *Homalozoon* (Kuhlmann et al. 1980), *Helicoprorodon* (Raikov and Kovaleva 1980), *Enchelydium* (Foissner and Foissner 1985), *Lepidotrachelophyllum* (Lynn and Nicholls 1985), and *Fuscheria* (Foissner and Foissner 1988), and the trichostomes *Dasytricha* (Paul et al. 1989), *Epidinium* (Furness and Butler 1983), *Eudiplodinium* (Furness and Butler 1985 a), and *Entodinium* (Furness and Butler 1985 b). There is diversity in oral kinetid structure, although the ancestral state is probably a dikinetid with oral monokinetids being either derived from dikinetids in haptorians and possibly lost altogether in trichostomes (Lipscomb and Riordan 1990).

The haptorian dikinetid is characterized as follows: a ciliated posterior kinetosome with a postciliary ribbon and sometimes a transverse ribbon; and a non-ciliated anterior kinetosome that bears a transverse ribbon, a nematodesma, and sometimes a single postciliary microtubule (Fig. 6 e–g). An exception to this pattern is *Enchelydium*, whose oral kinetids are non-ciliated monokinetids (Fig. 6 d) with only a transverse ribbon and nematodesma (Foissner and Foissner 1985).

The oralized "somatic" kinetids of the trichostomes, both the vestibuliferidan *Dasytricha* and the entodiniomorphids *Epidinium, Entodinium,* and *Eudiplodinium*, are strikingly similar to the somatic kinetids: nematodesmata arise from the bases of those kinetosomes closest to the cytostomal region and the kinetodesmal fibrils may be considerably reduced or absent.

Taxa of uncertain affinities

The plagiopylids *Plagiopyla* and *Lechriopyla* (Berger and Lynn 1984, de Puytorac et al. 1985) and the schizocaryid *Schizocaryum* (Lynn and Frombach 1987) have oral structures, which, like their somatic kinetids, are not comparable to any of the taxa described above and so their homologues remain unidentified.

Development of cortical structures

In a definitive series of papers, Jerka-Dziadosz (1980, 1981 a, b, 1982) carefully outlined the development and movement of the somatic and oral kinetids of *Paraurostyla* and briefly discussed the significance of comparative ultrastructural morphogenesis (Jerka-Dziadosz 1981 b). These papers, together with a number of ultrastructural studies of cortical regeneration (e.g., Bohatier 1979, Bernard and Bohatier 1981, Fahrni 1985) and studies on stomatogenesis in *Tetrahymena* (e.g., Bakowska et al. 1982), have demonstrated to systematists that a substantial increase in character information can be obtained from this approach. Bardele (1987) has argued strongly for this approach, referring to the notions of von Baer and Haeckel regarding the phylogenetic implications of ontogeny. In their modern interpretation, care must be taken not to "over-interpret" the ontogeny as an *actual* recapitulation of phylogeny (Gould 1977). With this in mind, several important issues arise that will have some systematic implications (see below).

First, Huttenlauch and Bardele (1987) presented sobering evidence for all comparative "ultrastructuralists" that some structures are *not* what they appear to be. Their study of the stomatogenesis of *Coleps* demonstrated that what were initially interpreted a transverse ribbons associated with the oral dikinetids only *appeared* that way. Due to morphogenetic movements of the kinetids and their kinetosomes during stomatogenesis, the postciliary ribbons changed orientation with respect to the kinetosomal perimeter so that they appeared to be transverse ribbons. In a general sense, we can only hope that *Coleps* is an exception. Nevertheless, this must give grounds for caution at over-interpretation, especially when structures are somewhat unusual, as indeed the prostome dikinetids appeared to be from the beginning.

Secondly, after replication, kinetosomes can rotate with respect to each other so that their orientation with respect to the kinety axis and/or body might be quite different. Eisler (1989) has most clearly demonstrated this in the development of the paroral structure of *Furgasonia* whose dikinetids have kinetosomes with quite different orientations. This interpretation was possible for *Furgasonia* since both kinetosomes have postciliary microtubules (Eisler 1989). Where postciliaries are not associated with each kinetosome, caution must be exercised in interpreting triplet numbers and kinetosome orientation.

Thirdly, Furness and Butler (1986) have demonstrated the likelihood that entodiniomorphids indeed derived from haptorian ancestors: by studying kinetid replication and development in *Eudiplodinium*, a second transverse ribbon microtubule appears transiently near triplet 5 (Fig. 4 k). This further emphasizes the systematic importance of morphogenetic studies using electron microscopy.

Fourthly, there is some dispute about the fundamental morphogenetic unit. For the stichotrichine hypotrich *Paraurostyla*, Jerka-Dziadosz (1980, 1981 b, 1982) showed that, following kinetosomal replication, the fundamental building unit for several cortical organellar complexes was the dikinetid. For the heterotrichs *Stentor, Condylostoma, Climacostomum*, and *Eufolliculina*, both in regeneration (Bernard and Bohatier 1981, Bohatier 1979, Fahrni 1985) and in stomatogenesis (Mulisch and Hausmann 1988), dikinetid formation is prerequisite to building more complex organelles. Bakowska et al. (1982) observed this also in stomatogenesis of the oral polykinetids of the hymenostome *Tetrahymena*.

On the other hand, Hoffmann (1987) and Eisler (1989) argued that, because the monokinetid units of cyrtophorids and nassulids have all the fibrillar associates, the monokinetid instead of the dikinetid must be the fundamental unit of stomatogenesis and therefore the ancestral state for the ciliate kinetid. Evenso, all the oral structures in these ciliates develop as associations of dikinetids or triads that have initially arisen from monokinetids. Huttenlauch and Bardele (1987) also demonstrated this to be the case for the development of oral structures of the prostome *Coleps*.

Hofmann (1987: 183) appeals to the biogenetic law, stating that "if this statement (= law) is true for stomatogenesis of ciliates, then the monokinetid turns out to be more primitive than the dikinetid". This is argued because, in some species, monokinetids give rise to dikinetids and then to trikinetids – a recapitulation of the evolutionary process (Bakowska et al. 1982, Eisler 1989, Hofmann 1987, Huttenlauch and Bardele 1987). However, in the hypotrichs and heterotrichs, dikinetids give rise to monokinetids, which then give rise to dikinetids. Using this example of ontogeny, one would infer dikinetids to be ancestral with monokinetids derived from them.

A further argument for the ancestry of the monokinetid has been proposed by Leipe and Hausmann (1989). Since monokinetids are found on ciliates judged to be of ancestral body type, having an apical cytostome, oral apparatus of proter retained during telokinetal stomatogenesis, often uniform and holotrichous ciliation, and toxicysts, monokinetids must also be ancestral. However, there really is no independent evidence to suggest that these ciliates are ancestral. Many that fit this morphotype are planktonic, highly adapted and almost radially symmetrical forms, whose body form could be argued to be highly derived, since "radial" symmetry itself is a rare condition among protists in general. Both Orias (1976) and Small (1984) envision the ancestral ciliate to be initially flattened and not holotrichously ciliated.

Grain et al. (1988) develop an argument that the kinetosomes in ciliates are not homologous to the dikinetids of flagellates. They consider ciliate kinetosomes to belong to a "second generation" of kinetosomes, similar to those developed in some multiciliated metazoan cells.

With such different hypotheses, we must use other criteria to assess which state is ancestral. Lynn and Small (1981) argued that both the outgroup comparison of ciliate kinetids to other protist taxa and the fact that "dikinetids" are widely distributed among eukaryotes indicate the dikinetid as the ancestral state for the ciliates. Small (1984) has added a further criterion for the ciliates: ciliates with the presumed ancestral nuclear condition have somatic dikinetids; thus, these karyorelicts might also be thought of as corticorelicts.

There is a further ontogenetic argument that one can make for the ancestral nature of the dikinetid. Since most flagellate groups have dikinetids that are composed of two kinetosomes linked by desmoses, the developmental genetic program for dikinetid construction must be present in all flagellate genomes and presumably in the genome of the flagellate ancestor of the ciliates. Ciliate dikinetids follow the steps of separate,

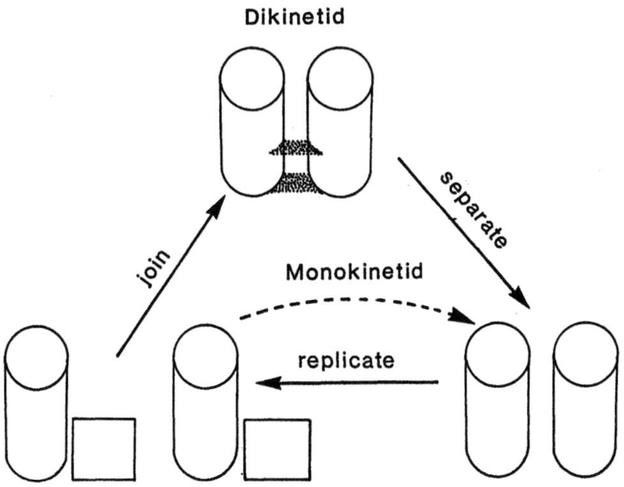

Fig. 7. A schematic representation of the replication of a somatic kinetid of a ciliate. Initially the kinetosomes separate as the connectives linking them disassemble. Replication occurs followed by the joining of the newly formed pairs of kinetosomes. A monokinetid pattern could arise from a dikinetid ontogeny by loss or repression of the "joining genes" and their gene products

replicate, and join during their development (e.g., Bohatier 1979). Elimination of desmoses and evolution of monokinetids from dikinetids could be accomplished by repression or deletion of the "desmose genes" (Fig. 7). On the other hand, if one assumes the monokinetid as ancestral for the ciliates, "desmose genes" must arise anew in each lineage that exhibits dikinetids. From this argument, assuming dikinetids as ancestral is more parsimonious.

Examination of the comparative morphogenesis of ciliates using electron microscopy has already provided some substantial results (see below). As more information accumulates, we will undoubtedly get closer to the goal of establishing a natural classification for this group (e.g., Eisler 1989).

Systematic implications

Most of the papers cited above briefly discuss some phylogenetic and/or taxonomic implications of the data that are presented. In this section, I will examine only those papers that, in my opinion, have made substantial contributions or are controversial. The treatment will not be exhaustive, but more illustrative of the systematic problems that will require resolution in the future.

De Puytorac et al. (1987) recognized the diversity of structure in the class Karyorelictea sensu Small and Lynn by elevating the class to a subphylum and its included orders to classes. If Orias (pers. comm.) is correct in his argument that nuclear dualism may have arisen independently in several ciliate lineages, then the cortical diversity of the karyorelicteans may be rationalized by relating them to other taxa: for example, the protoheterotrichid somatic dikinetid (see Lynn 1981) and cortical myonemes show strong similarities to those of the heterotrichs, suggesting these two groups may indeed belong together.

Both de Puytorac et al. (1987) and Small and Lynn (1985) recognized the folliculinids as a suborder within the heterotrich spirotrichs. Mulisch (1987) argued that strong similarities in the somatic cortex, cortical pigments, paroral structures, adoral polykinetids, and stomatogenesis suggest a close relationship between folliculinids and stentorids. She questioned the systematic validity of the life cycle differences that distinguish the folliculinids. The same argument could be raised against the tintinnines as a distinct taxon within the Oligotrichia.

There is still a great deal of controversy about relationships among the hypotrich spirotrichs. Molecular evidence indicated that Small and Lynn (1985) placed too much weight on similarities of the dorsal kinetid of the euplotines and discocephalines to that of the nassophoreans (Lynn and Sogin 1988) over similarities in the macronuclear organization and development, the presence of cirri, and the coupling of macronuclear reorganization and morphogenesis (Fleury 1988). The hypotrichs do form a monophyletic assemblage, but relationships within the class are still unclear. Fleury et al. (1985 a) and Tuffrau (1987) have presented two recent revisions based on ultrastructural and morphogenetic data. Fleury (1988) concluded that the Euhypotrichina Fleury et al. 1985 is a monophyletic group sharing such features as kinetosome-resorbing cysts, dorsal dikinetids that lack kinetodesmal fibrils, a pellicular monolayer of microtubules, and a morphogenesis that involves full turn-over of the whole somatic ciliature. The Pseudohypotrichina Fleury et al. 1985 is paraphyletic, and further features are required to demonstrate affinities (Fleury 1988).

The colpodeans are now generally recognized as a class (de Puytorac et al. 1987, Small and Lynn 1985), although their relationship to other higher taxa is still controversial: Small and Lynn (1985) placed the class in their subphylum Cyrtophora while de Puytorac et al. (1987) placed them in their subphylum Prostomata. De Puytorac et al. (1983), Foissner (1985), Small and Lynn (1985), and de Puytorac et al. (1987) presented slightly different classifications of this group based on differing emphases of various features. For example, as Wirnsberger et al. (1985) pointed out, the inclusion of the micronucleus within the macronuclear membrane is not likely to have arisen twice within the group although the classification of de Puytorac et al. (1987) suggests this. Further analysis is required on the relationships within this class.

Grell and Meister (1982 c) have presented a cladistic analysis of phyllopharyngean taxa in which they disagree with Small and Lynn (1985) in placing the hypocomids with the cyrtophorids: the latter have a heteromerous macronucleus while the former do not. Small and Lynn (1985) considered the presence of at least one dorsal kinetal fragment to be a more important character. However, Grell and Meister (1982 c, 1983) also demonstrated the presence of extrusomes in the hypocomids, very similar to those reported in the rhynchodids, and further evidence for the relationship that Grell and Meister (1982 c) prefer. De Puytorac et al. (1987) included the hypocomids in their order Rhynchodida.

Eisler and Bardele (1983) and Eisler (1989) have presented phylogenetic arguments for the relationships of the nassulids, which are united by the presence of paired alveolocysts and a cytopharyngeal basket with "X" lamellae associated with the nematodesmata. De Puytorac et al. (1987) united these taxa in the class Nassophorea in which Small and Lynn (1985) also included the peniculids. Eisler (1989) concluded that it is impossible to associate the nassulids sensu stricto to any other ciliate order at the present time.

The broad affinities among the oligohymenophoreans are largely undisputed. Small and Lynn (1985) removed the peniculids from this assemblage on the basis of kinetid features, fibrous trichocysts, and substantial oral nematodesmata while de Puytorac et al. (1987) retained the peniculids in the class Oligohymenophorea. Bradbury (1989) has argued that the apostomes may not be a subclass independent of the hymenostomes as indicated by both Small and Lynn (1985) and de Puytorac et al. (1987). Referring to similarities in the life cycle (i.e., a microstome-macrostome transformation) and oral ultrastructure, Bradbury (1989) concluded that the apostomes might more properly be considered a suborder within the Hymenostomatida, and derived from small, holotrichous bacterivorous forms with a capacity for macrostome transformation. Huttenlauch and Bardele (1987) have provided dramatic evidence that the ultrastructural study of stomatogenesis can yield significant phylogenetic information: their transfer of the prostomes sensu Small and Lynn from the subphylum Rhabdophora to the subphylum Cyrtophora was based on the demonstration that the microtubules supporting the cytopharynx are postciliary ribbons. The presence of radial transverse ribbons in the somatic kinetids suggests a close relationship to the oligohymenophoreans, as do some aspects of stomatogenesis and oral structure. De Puytorac et al. (1987) included the prostomates with litostomates sensu Small and Lynn (see below). Differences in both somatic and oral kinetid ultrastructure argue against this.

Discussion of relationships among the litostomates sensu Small and Lynn has been as controversial, and is as unsettled, as it has been for the hypotrichs sensu lato. As Foissner and Foissner (1988: 227) stated, establishing a natural system "is at present almost impossible because 'hard' information about the apomorphic and plesiomorphic state of the main characters is still lacking". Nevertheless, it is highly probable that the group is monophyletic, and includes both "prostomatous" and "trichostomatous" forms. This makes the suggestion (Leipe and Hausmann 1989) of a new subclass Ditransversalia Leipe and Hausmann, 1989 unnecessary for the following: (1) several taxa listed as not having the second transverse set probably do (e.g., *Dileptus*, Golinska pers. comm.); (2) in the case of those where a second set is not present the kinetids are densely aligned in a girdle, making secondary loss very probable (e.g., Didiniidae); and (3) a second transverse homologue has been identified during replication of entodiniomorphid kinetids (Furness and Butler 1986). Therefore, the revision of the litostomate kinetid by Lynn and Nicholls (1985) is justified (see also Paul et al. 1989).

Foissner and Foissner (1985) have presented convincing evidence that the oral kinetids of some haptorians (e.g., *Enchelydium*), that is litostomes with toxicysts, are monokinetids, and suggested that these represent the "missing-links" between contemporary archistomatids and their free-living ancestors. On this basis, they transferred the buetschliid archistomatids from the Entodiniomorphida to the Haptoria (see Foissner and Foissner 1988), and reorganized the Haptoria putting much weight on the presence of "oralized" kinetids bearing oral nematodesmata and the presence/absence of modified or clavate-like cilia of dorsal somatic kinetids. Lipscomb and Riordan (1990) have recently presented a detailed cladistic analysis of litostome genera that supports many of Foissner and Foissner's conclusions, but does not address the position of the entodiniomorphids.

Conclusions

A substantial body of literature has accumulated since my previous review (Lynn 1981). There are some clear resolutions to systematic relationships that have been provided by this ultrastructural research, the study of stomatogenesis of *Coleps* being the best example (Huttenlauch and Bardele 1987). Just as many other questions have been raised by this new information, especially problematic is the determination of ancestral and derived character states on which to base a phylogenetic analysis. While the study of ultrastructure continues, searches for molecular markers are providing encouraging prospects for unravelling some of the phylogenetic mysteries (Baroin et al. 1988; Lynn and Sogin 1988; Nanney et al. 1989 a, b; Preparata et al. 1989). For the next decade, the prospects for resolution of a natural system for the ciliates are very promising.

Acknowledgements

I am very grateful to the careful critical comments provided by Christian Bardele, Jean Grain, and Eugene Small whose suggestions considerably improved the manuscript. Any remaining errors or omissions remain my responsibility.

References

Bakowska J, Nelsen EM, Frankel J (1982) Development of the ciliary pattern of the oral apparatus of *Tetrahymena thermophila*. J Protozool 29: 366–382

Bardele CF (1987) Auf neuen Wegen zu alten Ziele – moderne Ansätze zur Verwandtschaftsanalyse der Ciliaten. Verh Dtsch Zool Ges 80: 59–75

Baroin A, Perasso R, Qu LH, Brugerolle G, Bachellerie JP, Adoutte A (1988) Partial phylogeny of the unicellular eukaryotes based on rapid sequencing of a portion of 28S ribosomal RNA. Proc Natl Acad Sci USA 85: 3473–3478

Berger J, Lynn DH (1984) Somatic and oral cortical ultrastructure of the plagiopylid ciliates *Lechriopyla mystax* Lynch, 1930 and *Plagiopyla minuta* Powers, 1933. J Protozool 31: 433–443

Bernard F, Bohatier J (1981) Ultrastructure et mise en place des organelles buccaux de la régénération oral chez *Stentor coeruleus* (cilié hétérotriche). Can J Zool 59: 2306–2318

Bohatier J (1979) Morphogenèse de régénération chez le cilié *Condylostoma magnum* (Spiegel): étude ultrastructurale. J Protozool 26: 404–414

Bonhomme A, Grain J, Collet JY (1989) Étude ultrastructurale de *Troglodytella gorillae*, cilié de l'intestin des gorilles. Considérations sur la systématique des ciliés entodiniomorphes. Eur J Protistol 24: 225–237

Bradbury PC (1989) Evidence for hymenostome affinities in an apostome ciliate. J Protozool 36: 95–103

– Olive LS (1980) Fine structure of the feeding stage of a sorogenic ciliate, *Sorogena stoianovitchae* gen. n., sp. n. J Protozool 27: 267–277

– Tyson GE (1982) The fine structure of *Conidophrys pitelkae* Bradbury related to its life cycle and taxonomic position in the Apostomatida. J Protozool 29: 184–194

Chapman GB, Kern RC (1983) Ultrastructural aspects of the somatic cortex and contractile vacuole of the ciliate, *Ichthyophthirius multifiliis* Fouquet. J Protozool 30: 481–490

Corliss JO (1974) The changing world of ciliate systematics: historical analysis of past efforts and a newly proposed phylogenetic scheme of classification for the protistan phylum Ciliophora. Syst Zool 23: 91–138

– (1979) The ciliated protozoa: characterization, classification, and guide to the literature, 2nd edn. Pergamon, London

Eisler K (1986) Licht- und elektronenmikroskopische Untersuchungen zur kortikalen Morphologie und Morphogenese nassulider Ciliaten. Ein Beitrag zur Klärung der Stellung der Nassulida (Ciliata, Cyrtophora) in einem natürlichen System der Ciliaten. PhD Thesis, University of Tübingen, Tübingen, Federal Republic of Germany

– (1988) Electron microscopical observations on the ciliate *Furgasonia blochmanni* Fauré-Fremiet, 1967. Part I. An update on morphology. Eur J Protistol 24: 75–93

– (1989) Electron microscopical observations on the ciliate *Furgasonia blochmanni* Fauré-Fremiet, 1967. Part II. Morphogenesis and phylogenetic conclusions. Eur J Protistol 24: 181–199

– Bardele CF (1983) The alveolocysts of the Nassulida: ultrastructure and some phylogenetic considerations. Protistologica 19: 95–102

Eperon S, Grain J (1983) Sur les fibres postciliaires de la polycinétie de *Thuricola folliculata* (O. F. Müller, 1786) (Ciliophora, Peritrichida) en bipartition. Protistologica 19: 123–129

Fahrni JF (1982) Morphologie et ultrastructure de *Spirochona gemmipara* Stein, 1852 (Ciliophora, Chonotrichida). I. Structures corticales et buccales de l'adulte. J Protozool 29: 170–184

– (1985) Morphogenesis in the heterotrich ciliate *Climacostomum virens*. II. Oral development during regeneration. J Protozool 32: 460–473

Fleury A (1988) The use of correlated ultrastructural and morphogenetic characters in evolutionary taxonomy of hypotrich ciliates. BioSystems 21: 309–316

– Iftode F, Deroux G, Fryd-Versavel G, Genermont J (1985a) Relation entre caractères ultrastructuraux et morphogénétique chez les Hypotriches (Protozoa, Ciliata): définition des Euhypotrichina n. s. o., et des Pseudohypotrichina n. s. o. C R Acad Sci Paris 300: 567–570

– – – – (1985b) Unité et diversité chez les hypotriches (protozoaires ciliés): II. – Éléments d'ultrastructure comparée chez divers représentants du sous-ordre des Euhypotrichina. Protistologica 21: 505–524

– – – – (1986) Unité et diversité chez les hypotriches (protozoaires ciliés): III. – Éléments d'ultrastructure comparée chez divers représentants du sous-ordre des Pseudohypotrichina et remarques générales. Protistologica 22: 65–87

Foissner W (1985) Klassifikation und Phylogenie der Colpodea (Protozoa: Ciliophora). Arch Protistenk 129: 239–290

– Foissner I (1985) Oral monokinetids in the free-living haptorid ciliate *Enchelydium polynucleatum* (Ciliophora, Enchelyidae): ultrastructural evidence and phylogenetic implications. J Protozool 32: 712–722

– – (1988) The fine structure of *Fuscheria terricola* Berger et al., 1983 and a proposed new classification of the subclass Haptoria Corliss, 1974 (Ciliophora, Litostomatea). Arch Protistenk 135: 213–235

Furness DN, Butler RD (1983) The cytology of sheep rumen ciliates. I. Ultrastructure of *Epidinium caudatum* Crawley. J Protozool 30: 676–687

– – (1985a) The cytology of sheep rumen ciliates. II. Ultrastructure of *Eudiplodinium maggii*. J Protozool 32: 205–214

– – (1985b) The cytology of sheep rumen ciliates. III. Ultrastructure of the genus *Entodinium* (Stein). J Protozool 32: 699–707

– – (1986) Morphogenesis of infraciliature in the entodiniomorphid ciliate *Eudiplodinium maggii*. J Ultrastruct Mol Struct Res 94: 16–29

Garcia-Rodriguez T, Perez-Paniagua F, Perez-Silva J (1981) Morphogenèse de division et ultrastructures corticale et buccale chez *Bresslaua vorax* Kahl (cilié, Colpodidae). Protistologica 17: 533–542

Gerassimova ZP (1981) The ultrastructure and systematic position of the rumen ciliate *Isotricha intestinalis* Stein, 1859. Tsitologiya 23: 861–866 (in Russian with English summary)

Görtz H-D (1982) The behavior and fine structure of the dorsal bristles of *Euplotes minuta*, *E. aediculatus*, and *Stylonychia mytilus* (Ciliata, Hypotrichida). J Protozool 29: 353–359

Gould SJ (1977) Ontogeny and phylogeny. Harvard University Press, London

Grain J (1966) Étude cytologique de quelques ciliés holotriches endocommensaux des ruminants et des équidés. II. Microscopie électronique. Protistologica 2: 2–51

– (1969) Le cinétosome et ses derivés chez les ciliés. Ann Biol Ser 4, 8: 53–97

– Didier P, Peck RK, Rodrigues de Santa Rosa M (1978) Étude ultrastructurale et position systématique des ciliés du genre *Cyclogramma* Perty, 1852. Protistologica 14: 225–240

– Iftode F, Fryd-Versavel G (1979) Étude des infraciliatures somatique et buccale de *Bryophrya bavariensis* et considérations systématiques. Protistologica 15: 581–595

– de Puytorac P, Grolière C-A (1982) Quelques précisions sur l'ultrastructure et la position systématique du cilié *Mesodinium rubrum*, et sur la constitution de ses symbiontes chloroplastiques. Protistologica 18: 7–21

– Mignot J-P, de Puytorac P (1988) Ultrastructures and evolutionary modalities of flagellar and ciliary systems in protists. Biol Cell 63: 219–237

Grell KG, Meister A (1982a) Die Ultrastruktur von *Acinetopsis rara* Robin (Suctoria). I. Tentakeln und Nahrungsaufnahme. Protistologica 18: 67–84

– – (1982b) Die Ultrastruktur von *Acinetopsis rara* Robin (Suctoria) II. Zellbau und Schwärmerbildung. Protistologica 18: 403–421

– – (1982c) Über die taxonomische Stellung der Hypocomidae (Ciliata). Z Naturforsch 37c: 1050–1052

– – (1983) Die Ultrastruktur von *Hypocoma acinetarum* Collin. Protistologica 19: 51–72

Grim JN (1988) A somatic kinetid study of the pycnotrichid ciliate *Vestibulongum corlissi* n. g., n. sp. (Class: Litostomatea), symbiont in the intestines of the surgeonfish, *Acanthurus xanthopterus*. J Protozool 35: 227–230

Grimes GW, Adler JA (1976) The structure and development of the dorsal birstle complex of *Oxytricha fallax* and *Stylonychia pustulata*. J Protozool 23: 135–143

Grolière C-A, de Puytorac P, Detcheva R (1980) À propos d'observations sur la stomatogenèse et l'ultrastructure du cilié *Protocruzia tuzeti* Villeneuve-Brachon, 1940. Protistologica 16: 453–466

Hausmann K, Hausmann E (1981) Structural studies on *Trichodina pediculus* (Ciliophora, Peritricha). I. The locomotor fringe and the oral apparatus. J Ultrastruct Res 74: 131–143

Hofmann AH (1987) Stomatogenesis in cyrtophorid ciliates. II. *Chilodonella cyprini* (Moroff, 1902): the kinetofragment as an anlagen-complex. Eur J Protistol 23: 165–184

– Bardele CF (1987) Stomatogenesis in cyrtophorid ciliates. I. *Trithigmostoma steini* (Blochmann, 1895): from somatic kineties to oral kineties. Eur J Protistol 23: 2–17

Huttenlauch I (1987) Ultrastructural aspects of the somatic and buccal infraciliature of *Coleps amphacanthus* Ehrenberg 1833. Protoplasma 136: 191–198

– Bardele CF (1987) Light and electron microscopical observations on the stomatogenesis of the ciliate *Coleps amphacanthus* Ehrenberg 1833. J Protozool 34: 183–192

Iftode F, Fryd-Versavel G, Wicklow BJ, Tuffrau M (1983) Le genre *Transitella*: stomatogenèse, ultrastructure. Affinités de la famille des Transitellidae. Protistologica 19: 21–39

– – Lynn DH (1984) New details of the oral structures of *Colpidium* and *Turaniella* and transfer of the genus *Colpidium* to the Turaniellidae Didier, 1971 (Tetrahymenina, Hymenostomatida). Protistologica 20: 463–474

Jankowski AW (1967) A new system of ciliate Protozoa (Ciliophora). Akad Nauk SSSR Trudy Zool Inst 43: 3–54 (in Russian)

Jerka-Dziadosz M (1980) Ultrastructural study on development of the hypotrich ciliate *Paraurostyla weissei*. I. Formation and morphogenetic movements of ventral ciliary primordia. Protistologica 16: 571–589

– (1981 a) Ultrastructural study on development of the hypotrich ciliate *Paraurostyla weissei*. II. Formation of the adoral zone of membranelles and its bearing on problems of ciliate morphogenesis. Protistologica 17: 67–81

– (1981 b) Ultrastructural study on development of the hypotrich ciliate *Paraurostyla weissei*. III. Formation of preoral membranelles and an essay on comparative morphogenesis. Protistologica 17: 83–97

– (1982) Ultrastructural study on development of the hypotrich ciliate *Paraurostyla weissei*. IV. Morphogenesis of dorsal bristles and caudal cilia. Protistologica 18: 237–251

Kuhlmann S, Patterson DJ, Hausmann K (1980) Untersuchungen zu Nahrungserwerb und Nahrungsaufnahme bei *Homalozoon vermiculare* Stokes 1887. 1. Nahrungserwerb und Feinstruktur der Oralregion. Protistologica 16: 39–55

Leipe DD, Hausmann K (1989) Somatic infraciliature in the haptorid ciliate *Homalozoon vermiculare* (Kinetophragminophora, Gymnostomata) Ditransversalia n. subcl. and phylogenetic implications. J Protozool 36: 280–289

Lipscomb DL, Riordan GP (1990) The ultrastructure of *Chaenea teres* and an analysis of the phylogeny of haptorid ciliates. J Protozool 37: 287–300

Lom J, Corliss JO (1971) Morphogenesis and cortical ultrastructure of *Brooklynella hostilis*, a dysteriid ciliate ectoparasitic on marine fishes. J Protozool 18: 261–281

Lynn DH (1976) Comparative ultrastructure and systematics of the Colpodida. Structural conservatism hypothesis and a description of *Colpoda steinii* Maupas. J Protozool 23: 302–314

– (1979) Fine structural specializations and evolution of carnivory in *Bresslaua* (Ciliophora: Colpodida). Trans Amer Microsc Soc 98: 353–368

– (1981) The organization and evolution of microtubular organelles in ciliated protozoa. Biol Rev 56: 243–292

– (1985) Cortical ultrastructure of *Coleps bicuspis* Noland, 1925, and the phylogeny of the class Prostomatea (Ciliophora). BioSystems 18: 387–397

– (1988) Cytoterminology of cortical components of ciliates: somatic and oral kinetids. BioSystems 21: 299–307

– Corliss JO (1991) Phylum Ciliophora. In: Harrison FR, Corliss JO (eds) Microscopic anatomy of invertebrates. Wiley-Liss, New York, pp 333–467

– Frombach S (1987) The ultrastructure and systematics of *Schizocaryum dogieli*, a ciliate endocommensal in Pacific echinoids. Can J Zool 65: 3133–3143

– Nicholls KH (1985) Cortical microtubular structures of the ciliate *Lepidotrachelophyllum fornicis* Nicholls and Lynn, 1984 and phylogeny of the litostomate ciliates. Can J Zool 63: 1835–1845

– Small EB (1981) Protist kinetids: structural conservatism, kinetid structure, and ancestral states. BioSystems 14: 377–385

– Sogin MLS (1988) Assessment of phylogenetic relationships among ciliated protists using partial ribosomal RNA sequences derived from reverse transcripts. BioSystems 21: 249–254

Maslin-Leny Y, Bohatier Y (1984) Cytologie ultrastructurale de *Trichodina* et *Tripartiella* (ciliés péritriches). Protistologica 20: 113–132

Matsusaka T, Nakamura T, Nagata K (1984) Ultrastructure, disintegration and formation of a cirrus in the vegetative, encysting and excysting ciliate, *Histriculus muscorum*. J Electron Microsc Japan 33: 217–229

Mogensen MM, Butler RD (1984) Cytological studies of *Trichophrya rotunda* (Hentschel). J Protozool 31: 101–111

Mulisch M (1987) Stomatogenesis during metamorphosis of *Eufolliculina uhligi* (Ciliophora, Heterotrichida). Eur J Protistol 23: 56–65

– Hausmann K (1984) Structure and ultrastructure of the oral apparatus of *Eufolliculina uhligi* Mulisch and Patterson 1983. Protistologica 20: 415–429

– – (1988) Transmission electron microscopical observations on stomatogenesis during metamorphosis of *Eufolliculina uhligi* (Ciliophora: Heterotrichida). J Protozool 35: 450–458

– Barthlott W, Hausmann K (1981) Struktur und Ultrastruktur von *Eufolliculina* spec. Schwärmer und sessiles Stadium. Protistologica 17: 285–312

Nanney DL, Meyer EB, Simon EM, Preparata RM (1989 a) Comparison of ribosomal and isozymic phylogenies of tetrahymenine ciliates. J Protozool 36: 1–8

– Preparata RM, Preparata FP, Meyer EB, Simon EM (1989 b) Shifting ditypic site analysis: heuristics for extending the phylogenetic range of nucleotide sequences in Sankoff analyses. J Mol Evol 28: 451–459

Orias E (1976) Ciliate architecture: evolution from a simple flagellate. Trans Amer Microsc Soc 95: 415–429

Paul RG, Butler RD, Williams AG (1989) Ultrastructure of the rumen ciliate *Dasytricha ruminantium*. Eur J Protistol 24: 205–215

Pitelka DR (1969) Fibrillar systems in protozoa. In: Chen TT (ed) Research in protozoology, vol 3. Pergamon, New York, pp 278–388

Preparata RM, Meyer EB, Preparata FP, Simon EM, Vossbrinck CR, Nanney DL (1989) Ciliate evolution: the ribosomal phylogenies of the tetrahymenine ciliates. J Mol Evol 28: 427–441

Puytorac P de, Grain J (1976) Ultrastructure du cortex buccal et évolution chez les ciliés. Protistologica 12: 49–67

– Njiné T (1980) À propos des ultrastructures corticale et buccale du cilié hypostome *Nassula tumida* Maskell, 1887. Prostistologica 16: 315–327

– Batisse A, Bohatier J, Corliss JO, Deroux G, Didier P, Dragesco J, Fryd-Versavel G, Grain J, Grolière C-A, Hovasse R, Iftode F, Laval M, Roque M, Savoie A, Tuffrau M (1974) Proposition d'une classification du phylum Ciliophora Doflein, 1901. C R Acad Sci Paris 278: 2799–2802

– Grain J, Rodrigues de Santa Rosa M (1976) À propos de l'ultrastructure corticale d'un cilié hypotriche *Stylonychia mytilus* Ehrbg. 1838: les caractéristiques du cortex buccal adoral et paroral des Polyhymenophora Jankowski, 1967. Trans Amer Microsc Soc 95: 327–345

– – Grolière C-A, Detcheva R (1980) À propos de l'ultrastructure d'un cilié Sphenophryidae: conservatisme ultrastructural du cortex chez les infusoires et rameau des phyllopharyngiens. C R Acad Sci Paris 290: 893–896

– – Legendre P, Devaux J (1984) Essai d'application de l'analyse phénétique à la classification du phylum des Ciliophora. J Protozool 31: 496–507

– Didier P, Detcheva R, Foissner W (1983) Sur l'ultrastructure du cilié Colpodida *Pseudoplatyophrya nana* (Kahl, 1926). Protistologica 19: 423–434

– Grolière JC, Detcheva R, Grain J (1985) Observations sur l'ultrastructure du cilié Vestibulifera, *Plagiopyla nasuta* Stein, 1860. Ann Sci Nat Zool Paris Ser 13, 7: 189–198

Puytorac P de, Grain J, Mignot J-P (1987) Précis de protistologie. Société Nouvelle des Éditions Boubée, Paris

Raikov IB, Kovaleva VG (1980) Fine structure of the circumoral ciliature and pharyngeal apparatus of the marine ciliate *Helicoprorodon gigas* (Gymnostomata). Protistologica 16: 177–198

Rosati G, Verni F, Bracchi P, Dini F (1987) An ultrastructural analysis of the ciliated protozoon *Aspidisca* sp. Trans Amer Microsc Soc 106: 31–52

Ruffolo JJ Jr (1976) Fine structure of the dorsal bristle complex and pellicle of *Euplotes*. J Morphol 148: 469–488

Small EB (1984) An essay on the evolution of ciliophoran oral cytoarchitecture based on descent from within a karyorelictean ancestry. Orig Life 13: 217–228

– Lynn DH (1981) A new macrosystem for the phylum Ciliophora Doflein, 1901. BioSystems 14: 387–401

– – (1985) Phylum Ciliophora. In: Lee JJ, Hutner SH, Bovee EC (eds) An illustrated guide to the protozoa. Allen Press, Lawrence, KS, pp 393–575 (Society of Protozoologists Special Publication)

Torres A, Martin J, Calvo P, Fedriani C, Rios RM (1986) Fine structure of *Laurentiella acuminata* (Ciliophora, Hypotrichida). Arch Protistenk 131: 225–237

Tuffrau M (1987) Proposition d'une classification nouvelle de l'ordre Hypotrichida (Protozoa, Ciliophora), fondée sur quelques données récentes. Ann Sci Nat Zool Paris 8: 111–117

Wicklow BJ (1981) Evolution within the order Hypotrichida (Ciliophora, Protozoa): ultrastructure and morphogenesis of *Thigmokeronopsis jahodai* (n. gen., n. sp.); phylogeny in the Urostylina (Jankowski, 1979). Protistologica 17: 331–351

– (1982) The Discocephalina (n. subord.): ultrastructure, morphogenesis and evolutionary implications of a group of endemic marine interstitial hypotrichs (Ciliophora, Protozoa). Protistologica 18: 299–330

– (1983) Ultrastructure and cortical morphogenesis in the euplotine hypotrich *Certesia quadrinucleata* Fabre-Domergue, 1885 (Ciliophora, Protozoa). J Protozool 30: 256–266

Williams DB, Williams BD, Hogan BK (1981) Ultrastructure of the somatic cortex of the gymnostome ciliate *Spathidium spathula* (O.F.M.). J Protozool 28: 90–99

Wirnsberger E, Foissner W, Adam H (1985) Morphogenesis, fine structure, and phylogenetic relationships of the "heterotrich" ciliate *Bryometopus atypicus* (Protozoa, Colpodea). Ann Sci Nat Zool Paris 7: 113–128

Wirnsberger-Aescht E, Foissner W, Foissner I (1989) Morphogenesis and ultrastructure of the soil ciliate *Engelmanniella mobilis* (Ciliophora, Hypotrichida). Eur J Protistol 24: 354–368

Protoplasma (1991) 164: 143–159

PROTOPLASMA

The cytoskeleton of chromophyte algae

R. A. Andersen*

Bigelow Laboratory for Ocean Sciences, West Boothbay Harbor, Maine

Received January 7, 1991
Accepted February 6, 1991

Summary. The cytoskeleton of flagellate chromophyte algae, zoospores and gametes is active during swimming, phototaxis, several types of phagotrophic feeding, the formation, secretion and deployment of silica-scales, and the abrupt movement of spine-scales. The flagellar basal bodies are anchored by microtubular roots and/or fibrous roots. The kinds, numbers, and paths of these roots are characteristic of different taxonomic groups within the chromophytes. There are more differences in flagellar apparatuses for taxonomic classes dominated by flagellates as compared to classes dominated by coccoid, filamentous, or parenchymatous forms. Swimming cells that exhibit phototaxis often contain an autofluorescent substance that is located at the base of one flagellum. Phagotrophy occurs in flagellates of several distantly related taxonomic classes, suggesting that phagotrophy evolved independently several times. The most complex phagotrophic process occurs in the Chrysophyceae where one microtubule of a flagellar root forms a feeding basket or pouch into which food particles are moved. The silica-scales of the Synurophyceae are formed, secreted and finally moved into position outside the cell by cytoskeletal components. The six spine-scales of *Apedinella* (Pedinellophyceae) lie outside the plasma membrane, but they are attached by microligaments and are repositioned almost instantly by a cytoskeletal complex of actin, centrin, and microtubules. A phylogenetic classification based upon a cladistic analysis suggests that aquatic fungi are natural members of the chromophyte group.

Keywords: Chrysophyceae; Flagellar apparatus; Microtubular root; Phaeophyceae; Phagotrophy; Photoreceptor; Phototaxis; Prymnesiophyceae; Rhizoplast; Scale; Striated root; Synurophyceae.

Introduction

Taxonomic groups

The chromophyte algae include the following classes: Bacillariophyceae, Chrysophyceae, Dictyochophyceae, Eustigmatophyceae, Pedinellophyceae, Phaeophyceae, Prymnesiophyceae, Raphidophyceae, Synurophyceae, and Xanthophyceae. The group is characterized by: (1) mitochondria with tubular cristae (Taylor 1976), (2) photosynthetically active light-harvesting pigments that usually include one or more of the chlorophyll *c* molecules and one or more carotenoids (Bjørnland and Liaaen-Jensen 1989, Jeffrey 1989), (3) an immature flagellum that usually bears tripartite tubular hairs (Bouck 1972), and (4) a polysaccharide storage product with β-1,3 (or -1,6)-linked glucans (Hellebust 1988). The Cryptophyceae, Dinophyceae, and Euglenophyceae have been included among the chromophytes (Chadefaud 1950, Cavalier-Smith 1986). However, their inclusion with the chromophytes is not supported by ultrastructural studies of the cytoskeleton (Roberts and Roberts 1991, Triemer and Farmer 1991) and molecular studies comparing rRNA sequences (Sogin et al. 1989 b, Perasso et al. 1989, Ariztia et al. 1991).

The recent amalgamation of the chromophyte algae, certain fungi (see Barr and Desaulniers 1989) and certain protozoa (Patterson 1986, 1989) into one group is now widely accepted. However, a name for this amalgamation has not received an equal acceptance. Suggested names include (1) the Heterokontophyceae, derived from Luther's (1899) Heterokontae (van den Hoek 1978, Leedale 1980), (2) the Chromophyta (Chadefaud 1950), and (3) the Stramenopiles (Patterson 1989). I strongly support this protistan approach even though I am limited to discussing a photosynthetic subgroup within these "Stramenopiles".

Our knowledge of chromophytes, including remarks on the cytoskeleton, has been summarized several times recently, either as an entire group (Cavalier-Smith 1986, Grain et al. 1988, Green et al. 1989, Patterson

* Correspondence and reprints: Bigelow Laboratory for Ocean Sciences, Provasoli-Guillard Center for Culture of Marine Phytoplankton, McKown Point, West Boothbay Harbor, ME 04575, U.S.A.

1989, Preisig 1989) or as classes (in Cox 1980; Kristiansen and Andersen 1986; Andersen 1987, 1989 b; Moestrup and Andersen 1991; Wetherbee et al. 1989, 1991). This article will emphasize the recent advances not included in previous reviews.

Cytoskeleton

The cytoskeleton consists of several structural components that may produce cellular movements, create shapes and organize organelles. Furthermore, the cytoskeleton can play a critical role in phototaxis, phagocytosis, scale formation-secretion-deployment, mitosis and cytokinesis, and cell-to-cell or cell-to-environment interactions. The flagellar apparatus is the dominating cytoskeletal component in the flagellate stages of chromophytes, and this article will emphasize the flagellate stages. The major topics will be the microtubular roots, fibrous roots, flagellum (tip, hairs, transition region, photoreceptor), flagellar transformation, nonflagellar fibers, scale secretion and deployment, phagocytosis, and phylogenetic relationships.

Microtubular roots

General description

Microtubular roots are a component of the basal apparatus in many chromophytes, but they appear to be absent in the Bacillariophyceae (Manton and von Stosch 1966, Heath and Darley 1972), Dictyochophyceae (Moestrup and Thomsen 1990) and Pedinellophyceae (Larsen 1985, Patterson 1986). Among the chromophytes with microtubular roots, one observes similarities in structure and/or function, suggesting they are homologous (Andersen 1987, 1989 a; Preisig 1989). Recently, the microtubular roots of several genera have been described: *Chrysosphaerella, Dinobryon, Epipyxis*, and *Uroglenopsis* (Chrysophyceae) (Andersen 1990; Owen et al. 1990 a, b; Andersen and Wetherbee 1991), *Vischeria* (Eustigmatophyceae) (Santos and Leedale 1991), *Prymnesium* (Prymnesiophyceae) (Green and Hori 1990), *Heterosigma* (Raphidophyceae) (Vesk and Moestrup 1987), *Mallomonas* (Synurophyceae) (Beech and Wetherbee 1990 b), and *Heterococcus* (Xanthophyceae) (O'Kelly 1989 b). The results are summarized below by class. Diagrammatical illustrations of the flagellar apparatuses of these and other chromophytes are found in Figs. 1–3.

Chrysophyceae

There are three important new discoveries concerning the structure of microtubular roots of the Chrysophyceae, and a fourth discovery involving the function of microtubular roots during phagotrophy (see below). One discovery is that *Chrysosphaerella, Dinobryon, Epipyxis*, and *Uroglenopsis* have a very distinct bifurcation in the R 3 microtubular root (Andersen 1990; Owen et al. 1990 a, b; Andersen and Wetherbee 1991). The proximal end of the R 3 root consists of six microtubules arranged as a trough-like structure and the individual microtubules can be identified and named (tubules a–f). A short distance from its proximal end near the basal bodies, the trough-like structure splits. The microtubules curve to form a large and a small loop under the short flagellum (Fig. 1 a–d). The bifurcation of the R 3 root is critical for feeding basket-type phagocytosis found in this group of chrysophytes (see phagotrophy below). The bifurcation is apparently present in many other chrysophytes including *Ochromonas* and *Poterioochromonas* (Bouck and Brown 1973, Schnepf et al. 1977). A bifurcating R 3 root is not found in all chrysophytes, however. *Hibberdia magna* completely lacks an R 3 root, and it is not able to phagocytize bacteria (Andersen 1989 a) (Fig. 1 f).

A second discovery concerns the descending rootlet of *Uroglenopsis* (Owen et al. 1990 b), a broad band of 11–14 microtubules overlying two additional microtubules (Fig. 1 d). The descending rootlet passes by the long flagellum basal body at the point where the proximal end of the R 1 root originates in many other chrysophytes. However, the proximal end of the descending rootlet is not attached to the basal body; rather, both ends of the descending root are in the posterior region of the cell. Unless a developmental process slides the remainder of the basal apparatus along the descending rootlet, it seems unlikely that the microtubular organizing center (MTOC) for the descending rootlet is on or near the basal bodies. Thus, its classification as a flagellar root is based upon position rather than origin. The descending root is superficially reminiscent of the R 1 (= X) and Y roots of *Chrysonebula* and *Hydrurus* (Hibberd 1977, Hoffman et al. 1986) (Fig. 1 e), but the Y "root" is actually a band of cytoplasmic microtubules nucleated from the R 1 root. No cytoplasmic microtubules are nucleated from the descending root in *Uroglenopsis* (Owen et al. 1990 b). The structure, origin, and occurrence of cytoplasmic microtubules do not support a homologous relationship for the descending rootlet of *Uroglenopsis* and the R 1 root of other chrysophytes.

A third discovery is the nucleation of cytoplasmic microtubules from the surface of the R 3 microtubular root in *Epipyxis* (Andersen and Wetherbee 1991). Prior

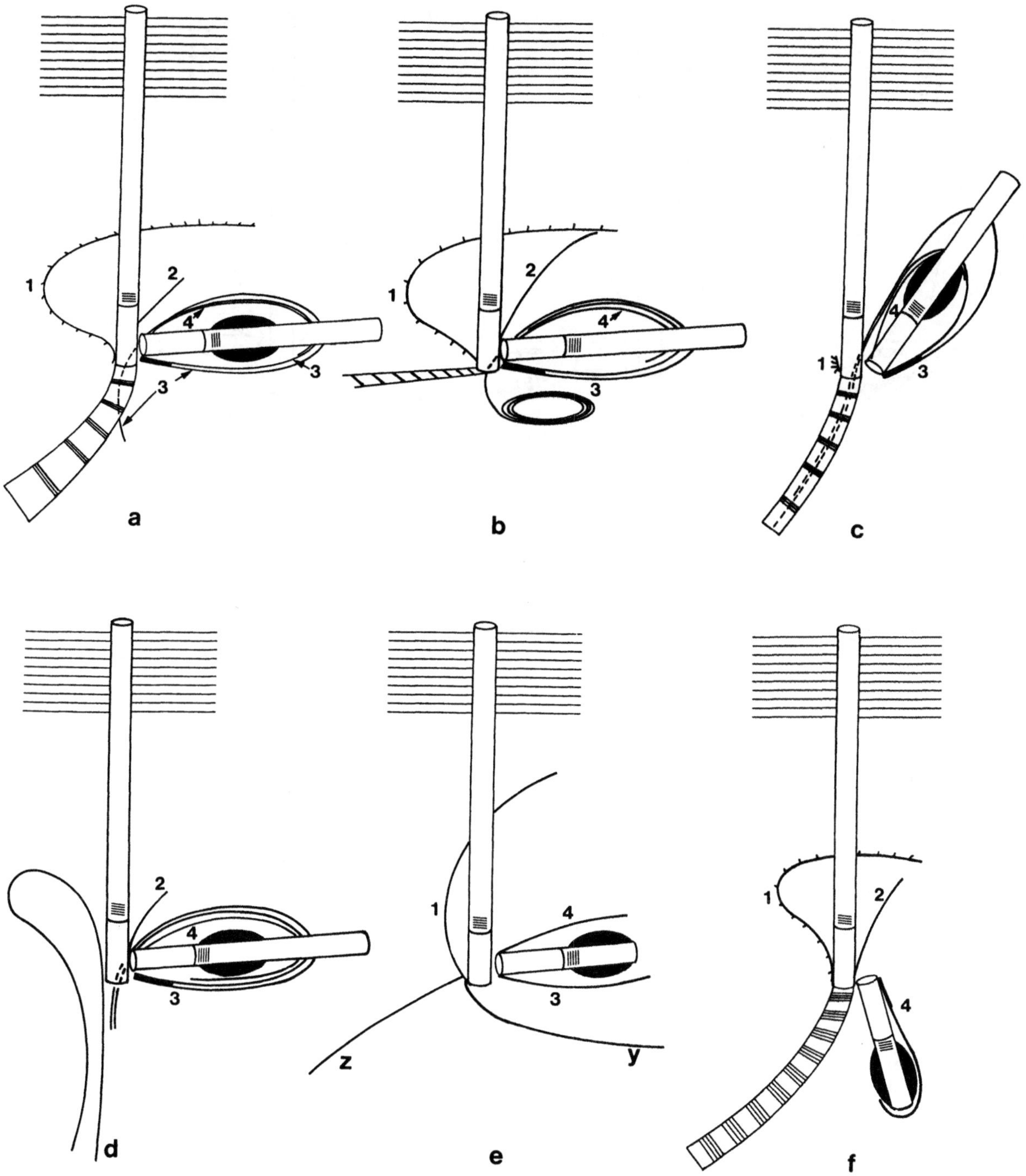

Fig. 1. Diagrams illustrating the flagellar apparatuses of the Chrysophyceae. **a** *Ochromonas*, **b** *Chrysosphaerella*, **c** *Epipyxis*, **d** *Uroglenopsis*, **e** *Hydrurus*, **f** *Hibberdia*. **a, e, f** After Andersen (1989 a), **b** after Andersen (1990), **c** after Andersen and Wetherbee (1991), **d** after Owen et al. (1990 b)

to this, only the R 1 root of typical chromophytes (i.e., those with tripartite flagellar hairs) was known to nucleate cytoplasmic microtubules. Because flagella transform from immature to mature forms (Melkonian et al. 1987, Wetherbee et al. 1988) (see below), it follows that flagellar roots transform also (Melkonian et al. 1987, Beech et al. 1991). In chromophytes, this suggests that the R 1 and R 3 roots are transformational homologues.

Questionable chrysophytes

The flagellar apparatus of *Sarcinochrysis marina* is described by O'Kelly (1989 a), although not in detail. The

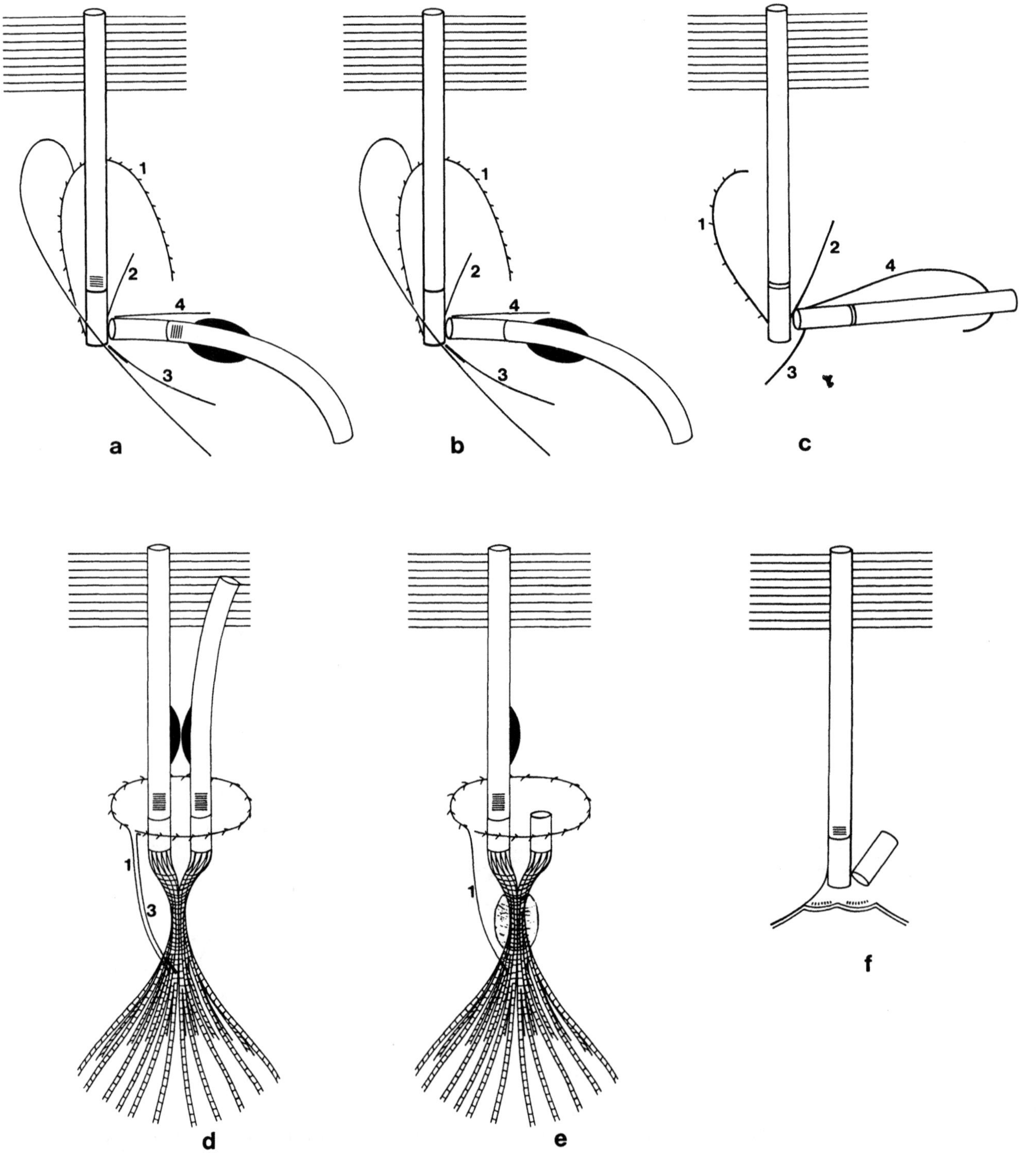

Fig. 2. Diagrams illustrating the flagellar apparatuses of the Sarcinochrysidales (Chrysophyceae), Phaeophyceae, Synurophyceae and Pedinellophyceae. **a** *Giraudyopsis*, **b** *Laminaria*, **c** *Sarcinochrysis*, **d** *Synura*, **e** *Mallomonas*, **f** *Pseudopedinella*. **a, b, d,** and **f** After Andersen (1987, 1989 a); **c** after O'Kelly (1989 a), **e** based upon Beech and Wetherbee (1990), Hibberd (1976) and unpubl. obs.

illustration shows four microtubular roots homologous with the R 1–R 4 roots typically found in chromophytes (Fig. 2 c). Of specific interest, the major posterior root (= R 3) has five microtubules arranged as trough-shaped structure near its proximal end. This root is reminiscent of the R 3 root in typical chrysophytes (Andersen 1990; Andersen and Wetherbee 1991; Owen et al. 1990 a, b), although it has five, rather than six,

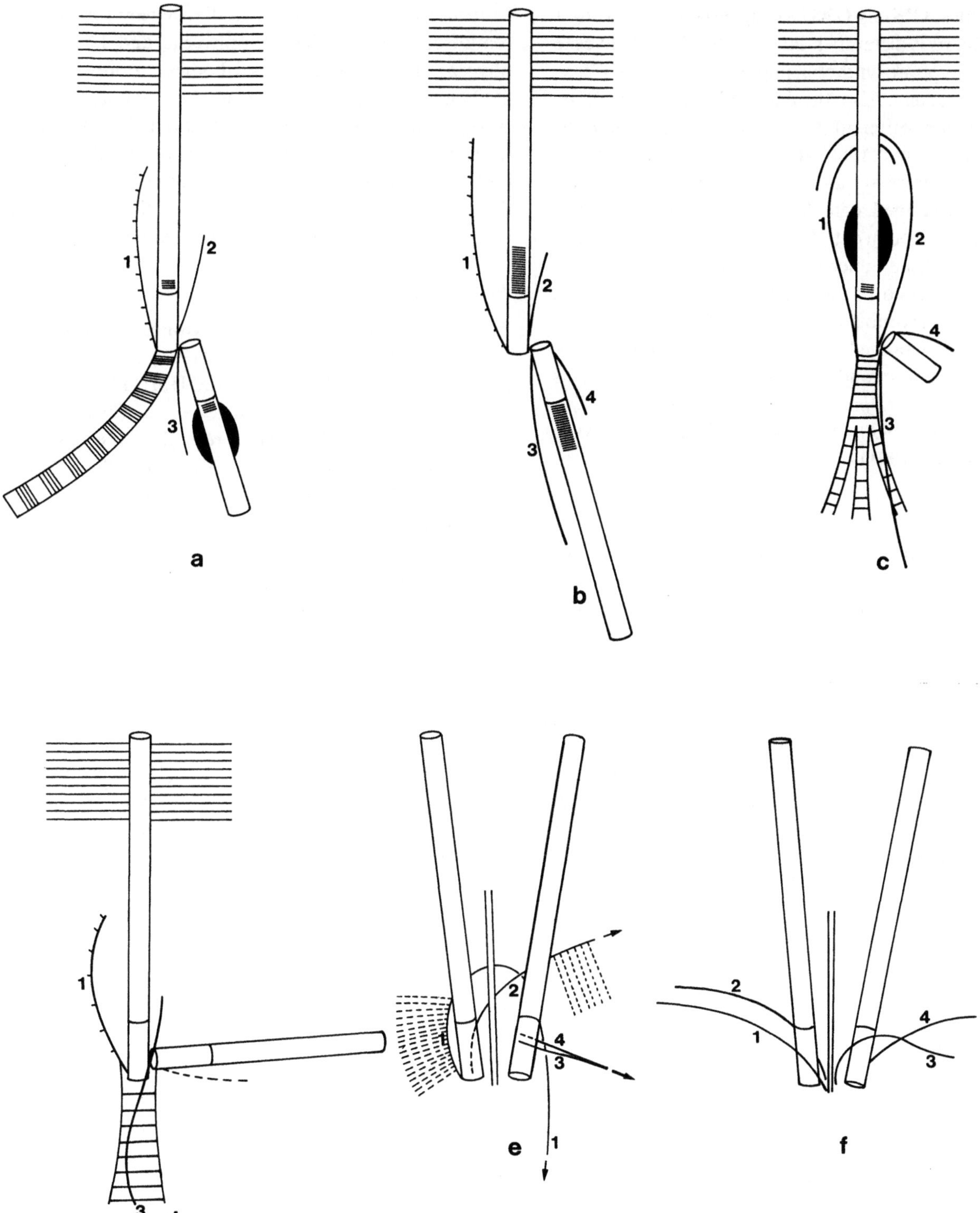

Fig. 3. Diagrams illustrating the flagellar apparatuses of the Xanthophyceae, Oomycetes, Eustigmatophyceae, Raphidophyceae and Prymnesiophyceae. **a** *Heterococcus*, **b** *Phytophthora*, **c** *Vischeria*, **d** *Heterosigma*, **e** *Pleurochrysis*. **a** After O'Kelly (1989 b), **b** and **e** after Andersen (1989 a), **c** after Santos and Leedale (1991), **d** after Preisig (1989)

microtubules. O'Kelly (1989 a) suggests, however, that *Sarcinochrysis* is distantly related to the Chrysophyceae based upon photosynthetic pigments and other taxonomic characters. Also, the microtubular roots of *Sarcinochrysis* are different from those of *Giraudyopsis* (O'Kelly and Floyd 1985) (Fig. 2 a), implying that the Sarcinochrysidales are an artifical group.

A recently described alga, *Reticulosphaera socialis*, has not been examined in detail regarding the cytoskeleton, but some published micrographs show structures similar to those found in the Chrysophyceae (Grell et al. 1990). One microtubular root (= R 3?) consists of six microtubules arranged as a trough. The root is associated with a mitochondrion, like chrysophytes such as *Chrysosphaerella, Dinobryon, Uroglenopsis*, and *Epipyxis* (Andersen 1990; Owen et al. 1990 a, b; Andersen and Wetherbee 1991).

Eustigmatophyceae

The flagellar apparatus of *Vischeria* zoospores (Eustigmatophyceae) is described (Santos and Leedale 1991). In zoospores of the Eustigmatophyceae, the flagellum bearing tripartite flagellar hairs also has the flagellar swelling that is associated with the eyespot (Hibberd and Leedale 1972, Lee and Bold 1973). Thus, as pointed out by Santos and Leedale (1991), a direct comparison to other chromophyte classes is difficult. Furthermore, the absence of cytoplasmic microtubules extending from the putative R 1 root (Santos and Leedale 1991) makes it even more difficult to establish flagellar root homologies. Yet, despite these reservations, there can be no doubt that the microtubular roots of *Vischeria* are the common chromophyte type (Fig. 3 c).

Phaeophyceae

No thorough studies describing the microtubular roots of brown algal zoospores have been published since the three original studies that characterized the class (Henry and Cole 1982 a, b; O'Kelly and Floyd 1984). However, micrographs and illustrations showing the microtubular roots in zoospores of *Hormosira banksii, Leathesia difformis, Notheia anomala*, and *Splachnidium rugosum* conform to the general model for brown algae (O'Kelly 1989 a) (Fig. 2 b). Apparently, the major evolutionary changes in the Phaeophyceae are associated with the thallus rather than the cytoskeleton of the zoospore (see phylogenetic comments below).

Also of interest is the recent discovery of flagellate eggs from *Laminaria angustata* (Motomura and Sakai 1988, Motomura 1990). The basal bodies lie near the nucleus, reminiscent of the Pedinellophyceae and Dictyochophyceae. However, the basal bodies are most distant at their proximal ends, and nearly touch at the transition region (Motomura and Sakai 1988). Thus, the basal bodies lie at an angle of approximately a *minus* 40°! Microtubular roots are not described, but root-like structures appear to be present in the micrographs (Figs. 8 and 9 of Motomura and Sakai 1988).

Prymnesiophyceae

The microtubular roots of the Prymnesiophyceae were thoroughly reviewed by Preisig (1989) and are not repeated here. Since then, Green and Hori (1990) describe the basal apparatus for *Prymnesium patellifera* (Prymnesiophyceae), finding that it resembles the basal apparatus of coccolithophores more than it does that of *Imantonia, Isochrysis*, or *Chrysochromulina*. As with other Prymnesiophytes, it is difficult to establish basal body homologies, and hence microtubular root homologies, because tripartite flagellar hairs are absent. We know the longer flagellum of *Prymnesium* contains a green autofluorescing substance (Kawai 1988), as does the shorter flagellum in other chromophytes (Coleman 1987, 1988; Kawai 1988). However, in *Prymnesium* it is not known whether the "left" or the "right" basal body produces the long flagellum (Green and Hori 1990). Beech and Wetherbee (1988) examined a thin-embedded cell of *Pleurochrysis* and established that for *Pleurochrysis* the left flagellum was the longer flagellum. Unfortunately, neither the long nor short flagellum of *Pleurochrysis* fluoresces (Kawai and Inouye 1989), and flagellar length is a questionable character for organisms in different orders.

Raphidophyceae

Vesk and Moestrup (1987) find that *Heterosigma akashiwo* has a well developed R 1 root (Fig. 3 d) and a layered-structure (= R 3 root?) similar to *Chattonella subsalsa* and *Gonyostomum semen* (Mignot 1967, 1976). An examination of the genuine *Olisthodiscus luteus* (not *Heterosigma akashiwo* that appears in most of the older literature as *Olisthodiscus*) suggests new variations in the flagellar apparatus of the Raphidophyceae (Inouye pers. comm.). *Olisthodiscus* has four microtubular roots, apparently homologous with the R 1–R 4 roots of other chromophytes; the R 1 consists of a broad band of microtubules and an associated layered-structure, and it nucleates cytoplasmic microtubules (Inouye pers. comm.). Thus, the Raphidophyceae may, like the

Chrysophyceae, show considerable variation of the microtubular roots resulting from evolutionary pressures on its vegetative state, the flagellate cell.

Synurophyceae

The microtubular roots of two species of *Mallomonas* and one species of *Synura* are described recently. *Mallomonas splendens* is the most thoroughly described of the three. It has a single three-membered microtubular root, the R 1 root, and it occurs in at least two absolute orientations (Beech and Wetherbee 1990 b) (Fig. 2 e). A three-membered R 1 root is also found in *Mallomonas tonsurata* var. *etortisetifera*, but the number of absolute orientations is not reported (Zhang et al. 1990 b). A second microtubular root consisting of two microtubules is reported for *M. tonsurata* var. *etortisetifera*, but its origin and path are not described. In the last study, one of the four absolute orientations of *Synura petersenii* is described, and both the three-membered R 1 root and the four-membered R 3 root are present (Andersen 1989 b) (Fig. 2 d).

Xanthophyceae

The flagellar apparatus of zoospores of *Heterococcus tectiformis* (Xanthophyceae) is described by O'Kelly (1989 b), the first thorough examination of a flagellar apparatus in the class. The zoospore has three microtubular roots: a major anterior root (= R 1) that nucleates cytoplasmic microtubules, a minor anterior root (= R 2) and a major posterior root (= R 3) (Fig. 3 a). Overall, the structure is similar to that of *Ophiocytium* (Hibberd 1980).

Bacillariophyceae, Dictyochophyceae, Pedinellophyceae

The striking feature of these three classes is the apparent lack of microtubular roots (Fig. 2 f). Microtubules do extend from the basal apparatus of sperm cells of *Biddulphia* (Bacillariophyceae) (Heath and Darley 1972). However, they are not organized like typical flagellar roots, and they resemble remnant spindle microtubules, perhaps remaining from the recent nuclear division. *Dictyocha speculum* (Dictyochophyceae) (Moestrup and Thomsen 1990) has a basal apparatus very similar to the Pedinellophyceae (Larsen 1985, Patterson 1986, Koutoulis et al. 1988). This and other similarities suggests the Dictyochophyceae and Pedinellophyceae should be aligned.

Fibrous roots

As in the past, fibrous roots of chromophytes continue to receive less attention than microtubular roots. None the less, one recent discovery is very significant. Melkonian et al. (1991) report that centrin, a Ca^{++} modulated contractile protein (Salisbury et al. 1984; Salisbury 1989 a, b), is present in the fibrous root (= rhizoplast, = system II fiber) of *Ochromonas* (Chrysophyceae) and *Synura* (Synurophyceae) as well as the basal body connecting bands of *Ectocarpus* zoospores. In addition, antibodies against centrin have reacted with material surrounding the basal bodies in *Apedinella radians* (Koutoulis et al. 1988), but the electron micrographs show a ring of fibrillar material, not striated roots. The noncontractile system I fiber of green algae (Melkonian et al. 1988, Lechtreck and Melkonian 1991) has as yet not been identified in chromophytes.

Although molecular studies of chromophytes are slow to develop, morphological discoveries continue to occur. The fibrous roots of chrysophytes and synurophytes have a typical cross-striated band, but a granular region is reported in several recent studies on *Synura* (Andersen 1989 b), *Mallomonas* (Beech and Wetherbee 1990 b), *Dinobryon* and *Uroglenopsis* (Owen et al. 1990 a, b). Granular material was previously described near the fibrous root of *Mallomonas caudata* (Synurophyceae) (Wujek 1978). However, it was not considered part of the fibrous root, and it may not be synonymous with the granular region mentioned here. The function or significance of the granular region is unknown.

A fibrous root was found in *Chrysosphaerella* (Chrysophyceae) using thick section stereo pairs even though it was not evident in thin sections (Andersen 1990). This suggests a delicate fibrous root may be found in some other chrysophytes where it is reported to be lacking. Of additional interest, the fibrous root of *Chrysosphaerella* is directed towards the plasma membrane and it attaches to a mitochondrion, not to the nucleus (Andersen 1990). The basal bodies are immediately above the nucleus and the distal end of the R 3 root coils over the nuclear surface (Fig. 1 b). Consequently, in *Chrysosphaerella*, the fibrous root cannot be considered a nucleus-basal body connector (sensu Salisbury et al. 1987).

Flagellum

Flagellar tip

The flagellar tip of green algae has been thoroughly studied (Moestrup 1982, Melkonian 1984), but little is known about the tips of chromophyte flagella. Recently, the short (smooth) flagellum of *Mallomonas guttata* (Synurophyceae) was found to split into several pseudopod-like fingers at its tip (Zhang et al. 1990 a). A similar feature is found in *M. papillosa* where each pseudopod apparently contains one or more of the outer doublet microtubules (P. L. Beech, pers. comm.). The function of the flagellar fingers is unknown.

Tripartite tubular hairs

The basic structure of tripartite flagellar hairs is thoroughly described (Bouck 1971, 1972) and the topic is reviewed recently (Leadbeater 1989). One new observation is that tripartite tubular hairs are absent on the flagella of flagellate eggs in *Laminaria angustata* (Phaeophyceae) (Motomura and Sakai 1988).
The presence of short and long lateral filaments has received recent attention, but still needs further study. Short and long lateral filaments are found on the tubular shafts of chrysophyte hairs, as well as on hairs of *Hemiselmis* (Cryptophyceae) (Bouck 1972). They are absent on flagellar hairs of zoospores in the Phaeophyceae and Sarcinochrysidales (Chrysophyceae) (Bouck 1972, Gayral and Morange 1970). Short and long lateral filaments are present on the tripartite hairs of the Synurophyceae (Andersen 1989 b) but are absent on the tripartite hairs of *Heterococcus* (Xanthophyceae) (O'Kelly 1989 b) and *Phytophthora* (Oomycetes) (unpubl. obs.). Several other classes remain to be examined carefully. A negative stain preparation is the best means for identifying the filaments but it is not widely used; shadowcast preparations are more popular, but they can obscure the filaments.

Transition region

The transitional helix is a widespread and consistent structure in many chromophyte and fungal groups (Hibberd 1979, Preisig 1989). One new observation is that the transitional helix of *Hibberdia magna*, an unusually pigmented chrysophyte (Alberte and Andersen 1986), has only four gyres (Andersen 1989 a). Also, the occurrence of a transitional helix is not consistent in *Mallomonas* (Synurophyceae). A helix is present in *Mallomonas oviformis* (= *Mallomonopsis elliptica* var. *oviformis*) (Hibberd 1976), *M. annulata, M. cratis, M. lychenensis*, and *M. papillosa* (Andersen unpubl. obs.), but it is absent in *M. splendens* (Beech and Wetherbee 1990 b), *M. tonsurata* var. *etortisetifera* (Zhang et al. 1990 b) and *M.* cf. *asmundae* (Andersen unpubl. obs.). Finally, it is worth noting that the newly described *Reticulosphaera socialis* (Grell 1989) has a transitional helix consisting of five gyres (Grell et al. 1990). Five gyres is the typical transitional helix structure for the Chrysophyceae, and together with the possible similarity of microtubular roots (see above), suggests this alga may be a chrysophyte.
An unusual helical structure consisting of at least seven gyres occurs in the transitional region of *Pleurochrysis carterae* (Beech and Wetherbee 1988). However, the spacing between gyres is much greater than the typical chromophyte transitional helix. *Pleurochrysis* also has rings (or a helix) below the transitional region (Inouye and Pienaar 1985, Beech and Wetherbee 1988).

Phototactic photoreceptor

Most flagellate cells of chromophyte algae are presumed to be phototactic. The flagellum swelling, the stacked membranes and/or the eyespot have been proposed as the photoreceptor. Foster and Smyth (1980) suggest that the typical chromophyte photoreceptor is a variation of the quarter-wave stack model. Some recent evidence may challenge this hypothesis.
An autofluorescent green flagellum is reported for many chromophytes (Coleman 1987, 1988; Müller et al. 1987; Kawai 1988; Kawai and Inouye 1989). These results are summarized in Table 1. Apparently, the phenomenon is limited to chromophytes, i.e., it is not found in the examined species of the Chlorophyceae, Prasinophyceae, Cryptophyceae, Dinophyceae, or Euglenophyceae (Coleman 1988, Kawai and Inouye 1989).
The absence of flagellar green autofluorescence in chromophytes is usually correlated with the absence of an eyespot, and sometimes the flagellar swelling (Müller et al. 1987, Kawai and Inouye 1989). The major exceptions are the Synurophyceae which lack an eyespot but possess the autofluorescence and the Eustigmatophyceae which possess an eyespot but lack the autofluorescence. Admittedly, the eyespot of the Eustigmatophyceae is different from that of many other chromophytes (Hibberd and Leedale 1972). Presumably, the eustigmatophyte zoospores, raphidophytes, coccolithophorid swimming cells, Pavlovales and pedinellophytes either are not phototactic or use a photoreceptor that does not require the green autofluorescent substance.
The compound responsible for the green autofluorescence is not known with certainty, but its spectral properties are similar to flavins (Müller et al. 1987, Kawai 1988). The peak emission (excited at 440, 450, 460, 480 nm) is approximately 530 nm in *Giffordia mitchellae*

Table 1. Summary of the occurrence of green autofluorescent flagella in chromophyte algae

Taxon	Autofluorescence	Source
Bacillariophyceae	–	Kawai and Inouye 1989
Chrysophyceae	+ [a]	Coleman 1987, 1988; Müller et al. 1987; Kawai 1988; Kawai and Inouye 1989
Dictyochophyceae	?	
Eustigmatophyceae	–	Kawai and Inouye 1989
Pedinellophyceae	–	Kawai and Inouye 1989
Phaeophyceae	+ [b]	Coleman 1987, 1988; Müller et al. 1987; Kawai 1988; Kawai and Inouye 1989
Prymnesiophyceae	+/– [c]	Coleman 1987, Coleman 1988, Kawai and Inoue 1989
Raphidophyceae	–	Coleman 1988, Kawai and Inouye 1989
Synurophyceae	+	Coleman 1988, Kawai 1988, Kawai and Inouye 1989
Xanthophyceae	+/– [d]	Kawai and Inouye 1989

[a] Absent in one species of *Ochromonas* (Kawai and Inouye 1989)
[b] Absent in species lacking an eyespot (Müller 1987, Kawai 1988, Kawai and Inouye 1989)
[c] Present in the Isochrysidales and Prymnesiales, absent in Coccosphaerales and Pavlovales (Kawai and Inouye 1989)
[d] Present in Mischococcales, absent in Vaucheriales (Kawai and Inouye 1989)

(Phaeophyceae) (Müller et al. 1987). Similarly, the peak emission (excited at 440 nm) is between 515 and 520 nm for *Synura petersenii* (Synurophyceae) and *Scytosiphon lomentaria* (Phaeophyceae) (Kawai 1988). Fluorescence excitation spectra (emission at 530 nm) show a broad maximum between 370 and 400 nm with a second peak (450 nm) and shoulder (470 nm) in the visible range for *Laminaria, Giffordia*, and *Sphacelaria* (Phaeophyceae) (Müller et al. 1987). In a clever experiment, Müller et al. (1987) show that living gametes of *Ectocarpus siliculosus* (Phaeophyceae) best respond phototactically to light between approximately 440 and 460 nm.

The autofluorescent material is concentrated near the base of the smooth flagellum where the swelling occurs, but is not limited to the swelling (Coleman 1988, Kawai 1988). Coleman (1988) states that after prolonged observation the other (hairy) flagellum may exhibit a faint fluorescence. Kawai and Inouye (1989) report that both flagella are autofluorescent in *Botrydiopsis intercedens* (Xanthophyceae).

A second new discovery for chromophyte phototactic photoreceptors is the reflective property of the eyespot in *Ectocarpus* (Kriemer et al. 1991). Using confocal laser scanning microscopy to produce a profile, the authors show that light is reflected by the eyespot and is focussed exactly on the flagellar swelling. The curved, dish-like shape of the eyespot provides the means for focussing the light. The reflective property also fulfills the barrier role assigned by Foster and Smyth (1980). When the eyespot is illuminated by light that first passes through the cell, the light is reflected away from the flagellar swelling (Kriemer et al. 1991).

Flagellar transformation

The recently discovered phenomenon that flagella and/or basal bodies require more than one cell cycle to attain maturity (Melkonian et al. 1987) is now known to occur in many if not all protists. *Pleurochrysis carterae* (Prymnesiophyceae), *Epipyxis pulchra* (Chrysophyceae), and *Mallomonas splendens* (Synurophyceae) are three chromophytes that played a role in the elucidation of the various patterns of flagellar transformation (Beech et al. 1988, Wetherbee et al. 1988, Beech and Wetherbee 1990 a). The "typical" chromophyte flagellates, i.e., with tripartite tubular hairs on the immature flagellum and a photoreceptor-related swelling on the mature flagellum, may portray a consistent pattern of transformation. It is likely that the flagellar developmental sequence can be inferred using characters such as tripartite hairs and the flagellar swelling, and perhaps also the rhizoplast as suggested by Beech et al. (1991). The Prymnesiophyceae are not so easily explained by a general chromophyte model because they lack tripartite hairs, a rhizoplast and (Pavlovales excepted) a photoreceptor-related flagellar swelling.

Flagellar transformation per se does not occur during the formation of zoospores for chromophytes such as the diatoms, brown seaweeds or xanthophytes. However, since one basal body should be older than the other, it will be interesting to see which basal body gives rise to the hairy flagellum.

The point in the cell cycle at which new basal bodies first form is variable for the chromophytes (Beech et al. 1991). *Dinobryon cylindricum* and *D. divergens* have two dormant basal bodies, perhaps first forming even before cytokinesis is complete (Owen et al. 1990 a). This is unusual (unique?) among chromophytes where often only flagellum-bearing basal bodies are observed. Several authors report new, nonflagellate basal bodies in some vegetative cells but not others (Slankis and Gibbs 1972; Schnepf et al. 1977; Andersen 1985, 1990), suggesting that next generation basal bodies are formed during the late stages of interphase.
Finally, *Pelagococcus subviridis* (? Chrysophyceae) may become an important experimental organism for the study of basal body development in chromophytes. Although it has no known flagellate stage, it produces two new centrioles de novo just prior to nuclear division, and the centrioles quickly disassemble as an extranuclear spindle forms (Vesk and Jeffrey 1987). The succinct development of the centrioles (and material associated with them, such as the MTOCs for the mitotic spindle) may provide a good model system for insights into the genetic regulation of basal body, flagellum and microtubular root development in flagellate chromophytes.

Phagotrophy

Phagotrophy was first discovered over a century ago for species of Chrysophyceae (James-Clark 1867, Cienkowsky 1870, Saville-Kent 1880) and Pedinellophyceae (Wyssotzki 1887). In *Epipyxis*, the flagella create water currents that draw the food particles toward the cell, and the flagella capture, hold and "taste" food particles prior to phagocytosis (Wetherbee and Andersen 1991). Chromophytes use four distinct methods for capturing food particles. The most widespread method is the feeding basket (Andersen and Wetherbee 1991, Wetherbee and Andersen 1991, Moestrup and Andersen 1991). Other methods are the use of tentacles by pedinellophytes, rhizopodia by chrysophytes (Moestrup and Andersen 1991) and haptonemata by prymnesiophytes (Kawachi and Inouye pers. comm.).
The cytoskeleton is directly involved in phagocytosis, but this is only now being elaborated in full detail. The most complete study is on *Epipyxis pulchra* (Chrysophyceae) where light and electron microscopy are used to describe the feeding process (Andersen and Wetherbee 1991, Wetherbee and Andersen 1991). First, the cell undergoes a complete morphogenesis changing its shape to form an elongate cell with a contractile posterior stalk. The flagellar apparatus moves about 4 µm from its nonfeeding position near the contractile vacuole, Golgi body, and nucleus. A complex array of root microtubules and cytoplasmic microtubules, a rhizoplast and a mitochondrion serve to maintain the cell in the feeding position. A feeding basket forms around a large microtubular loop and provides a quick means of engulfment of food particles. The microtubular loop forms by the upwards sliding of the f tubule of the R 3 microtubular root once a food particle is captured.
Food particle capture by flagella is also demonstrated for *Ochromonas danica* (Chrysophyceae) (Wetherbee and Andersen unpubl. obs.), and presumably the flagella capture food in most other Chrysophyceae (Moestrup and Andersen 1991). The dynamic function of the R 3 microtubular root during phagocytosis is known for *Chrysophaerella* also (Andersen 1990). Rather than forming a large feeding basket, however, it forms a "pocket" or "pouch" that is not as obvious as the *Epipyxis* feeding basket. It is likely that a similar "feeding pouch" is used during phagocytosis in *Ochromonas* and many other chrysophytes.

Nonflagellar fibers

Centrin, which is becoming a well known component of basal apparatus fibrous fibers (see above), is also active in spine-scale movement in *Apedinella radians* (Pedinellophyceae) (Koutoulis et al. 1988). A complex, star-shaped apparatus with interlinking bands of centrin and actin is attached to special plaques at the plasma membrane and to the extracellular spine-scales. The nearly instantaneous reorientation of the spine-scales occurs as centrin and actin contract. Along with the flagellum and haptonema, this is one of the most complex, moving-appendage systems in the chromophyte algae. It is not known if the *Apedinella* apparatus evolved from a homologous system found in other pedinellids which lack spine-scales. A simpler network, not yet characterized to identify the fibrillar components, exists in *Actinomonas pusilla* and *Pteridomonas danica* (Pedinellophyceae) (Larsen 1985). This may suggest a somewhat wider occurrence.
The microligament, a striated band found *outside* the plasma membrane, is another unusual nonflagellar fiber found in *Apedinella* (Koutoulis et al. 1988). Although the microligament is not known to contract, it is still remarkable that such a striated band would be found outside the plasma membrane. This represents an excellent example of the evolutionary experimentation that exists among the protists.

Scale formation, secretion, and deployment

The scales and bristles of the Synurophyceae do not reorient like the spine-scales of *Apedinella*, but they are produced, translocated, and finally assembled extracellularly by a complex and dynamic cytoskeletal system. The role of the cytoskeleton (microtubules and actin) during synurophyte scale formation is well established (Schnepf and Deichgräber 1969; Wujek and Kristiansen 1978; Mignot and Brugerolle 1982; Brugerolle and Bricheux 1984; Andersen 1985, 1987). The model can be generalized as follows. The silica deposition vesicle forms along the outer surface of one chloroplast as vesicles are transported anteriorly from the Golgi body along the proximal end of the R 1 microtubular root and then posteriorly along a special band of cytoplasmic microtubules nucleated from the loop region of the R1 root. The band of cytoplasmic microtubules forms an anticlockwise spiral down and along the chloroplast. The microtubular band, the chloroplast ER and actin filaments mold the silica deposition vesicle into the shape of a scale. Silica is added and the scale forms.

The formation of bristles occurs separately from the formation of scales (Wujek and Kristiansen 1978), and this raises questions as to how bristles are secreted and how bristles are attached to scales once they are outside the plasma membrane. In addition to the deployment problem, bristles form with the tip directed anteriorly. Thus, when a bristle is secreted posteriorly, it is not only spatially and temporally disjunct from the scale, it is also released with the attachment base being pushed away from the cell. This is so unbelievable that Siver and Glew (1990) reinterpret the orientations of Wujek and Kristiansen (1978) suggesting that the bristles are formed and secreted in the opposite orientation. However, in an elegant study of *Mallomonas splendens*, the unbelievable is shown to happen (Wetherbee et al. 1989, Beech et al. 1990). A long cellular protuberance accompanies the bristles as they are extruded, reoriented 180° outside the cell and pulled back to the respective baseplate scales. This protuberance, which is visible with image enhanced light microscopy, appears as a thread-like, cytoplasmic extension containing one or more microtubules when viewed using an electron microscope (Beech et al. 1990).

Siliceous scales cover all synurophytes, either encasing individual cells (e.g., *Mallomonas, Synura, Chrysodidymus*) or surrounding the entire colony (*Synura lapponica, Tesselaria volvocina*). Most attention has focused on the scale cases that surround individual cells. Two recent papers provide new theories on scale case formation (Leadbeater 1990, Siver and Glew 1990). Leadbeater (1990) suggests *Synura* scales are added in rows, and he provides micrographs with scales arranged in a row just beneath the plasma membrane. In this model the scale case separates and spreads along a seam, and the new row of scales is secreted into the seam. The mechanism for ungluing/regluing the scales is not known, but enzymes can digest the glues (Leadbeater 1986). Recent evidence suggest different glues are used for gluing: (1) scales to scales, (2) bristles to scales and (3) scales to plasma membrane (Ludwig and Wetherbee 1990).

Siver and Glew (1990) put forth a hypothesis for scale case formation in *Mallomonas*. The model proposes a single site for scale and bristle release, and a highly regimented movement of scales over the cell surface to produce the scale case. The hypothesis is designed in part for naked cells that are excysting. No data (light or electron micrographs) are provided to support the hypothesis, and it is difficult to apply this model of scale deployment to normal vegetative cells.

By combining the observations and ideas from the various studies, a model for scale case assembly is beginning to emerge. Scales can and do change position outside the cell. This is most clearly demonstrated during cytokinesis in *Mallomonas splendens* (Beech et al. 1990). The scales separate along the cleavage furrow and then reassemble into the two scale cases of the daughter cells. The scales are slightly more overlapped in the mother cell, and then spread slightly to cover the increased surface area of the daughter cell (see discussion in Siver and Glew 1990). And, obviously, the scale number must double during each cell cycle. In *Mallomonas splendens* base plate scales and posterior bristles are secreted before cytokinesis and are distributed equally to the daughter cells. The anterior bristles are added after cytokinesis. The large scales of synurophytes are proving to be excellent markers for studying extracellular structures manipulated by intracellular cytoskeletal components (Wetherbee et al. 1991). Several other observations are reported recently. The role of mitochondria during scale morphogenesis is shown for *Synura glabra* and *S. petersenii* (Andersen 1989 b). Silica scales also occur in the Chrysophyceae, and a recent study of *Chrysosphaerella* suggests a mitochondrion may be involved in its scale morphogenesis (Andersen 1990). On a different level, the scales of marine phytoplankton (Parmales, ? Chrysophyceae) are discussed in relationship to the evolution of the Bacillariophyceae (Mann and Marchant 1989), and more specifically, bristles of *Corethron* (Bacillariophyceae) are compared to bristles of *Mallomonas* (Crawford and Round 1989).

Phylogenetic relationships

The cytoskeleton has played an important role in the study of evolutionary relationships among chromophyte algae. Flagellar hairs were used much earlier to show evolutionary relatedness of protists (Vlk 1931), and they continue to be an important feature in chromophyte phylogenetic discussions (Moestrup 1982, Leadbeater 1989). Short and long lateral filaments on the tubular shaft is a character shared by the Chrysophyceae, Pedinellophyceae, and Synurophyceae – three classes previously united as one (Bourrelly 1981). The absence of tripartite hairs on the flagella of the Prymnesiophyceae leads some workers to exclude or isolate the class from the other chromophytes (Cavalier-Smith 1986).

Microtubular root comparisons of chromophytes for phylogenetic purposes are more recent (Andersen 1987, 1989 a, b; Preisig 1989). The absence of microtubular roots in the Bacillariophyceae, Dictyochophyceae, and Pedinellophyceae is as troubling as the absence of flagellar hairs in the Prymnesiophyceae. The loss of microtubular roots is almost certainly a secondary loss because these roots are common to most protists, including those like *Giardia* which diverged very early from the lineage giving rise to other eukaryotes (Sogin et al. 1989 a, Brugerolle 1991). Of those chromophytes with microtubular roots, all (except *Uroglenopsis*?) have an R 1 root, and it typically gives rise to cytoplasmic microtubules. These cytoplasmic microtubules, together with the R 1 root, determine the shape of the cell body. Thus, evolutionary pressures on the cell shape will necessarily act on the R 1 root complex.

The trough-like structure of the R 3 microtubular root is also emerging as a shared character among several groups of chromophytes. Although its function varies widely, from a posteriorly directed rhizostyle in *Heterosigma* (Vesk and Moestrup 1987) to a bifurcating phagotrophic organelle in *Epipyxis* (Andersen and Wetherbee 1991), the structure is remarkably conserved.

The fibrous root is a common feature in chromophyte algae, but it too is troubling by its absence in certain groups. One insight may be evident following careful study of *Chrysosphaerella*. Although it is a chrysophyte (and fibrous roots are more typical of chrysophytes than any other chromophyte class), its basal bodies sit on the nucleus and its fibrous root is delicate and directed away from the nucleus (Andersen 1990). This suggests that the structure, position, and presence of the fibrous root may not be a conservative character. The transitional helix is a good taxonomic character (Hibberd 1979), but again, its absence in several groups is troubling. Furthermore, its doubled appearance in *Heterococcus* (O'Kelly 1989 b) is difficult to interpret: is it a double helix, or is it a separate, independently derived character?

It has become apparent to me during the past two years that evolutionary change in the flagellar apparatus probably occurs at a substantially different rate in flagellates as compared to zoospores and gametes. This should not seem illogical or counterintuitive; the life stage that occurs during the vast majority of the life cycle should be the stage that evolves to fill new niches while giving rise to new species. In support of this idea, we find considerable variation in the flagellar apparatus of the Prymnesiophyceae (Preisig 1989), and similarly we find considerable variation in that of the Chrysophyceae (Andersen 1989 a) (see also Fig. 1); both classes are dominated by flagellate forms. Conversely, all of the Phaeophyceae zoospores are virtually identical in flagellar structure (O'Kelly 1989 a) even though the morphology of their thalli is very diverse (Clayton 1989).

Table 2. Data matrix used in the cladistic analysis. See Table 3 for explanation of codes

Taxon	Characters											
	1	2	3	4	5	6	7	8	9	10	11	12
Amphidium	1	1	3	1	3	3	1	1	2	2	1	1
Biddulphia	2	3	3	1	3	3	2	1	1	1	1	1
Ochromonas	3	1	4	3	1	1	1	1	1	1	2	2
Chrysosphaerella	3	1	4	3	1	2	2	1	1	1	2	2
Hibberdia	3	1	1	2	1	1	1	1	1	1	2	2
Dictyocha	2	3	3	1	3	3	2	1	1	1	1	9
Vischeria	2	2	2	2	4	1	1	1	2	2	2	1
Pseudopedinella	3	3	3	1	3	3	1	1	1	9	9	1
Laminaria	2	1	2	1	1	3	1	1	1	1	2	2
Prymnesium	1	2	1	1	3	3	1	2	2	1	1	2
Pleurochrysis	1	1	1	1	3	3	1	2	2	1	1	2
Heterosigma	2	1	2	1	3	1	1	1	1	1	2	1
Synura	3	1	2	4	2	1	2	1	1	1	2	2
Mallomonas	3	1	3	4	2	1	2	1	1	1	2	2
Heterococcus	2	1	2	5	1	1	1	1	1	2	1	2
Phytothphora	2	1	2	5	3	3	1	1	3	3	3	9

9 Unknown value

Accepting the assumption that flagellar apparatuses evolve more rapidly in flagellates, we are both blessed and damned in our phylogenetic pursuits. The evolutionary changes in the cytoskeleton of flagellates are bountiful and that is a blessing; conversely, flagellar apparatuses of the zoospores or gametes have changed very little, and the flagellar apparatus is a poor measure of evolutionary change for these organisms.

The chromophyte algae were originally grouped based upon chloroplast pigmentation. Recent molecular studies support this concept (Perasso et al. 1989, Ariztia et al. 1991), and I remain convinced that the chromophyte algae have one common, pigmented ancestor. I combine pigment data with cytoskeletal features (Tables 2 and 3) in a cladistic analysis [PAUP (Swofford 1985)]. Chromophyte genera for which substantial data exist are used, along with *Phytophthora* (Oomycetes) and *Amphidinium* (Dinophyceae). Twelve phylogenetic trees are constructed in the analysis, and these are subsequently analyzed using the CONTREE program of PAUP to produce a single consensus tree (Fig. 4). Assuming the genera are representative of their respective classes, the Prymnesiophyceae form a clade that diverges deep in the tree, before the occurrence of tripartite hairs and girdle lamellae. The Bacillariophyceae, Dictyochophyceae, and Pedinellophyceae are grouped, based on the absence of microtubular roots. The analysis suggests this loss occurred only once, but I suspect the loss in diatoms is independent of the loss in the Dictyochophyceae and Pedinellophyceae. The Chrysophyceae and Synurophyceae group together, but the position of *Hibberdia* is unresolved. This is not surprising considering the unusual features of *Hibberdia* (Alberte and Andersen 1986, Andersen 1989 a). The

Table 3. Characters used to establish the data matrix (Table 2) for the cladistic analysis

Character 1	tripartite hairs	1	absent
		2	present
		3	present with filaments
Character 2	R 1 root	1	R 1 root with cytoplasmic microtubules
		2	R 1 root without cytoplasmic microtubules
		3	R 1 root absent
Character 3	R 3 root	1	present
		2	present, U-shaped
		3	absent
		4	present, U-shaped, split
Character 4	transitional helix	1	absent
		2	present, < 5 gyres
		3	present, 5 gyres
		4	present, > 5 gyres
		5	double helix
Character 5	photoreceptor	1	short flagellum, eyespot-type
		2	Synurophyte type
		3	absent
		4	Eustigmatophyte type
Character 6	fibrous root	1	present, basal body-nucleus connector
		2	present, not basal body-nucleus connector
		3	absent
Character 7	silica deposition vesicle	1	absent
		2	present
Character 8	haptonema	1	absent
		2	present
Character 9	girdle lamella	1	present
		2	absent
Character 10	flucoxanthin	1	present
		2	absent
Character 11	carotenoids	1	diatoxanthin present
		2	violoxanthin/antheroxanthin present
Character 12	green flagellum	1	autofluorescence absent
		2	autofluorescence present

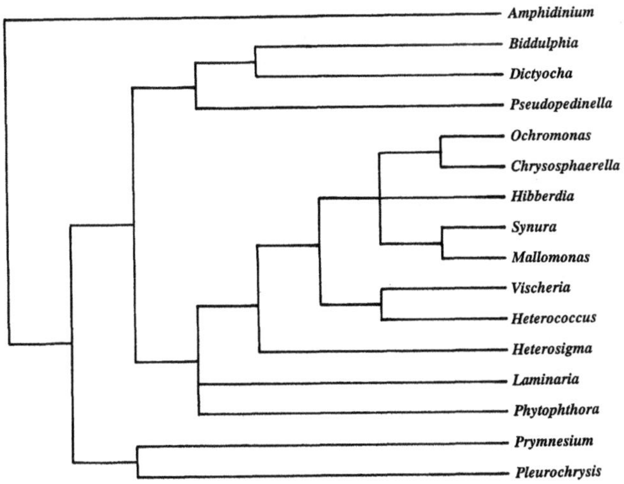

Fig. 4. A consensus tree analysis of selected chromophyte genera, *Phytophthora* (Oomycetes) and *Amphidinium* (Dinophyceae) constructed using the PAUP CONTREE program (adams consensus). The consensus tree is based upon 12 trees formed using the data matrix in Table 2 (characters 4–6 are unordered) and analyzed using PAUP with the MULPARS and SWAP = GLOBAL options. The Mikevich's consensus information value is 0.607, the weighted consensus fork value is 0.59, and Rohlf's consistency index value is 0.843

Eustigmatophyceae and Xanthophyceae form a clade, and they are sister taxa to the Chrysophyceae and Synurophyceae. The Raphidophyceae (*Heterosigma*, not *Gonyostomum* or *Vacuolaria*), Phaeophyceae and Oomycetes are included in a large grouping of "core" chromophytes.

The consensus tree in Fig. 4 is similar to phylogenetic trees constructed from sequence comparisons of rRNA (Gunderson et al. 1987, Perasso et al. 1989, Ariztia et al. 1991), but unlike the trees based upon sequence comparisons, it contains at least one representative from each chromophyte class. The similarity of trees suggests a generally accepted phylogeny may be emerging for the various classes of chromophyte algae. Many problems still remain, however. For example, the inclusion of *Hydrurus, Giraudyopsis, Sarcinochrysis, Pelagococcus*, and other "unusual" chrysophytes complicate rather than clarify the definition of the Chrysophyceae. Similar situations, created by "unusual" taxa, exist in other chromophyte classes. To be certain, some of these taxa may not belong to the class in question, and ample work remains.

Although the exact evolutionary relationships within the chromophytes remain somewhat unclear, the phylogenetic relationships of chromophytes to other protists is even more uncertain. The best understood relationship may be with the Oomycete fungi (Fig. 4) (Barr and Désaulniers 1989). The recognition of a possible relationship between these fungi and the chromophytes dates from the mid-nineteenth century and is supported by light and electron microscopic data (see Barr and Désaulniers 1989). Molecular studies support this relationship also (Gunderson et al. 1987, Ariztia et al. 1991).

The chromophytes and Oomycetes are part of a larger group, the Stramenopiles (Patterson 1989). Tripartite flagellar hairs, tubular mitochondrial cristae and β-1,3 linked polysaccharides are commonly shared characters. However, which other protistan group is most closely related to the Stramenopiles is not obvious; probably, it is a group with tubular mitochondrial cristae such as the dinoflagellates (Farmer and Roberts 1989, Roberts and Roberts 1991) or the ciliates (Lynn 1991). This appears to be supported by molecular comparisons of rRNA sequences, although the level of statistical confidence is low (Gunderson et al. 1987, Sogin 1989 b).

The chromophyte cytoskeleton still holds important information that will help resolve the phylogenetic relationships. Future studies, including molecular sequence comparisons of cytoskeletal components, will be powerful tools for phylogenetic studies. And, the unusual cytoskeletal functions of chromophytes (i.e., feeding basket phagotrophy, extra-cellular spine-scale movement, scale formation and deployment) are excellent systems for cell biology.

Acknowledgements

I thank Dr. Peter Beech for critically reading the manuscript and for drawing attention to the variability of the transitional helix in *Mallomonas*. I thank Gina Ross for preparing Figs. 1 a, 2 b, d, f, and 3 a, e. This research was supported by NSF grant BSR 8709222 and is Bigelow Laboratory for Ocean Sciences contribution number 91013.

References

Alberte RS, Andersen RA (1986) Anteraxanthin, a light-harvesting carotenoid found in a chromophyte alga. Plant Physiol 80: 583–587

Andersen RA (1985) The flagellar apparatus of the golden alga *Synura uvella*: four absolute orientations. Protoplasma 128: 94–106

– (1987) *Synurophyceae* classis nov., a new class of algae. Amer J Bot 74: 337–353

– (1989 a) Absolute orientation of the flagellar apparatus of *Hibberdia magna* comb. nov. (Chrysophyceae). Nord J Bot 8: 653–669

– (1989 b) The Synurophyceae and their relationship to other golden algae. Beih Nova Hedwigia 95: 1–26

– (1990) The three-dimensional structure of the flagellar apparatus of *Chrysosphaerella bevispina* (Chrysophyceae) as viewed by high voltage electron microscopy stereo pairs. Phycologia 29: 86–97

– Wetherbee R (1991) Microtubules of the flagellar apparatus are active during prey capture in the chrysophycean alga *Epipyxis pulchra*. Protoplasma (in press)

Ariztia E, Andersen RA, Sogin ML (1991) A new phylogeny for chromophyte algae using 16S-like rRNA sequences from *Mallomonas papillosa* (Synurophyceae) and *Tribonema aequale* (Xanthophyceae). J Phycol 27: 428–436

Barr DJS, Désaulniers NL (1989) The flagellar apparatus of the Oomycetes and Hyphochytridiomycetes. In: Green JC, Leadbeater BSC, Diver WL (eds) The chromophyte algae: problems and perspectives. Clarendon Press, Oxford, pp 343–355

Beech PL, Wetherbee R (1988) Observations on the flagellar apparatus and peripheral endoplasmic reticulum of the coccolithiophorid *Pleurochrysis carterae* (Prymnesiophyceae). Phycologia 27: 142–158

– – (1990 a) Direct observations on flagellar transformation in *Mallomonas splendens* (Synurophyceae). J Phycol 26: 90–95

– – (1990 b) The flagellar apparatus of *Mallomonas splendens* (Synurophyceae) at interphase and its development during the cell cycle. J Phycol 26: 95–111

– – Pickett-Heaps JD (1988) Transformation of the flagella and associated flagellar components during cell division in the coccolithophorid *Pleurochrysis carterae*. Protoplasma 145: 37–46

– – – (1990) Secretion and deployment of bristles in *Mallomonas splendens* (Synurophyceae). J Phycol 26: 112–122

– Heimann K, Melkonian M (1991) Development of the flagellar apparatus during the cell cycle in unicellular algae. Protoplasma 164: 23–37

Bjørnland T, Liaaen-Jensen S (1989) Distribution patterns of carotenoids in relation to chromophyte phylogeny and systematics. In: Green JC, Leadbeater BSC, Diver WL (eds) The chromophyte algae: problems and perspectives. Clarendon Press, Oxford, pp 37–60

Bouck GB (1971) The structure, origin, isolation and composition of the tubular mastigonemes of the *Ochromonas* flagellum. J Cell Biol 50: 362–384

– (1972) Architecture and assembly of mastigonemes. In: DuPraw EJ (ed) Advances in cell and molecular biology, vol 2. Academic Press, New York, pp 237–271

– Brown DL (1973) Microtubule biogenesis and cell shape in *Ochromonas*. I. The distribution of cytoplasmic and mitotic microtubules. J Cell Biol 56: 340–359

Bourrelly P (1981) Les algues d'eau douce, vol 2, les algues jaunes et brunes. N Boubeé, Paris

Brugerolle G (1991) Flagellar and cytoskeletal systems in amitochondrial flagellates: *Arachamoeba, Metamonadea* and *Parabasalia*. Protoplasma 164: 70–90

– Bricheux G (1984) Actin microfilaments are involved in scale formation of the chrysomonad cell *Synura*. Protoplasma 123: 203–212

Cavalier-Smith T (1986) The kingdom Chromista: origin and systematics. In: Round FE, Chapman DJ (eds) Progress in phycological research, vol 4. BioPress, Bristol, pp 309–347

Chadefaud M (1950) Les cellules nageuses des algues dans l'embranchement des chromophycées. C R Acad Sci Paris 231: 788–790

Cienkowski L (1870) Ueber Palmellaceen und einige Flagellaten. Arch Mikrosk Anat 8: 421–438

Clayton MN (1989) Brown algae and chromophyte phylogeny. In: Green JC, Leadbeater BSC, Diver WL (eds) The chromophyte algae: problems and perspectives. Clarendon Press, Oxford, pp 229–253

Coleman AW (1987) A cytoskeletal mechanism for plastid assortment at cell division. J Cell Biol 105: 35 a

– (1988) The autofluorescent flagellum: a new phylogenetic enigma. J Phycol 22: 118–120

Cox ER (ed) (1980) Phytoflagellates. Developments in marine biology, vol 2. Elsevier/North-Holland, New York

Crawford RM, Round FE (1989) *Corethron* and *Mallomonas* – some striking morphological similarities. In: Green JC, Leadbeater BSC, Diver WL (eds) The chromophyte algae: problems and perspectives. Clarendon Press, Oxford, pp 295–305

Farmer MA, Roberts KR (1989) Comparative analyses of the dinoflagellate flagellar apparatus. III. Freeze substitution of *Amphidinium rhynchocephalum*. J Phycol 25: 280–292

Foster KW, Smyth RD (1980) Light antennas in phototactic algae. Microbiol Rev 44: 572–630

Gayral P, Morange J (1970) Structure fine du fouet pleuronématé chez quelques Chrysophycées. In: Septiéme Congrès Internationale Microscopie Electronique, Grenoble, 3: 387–388

Grain J, Mignot J-P, Puytorac P (1988) Ultrastructures and evolutionary modalities of flagellar and ciliary systems in protists. Biol Cell 63: 219–237

Grell KG (1989) *Reticulosphaera socialis* n. gen., n. sp., ein plasmodialer und phagotropher Vertreter der heterokonten Algen. Z Naturforsch 44 c: 330–332

– Heini A, Schüller S (1990) The ultrastructure of *Reticulosphaera socialis* Grell (Heterokontophyta). Eur J Protistol 26: 37–54

Green JC, Hori T (1990) The architecture of the flagellar apparatus of *Prymnesium patellifera* (Prymnesiophyta). Bot Mag Tokyo 103: 191–207

– Leadbeater BSC, Diver WL (eds) (1989) The chromophyte algae: problems and perspectives. Clarendon Press, Oxford

Gunderson JH, Elwood H, Ingold A, Kindle K, Sogin ML (1987) Phylogenetic relationships between chlorophytes, chrysophytes and Oomycetes. Proc Natl Acad Sci USA 84: 5823–5827

Heath IB, Darley WM (1972) Observations on the ultrastructure of the male gametes of *Biddulphia levis* Ehr. J Phycol 18: 51–59

Hellebust JA (1988) Polysaccharides produced by chromophyte microalgae. In: Ramus J, Jones MJ (eds) Polysaccharides from microalgae: a new agroindustry. Proceedings of the International Workshop, Duke University Marine Laboratory, Beaufort, NC, pp 13–19

Henry EC, Cole KM (1982 a) Ultrastructure of swarmers in the *Laminariales* (Phaeophyceae). I. Zoospores. J Phycol 18: 550–569

– – (1982 b) Ultrastructure of swarmers in the *Laminariales* (Phaeophyceae). II. Sperm. J Phycol 18: 570–579

Hibberd DJ (1976) The ultrastructure and taxonomy of the Chrysophyceae and Prymnesiophyceae (Haptophyceae): a survey with some new observations on the ultrastructure of the Chrysophyceae. Bot J Linn Sic 72: 55–80

– (1977) The cytology and ultrastructure of *Chrysonebula holmesii* Lund (Chrysophyceae) with special reference to the flagellar apparatus. Br Phycol J 12: 369–383

– (1979) The structure and phylogenetic significance of the flagellar transition region in the chlorophyll *c*-containing algae. BioSystems 11: 243–261

– (1980) Xanthophytes. In: Cox ER (ed) Phytoflagellates. Developments in marine biology, vol 2. Elsevier/North-Holland, New York, pp 243–271

– Leedale GF (1972) Observations on the cytology and ultrastructure of the new algal class, Eustigmatophyceae. Ann Bot 36: 49–71

Hoffman LR, Vesk M, Pickett-Heaps JD (1986) The cytology and ultrastructure of zoospores of *Hydrurus foetidus* (Chrysophyceae). Nord J Bot 6: 105–122

Inouye I, Pienarr RN (1985) Ultrastructure of the flagellar apparatus in *Pleurochrysis* (Class Prymnesiophyceae). Protoplasma 125: 24–35

James-Clark H (1867) On the Spongiae Ciliatae as Infusoria Flagellata; or, observations on the structure, animality and relationship of *Leucosolenia botryoides* Bowerbank. Mem Boston Soc Nat Hist 1: 305–340

Jeffrey SW (1989) Chlorophyll *c* pigments and their distribution in the chromophyte algae. In: Green JC, Leadbeater BSC, Diver WL (eds) The chromophyte algae: problems and perspectives. Clarendon Press, Oxford, pp 13–36

Kawai H (1988) A flavin-like autofluorescent substance in the posterior flagellum of golden and brown algae. J Phycol 24: 114–117

– Inouye I (1989) Flagellar autofluorescence in forty-four chlorophyll *c*-containing algae. Phycologia 28: 222–227

Koutoulis A, McFadden GI, Wetherbee R (1988) Spine-scale reorientation in *Apedinella radians* (Pedinellales, Chrysophyceae): the microarchitecture and immunocytochemistry of the associated cytoskeleton. Protoplasma 147: 25–41

Kreimer G, Kawai H, Müller DG, Melkonian M (1991) Reflective properties of the stigma in male gametes of *Ectocarpus siliculosus*

(Phaeophyceae) studied by confocal laser scanning microscopy. J Phycol 27: 268–276

Kristiansen J, Andersen RA (eds) (1986) Chrysophytes: aspects and problems. Cambridge University Press, New York

Larsen J (1985) Ultrastructure and taxonomy of *Actinomonas pusilla*, a heterotrophic member of the Pedinellales (Chrysophyceae). Br Phycol J 20: 341–355

Leadbeater BSC (1986) Scale case construction in *Synura petersenii* Korsch. (Chrysophyceae). In: Kristiansen J, Andersen RA (eds) Chrysophytes: aspects and problems. Cambridge University Press, New York, pp 121–131

– (1989) The phylogenetic significance of flagellar hairs in the Chromophyta. In: Green JC, Leadbeater BSC, Diver WL (eds) The chromophyte algae: problems and perspectives. Clarendon Press, Oxford, pp 145–165

– (1990) Ultrastructure and assembly of the scale case in *Synura* (Synurophyceae Andersen). Br Phycol J 25: 117–132

Lechtreck K-F, Melkonian M (1991) An update on fibrous flagellar roots in green algae. Protoplasma 164: 38–44

Lee KW, Bold HC (1973) *Pseudocharaciopsis texensis* gen. et sp. nov., a new member of the Eustigmatophyceae. Br Phycol J 8: 31–37

Leedale GF (1980) Algal classes: delimitation, interrelationships and phylogeny. In: Desikachary TV, Rajo Rao VN (eds) Taxonomy of algae: 1974 Symposium. University of Madras, Madras, pp 1–19

Ludwig M, Wetherbee R (1990) Bristle/scale deployment and attachment in *Mallomonas splendens*: antibodies against cell surface molecules. J Phycol 26: 9 a

Luther A (1899) Ueber *Chlorosaccus* eine neue Gattung der Süßwasseralgen, nebst einigen Bemerkungen zur Systematik verwandter Algen. Bih Kungl Svenska Vetensk Handl 24, Afd 3, 13: 1–22

Lynn DH (1991) The implications of recent descriptions of kinetid structure to the systematics of the ciliated protists. Protoplasma 164: 123–142

Mann DG, Marchant HJ (1989) The origins of the diatom and its life cycle. In: Green JC, Leadbeater BSC, Diver WL (eds) The chromophyte algae: problems and perspectives. Clarendon Press, Oxford, pp 307–323

Manton I, von Stosch HA (1966) Observations on the fine structure of the male gamete of the marine centric diatom *Lithodesmium undulatum*. J Roy Microsc Soc 85: 119–134

Melkonian M (1984) Flagellar apparatus ultrastructure in relation to green algal classification. In: Irvine DE, John DM (eds) Systematics of the green algae. Academic Press, New York, pp 73–120

– – Beech PL, Katsaros C, Schulze D (1991) Centrin mediated cell motility. In: Melkonian M (ed) Algal cell motility. Chapman and Hall, New York (in press)

– – Reize IB, Preisig HR (1987) Maturation of a flagellum/basal body requires more than one cell cycle in algal flagellates: studies on *Nephroselmis olivacea* (Prasinophyceae). In: Wiessner W, Robinson DG, Starr RC (eds) Algal development, molecular and cellular aspects. Springer, Berlin Heidelberg New York Tokyo, pp 102–113

– Schulze D, McFadden GI, Robenek H (1988) A polyclonal antibody (Anticentrin) distinguishes between two types of fibrous flagellar roots in green algae. Protoplasma 144: 56–61

Mignot JP (1967) Structure et ultrastructure de quelques Chloromonadines. Protistologica 3: 5–23

– (1976) Compléments à l'étude des Chloromonadines. Ultrastructure de *Chattonella subsalsa* Biecheler, Flagellé d'eau Saumâtre. Prostistologica 12: 279–293

– Brugerolle G (1982) Scale formation in chrysomonad flagellates. J Ultrastruct Res 81: 13–26

Moestrup Ø (1982) Flagellar structure in algae: a review, with new observations particularly on the Chrysophyceae, Phaeophyceae, (Fucophyceae), Euglenophyceae, and *Reckertia*. Phycologia 21: 427–528

– Andersen RA (1991) Organization of heterotrophic heterokonts. In: Patterson DJ, Larsen J (eds) The biology of free-living heterotrophic flagellates. Clarendon Press, London, (in press)

– Thomsen H (1990) *Dictyocha speculum* (Silicoflagellate, Dictyochophyceae), studies on armoured and unarmoured stages. Det Kongel Danske Vidensk Selskab, Biol Skri 37: 1–57

Motomura T (1990) Ultrastructure of fertilization in *Laminaria angustata* (Phaeophyta, Laminariales) with emphasis on the behavior of centrioles, mitochondria and chloroplasts of the sperm. J Phycol 26: 80–89

– Sakai Y (1988) The occurrence of flagellated eggs in *Laminaria angustata* (Phaeophyta, Laminariales). J Phycol 24: 282–285

Müller DG, Maier I, Müller H (1987) Flagellum autofluorescence and photoaccumulation in heterokont algae. Photochem Photobiol 46: 1003–1008

O'Kelly CJ (1989 a) The evolutionary origin of the brown algae: information from studies of motile cell ultrastructure. In: Green JC, Leadbeater BSC, Driver WL (eds) The chromophyte algae: problems and perspectives. Clarendon Press, Oxford, pp 255–278

– (1989 b) Reconstruction from serial sections of *Heterococcus tectiformis* (Tribophyceae = Xanthophyceae) zoospores, with emphasis on the flagellar apparatus. Crypt Bot 1: 58–69

– Floyd GL (1984) The absolute configuration of the flagellar apparatus in zoospores from two species of Laminariales (Phaeophyceae). Protoplasma 123: 18–25

– – (1985) Absolute configuration analysis of the flagellar apparatus in *Giraudyopsis stellifer* (Chrysophyceae, Sarcinochrysidales) zoospores and its significance in the evolution of the Phaeophyceae. Phycologia 24: 263–274

Owen HA, Mattox KR, Stewart KD (1990 a) Fine structure of the flagellar apparatus of *Dinobryon cylindricum* (Chrysophyceae). J Phycol 26: 131–141

– Stewart KD, Mattox KR (1990 b) Fine structure of the flagellar apparatus of *Uroglena americana* (Chrysophyceae). J Phycol 26: 142–149

Patterson DJ (1986) The actinophryid heliozoa (Sarcodina, Actinopoda) as chromophytes. In: Kristiansen J, Andersen RA (eds) Chrysophytes: aspects and problems. Cambridge University Press, New York, pp 49–67

– (1989) Stramenopiles: chromophytes from a protistan perspective. In: Green JC, Leadbeater BSC, Diver WL (eds) The chromophyte algae: problems and perspectives. Clarendon Press, Oxford, pp 357–379

Perasso R, Baroin A, Qu LH, Bachellerie JP, Adoute A (1989) Origin of the algae. Nature 339: 142–144

Preisig HR (1989) The flagellar base ultrastructure and phylogeny of chromophytes. In: Green JC, Leadbeater BSC, Diver WL (eds) The chromophyte algae: problems and perspectives. Clarendon Press, Oxford, pp 167–187

Roberts KR, Roberts JE (1991) The flagellar apparatus and cytoskeleton of the dinoflagellates: a comparative overview. Protoplasma 164: 105–122

Salisbury JL (1989 a) Algal centrin: calcium-sensitive contractile organelles. In: Coleman AW, Goff LJ, Stein-Taylor JR (eds) Algae as experimental systems. AR Liss, New York, pp 19–37

– (1989 b) Centrin and the algal flagellar apparatus. J Phycol 25: 201–206

– Baron A, Surek B, Melkonian M (1984) Striated flagellar roots: isolation and partial characterization of a calcium-modulated contractile organelle. J Cell Biol 99: 962–970

– Sanders MA, Harpst L (1987) Flagellar root contraction and nuclear movement during flagellar regeneration in *Chlamydomonas reinhardtii*. J Cell Biol 105: 1799–1805

Santos LMA, Leedale GF (1991) *Vischeria stellata* (Eustigmatophyceae): ultrastructure of the zoospores, with special reference to the flagellar apparatus. Protoplasma 164: 160–167

Saville-Kent W (1880) A manual of Infusoria, 1. London

Schnepf E, Deichgräber G (1969) Über die Feinstruktur von *Synura petersenii* unter besonderer Berücksichtigung der Morphogenese ihrer Kieselschuppen. Protoplasma 68: 85–106

– Deichgräber G, Röderer G, Herth W (1977) The flagellar root apparatus, the microtubular system and associated organelles in the chrysophycean flagellate *Poterioochromonas malhamensis* Peterfi (syn. *Poteriochromonas stipitata* (Scherffel and *Ochromonas malhamensis* Pringsheim). Protoplasma 92: 87–107

Siver PA, Glew JR (1990) The arrangement of scales and bristles on *Mallomonas* (Chrysophyceae): a proposed mechanism for the formation of the cell covering. Can J Bot 68: 374–380

Slankis T, Gibbs SP (1972) The fine structure of mitosis and cell division in the chrysophycean alga *Ochromonas danica*. J Phycol 8: 243–256

Sogin ML, Gunderson JH, Elwood HJ, Alonso RA, Peattie DA (1989 a) Phylogenetic significance of the kingdom concept: an unusual eukaryotic 16S-like ribosomal RNA from *Giardia lamblia*. Science 243: 75

– Edman U, Elwood HE (1989 b) A single kingdom of eukaryotes. In: Fernholm B, Bremer K, Jornvall H (eds) The hierarchy of life. Elsevier, New York, pp 133–143

Swofford DL (1985) PAUP. Phylogenetic analysis using parsimony. Version 2.4. Ill Nat Hist Surv, Champagne, IL

Taylor FJR (1976) Flagellate phylogeny: a study in conflicts. J Protozool 23: 28–40

Triemer RE, Farmer MA (1991) An ultrastructural comparison of the mitotic apparatus, feeding apparatus, flagellar apparatus and cytoskeleton in Euglenoids and Kinetoplastids. Protoplasma 164: 91–104

van den Hoek C (1978) Algen. Thieme, Stuttgart

Vesk M, Jeffrey SW (1987) Ultrastructure and pigments of two strains of the picoplanktonic alga *Pelagococcus subviridis* (Chrysophyceae). J Phycol 23: 322–336

– Moestrup Ø (1987) The flagellar root system in *Heterosigma akashiwo* (Raphidophyceae). Protoplasma 137: 15–28

Vlk W (1931) Über die Struktur der Heterokontengeisseln. Beih Bot Zentralbl 48: 214–228

Wetherbee R, Andersen RA (1991) Flagella of chrysophycean algae play an active role in prey capture and selection: direct observations on *Epipyxis pulchra* and *Ochromonas danica* using image enhanced video microscopy. Protoplasma (in press)

– Koutoulis A, Andersen RA (1991) The microarchitecture of the chrysophycean cytoskeleton. In: Menzel D (ed) The cytoskeleton of the algae. CRC Press, Boca Raton (in press)

– – Beech PL (1989) The role of the cytoskeleton during the assembly, secretion and deployment of scales and spines. In: Coleman AW, Goff LJ, Stein-Taylor JR (eds) Algae as experimental systems. AR Liss, New York, pp 93–108

– Platt SJ, Beech PL, Pickett-Heaps JD (1988) Flagellar transformation in the heterokont *Epipyxis pulchra* (Chrysophyceae): direct observations using image enhanced light microscopy. Protoplasma 145: 47–54

Wujek DE (1978) Ultrastructure of flagellated chrysophytes. III. *Mallomonas caudata*. Trans Kansas Acad Sci 81: 327–335

– Kristiansen J (1978) Observations on bristle- and scale-production in *Mallomonas caudata* (Chrysophyceae). Arch Protistenk 120: 213–221

Wyssotski A (1887) Les Mastigophores et Rhizopodes, trouvés dans les lacs scalés. Trav Soc Nat Univ Imp Kharkow 21: 119–140 (in Russian)

Zhang X, Inouye I, Chihara M (1990 a) An unusual short flagellum in *Mallomonas guttata* (Synurophyceae, Chrysophyta). Bot Mag Tokyo 103 97–101

– – – (1990 b) Taxonomy and ultrastructure of a freshwater scaly flagellate *Mallomonas tonsurata* var. *etortisetifera* var. nov. (Synurophyceae, Chromophyta). Phycologia 29: 65–73

Protoplasma (1991) 164: 160–167

PROTOPLASMA

Vischeria stellata (Eustigmatophyceae): ultrastructure of the zoospores, with special reference to the flagellar apparatus

Lilia M. A. Santos* and **G. F. Leedale**

Department of Pure and Applied Biology, University of Leeds, Leeds

Received September 5, 1990
Accepted December 11, 1990

Summary. Ultrastructure of the zoospores of *Vischeria stellata* (R. Chodat ex Poulton) Pascher is investigated, with particular reference to the system of flagellar roots. Microtubular roots and a rhizoplast are present and a model showing their distribution is proposed. Four microtubular roots attach to the basal bodies in a system basically similar to that displayed by the heterokont algae and fungi. The rhizoplast is also similar to that of other heterokont algae. We conclude from these observations that the class Eustigmatophyceae should be placed within the division Heterokontophyta.

Keywords: Flagellar roots; Eustigmatophyceae; Heterokontophyta; *Vischeria*; Zoospores.

Abbreviations: C chloroplast; B′ basal body of the emergent flagellum; B″ second basal body; E eyespot; F emergent flagellum; FS flagellar swelling; LV lamellate vesicle; M mastigonemes; MTs microtubules; N nucleus; R 1–R 4 microtubular roots; Rh rhizoplast; SB striated band; SV spiral vesicle.

* Correspondence and reprints: Instituto Botânico, Universidade de Coimbra, P-3049 Coimbra Codex, Portugal.

Introduction

The separation of the Eustigmatophyceae from the Tribophyceae as a new class of yellow-green algae was based, essentially, on the unique ultrastructure displayed by the zoospores (Hibberd and Leedale 1970, 1971 a, 1972). However, despite the general structure of the uniflagellate and biflagellate zoospores being relatively well known (Hibberd 1980, Hibberd and Leedale 1972, Lee and Bold 1973), the system of flagellar roots has not yet been documented in detail.
Ultrastructural aspects of the flagellar apparatus have been considered to be of taxonomic and phylogenetic significance in green algae (for reviews, see Mattox and Stewart 1984, Melkonian 1984, O'Kelly and Floyd 1984) and are also receiving increased attention in phylogenetic considerations of other algal groups and heterokont protists (for reviews, see Hibberd 1979, Moestrup 1982, Preisig 1989). This paper presents a recon-

Figs. 1–10. Transmission electron micrographs of *Vischeria stellata*.

Fig. 1. Longitudinal section of the zoospore, showing its general structure. View from left side of the zoospore. × 23,000

Fig. 2. Mastigonemes on the emergent flagellum. × 8,500

Fig. 3. Longitudinal section of the base of the emergent flagellum, showing the transitional helix. Note the number of gyres in section (arrow), four on one side and five on the other. *E* Eyespot, *FS* flagellar swelling. × 42,000

Figs. 4–10. Seven selected transverse sections of two different series (Figs. 4–7 and Figs. 8–10) through the anterior end of the cell, to show the transition region of the long flagellum, the two flagellar basal bodies and their associated roots. Dorsal side of the zoospore to the top of the page. × 72,000

Fig. 4. Axonemal MTs, transitional helix and roots R 1 and R 2

Fig. 5. Transitional fibres connecting axoneme with the plasmalemma (arrows), roots R 1 and R 2

Fig. 6. Triplet MTs in basal body B′, roots R 1, R 2, and one MT of root R 4

Fig. 7. Basal bodies, roots R 1 and R 2. Note a striated fibrous band connecting both basal bodies

Fig. 8. Roots R 1, R 4, and striated band

Fig. 9. Four MTs of root R 3 between the basal bodies. Root R 1 is still visible

Fig. 10. Cartwheel pattern in basal body B′

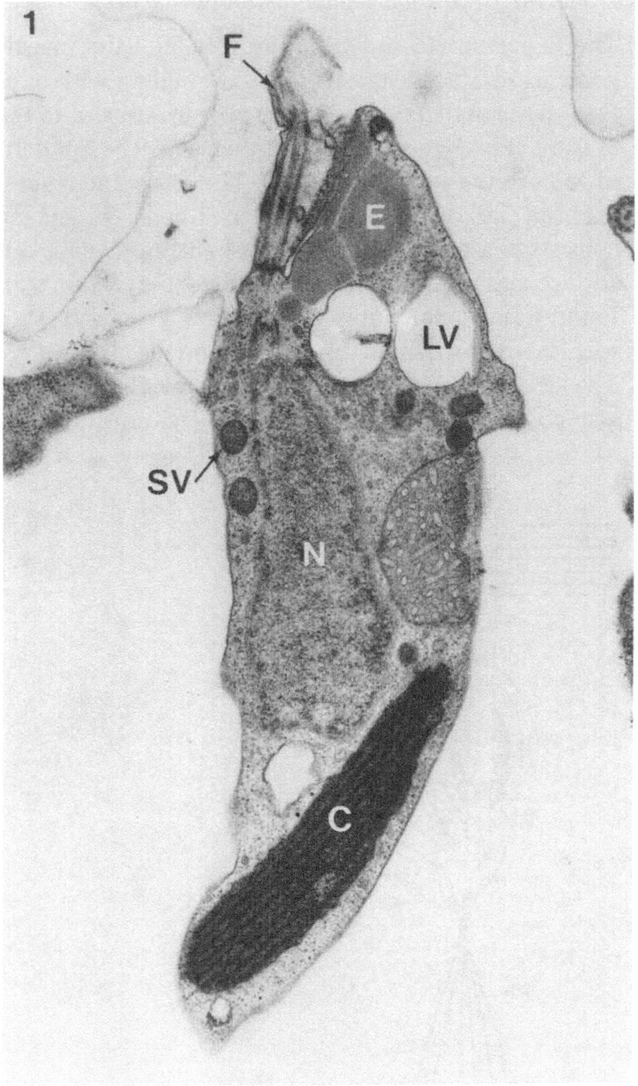
1
F
E
LV
SV
N
C

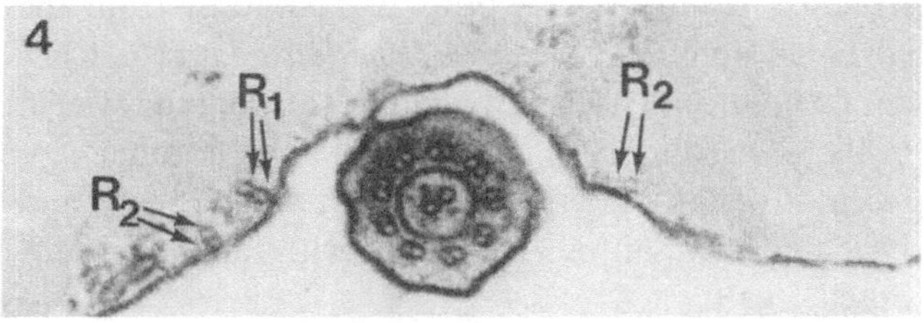
4
R_1
R_2
R_2

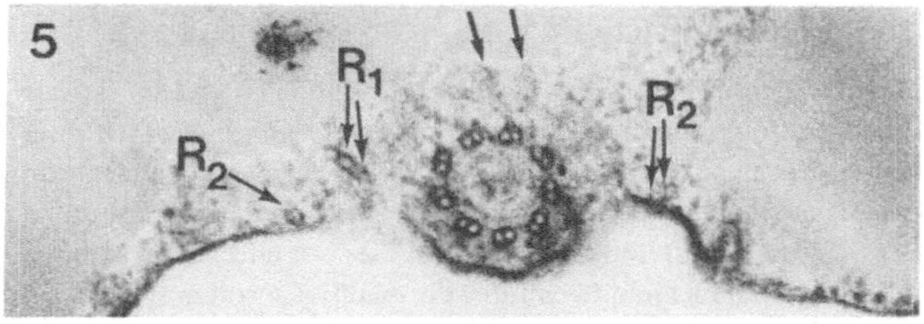
5
R_1
R_2
R_2

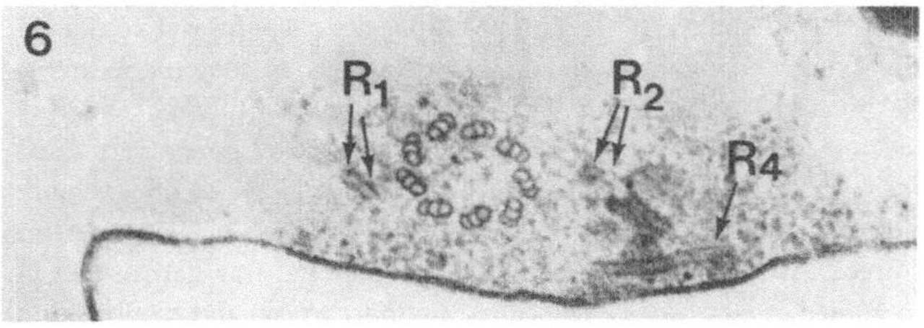
6
R_1
R_2
R_4

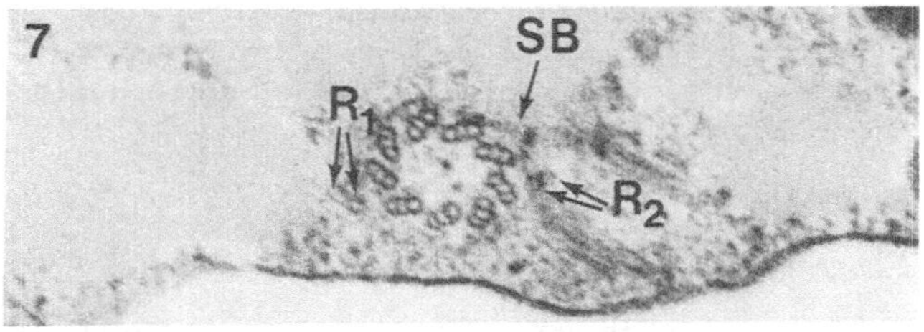
7
SB
R_1
R_2

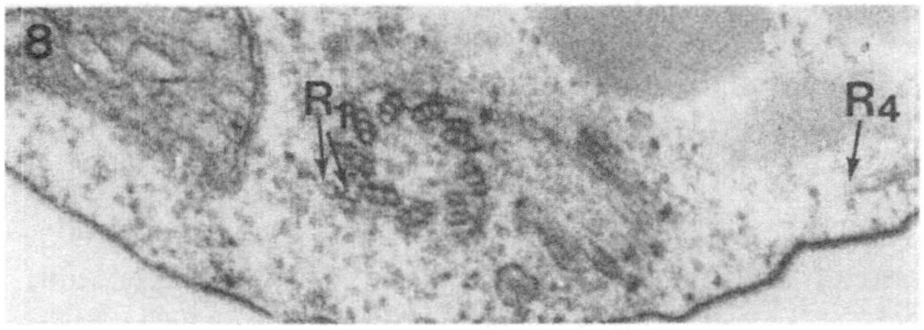
8
R_1
R_4

2

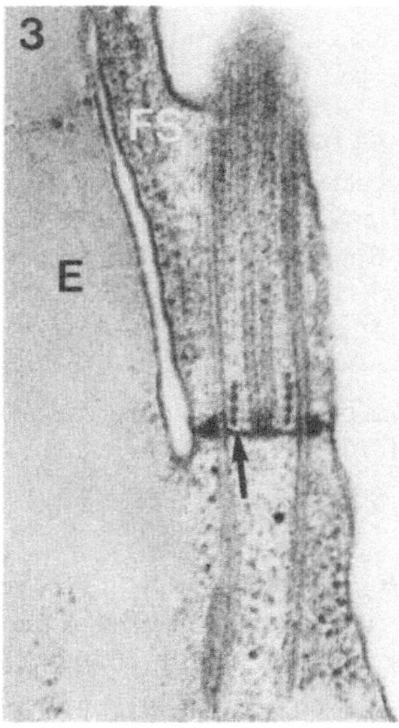
3
FS
E

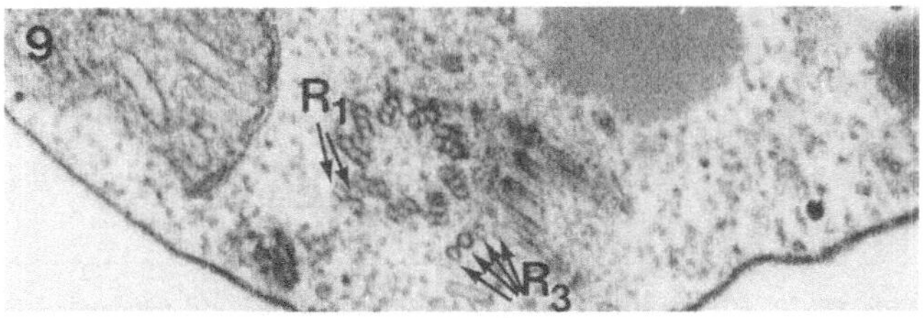
9
R_1
R_3

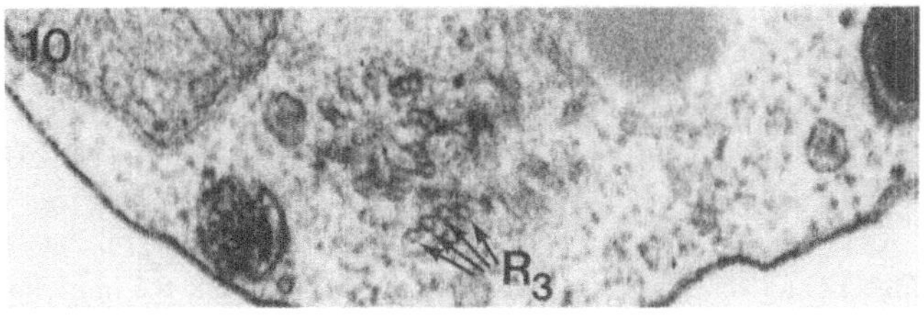
10
R_3

struction from serial sections of zoospores from the eustigmatophycean species *Vischeria stellata* (R. Chodat ex Poulton) Pascher of a model showing the system of flagellar roots. So far, data available on other representatives of this class (Hibberd 1969, Hibberd and Leedale 1972, Santos 1990) agree with the model proposed.

Materials and methods

Unialgal cultures of *Vischeria stellata* (strains 33.83 and 887.2) were obtained from the culture collection of algae at Göttingen (Schlösser 1982). Cultures were mantained at 12 °C in constant light, at an irradiance of 5,000 lux on medium 1 or 1 b (for chemical composition, see Schlösser 1982). Induction of zoospores was effected by addition of fresh liquid medium to cultures on solidified medium (more than 1 month old), followed by subjecting the culture to a long period of darkness (2–3 days). Direct preparations of zoospores were obtained by fixing a drop of material on a formvar/carbon-coated grid in 2% osmium tetroxide vapour, drying at room temperature and shadow-casting with gold/palladium. For sectioning, zoospores were fixed in a mixture of equal parts of 5% glutaraldehyde and 4% osmium tetroxide in 0.2 M cacodylate buffer, pH 7.2, for 1–2 h at 4 °C. After three 5–10 °C min rinses in buffer, the material was dehydrated in a graded ethanol series, embedded in Spurr's resin and sectioned on a Reichert OMU 4 ultramicrotome. Sections were stained with 0.5% uranyl acetate in 20% ethanol and Reynolds' lead citrate. Observation of direct preparations and sections was carried out on a JEOL 1200 EX electron microscope.

Results

General structure of the zoospores

The general morphology of the zoospores of *Vischeria stellata* (Fig. 1) has been previously described (Hibberd 1969, Hibberd and Leedale 1972). The zoospores are flask-shaped, usually 10–15 µm long and about 5 µm at maximum width. They are naked cells, with a single emergent flagellum, almost as long as the cell body, inserted subapically. A large extraplastidial eyespot and a pyriform nucleus occupy the anterior part of the cell, while a single chloroplast lacking a pyrenoid is situated at the posterior end. The rest of the cell is filled with the other organelles and different sorts of vesicles, in particular the lamellate and spiral vesicles typical of eustigmatophycean zoospores.

Flagellum, transition region, and basal bodies

The flagellum (F) has tripartite tubular hairs (mastigonemes) on its surface (Fig. 2), a swelling with electron-dense material close to the cell surface and to the eyespot (Fig. 3), and consists of the usual 9 + 2 pattern of MTs in the axoneme (Fig. 4). The transition region, between the axoneme and the basal body (Figs. 3–5), consists of a transverse partition slightly above the cell surface and a transitional helix with 4–5 gyres (Fig. 3). Faint transitional fibres connect the axonemal MTs with the plasmalemma in this region (Fig. 5).

The basal body (B′) consists of nine triplet MTs (Fig. 6) also connected to the plasmalemma by transitional fi-

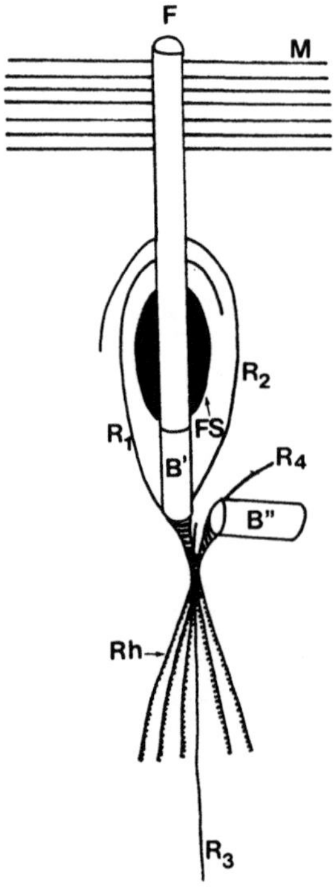

Fig. 11. Schematic drawing showing the relative positions of basal bodies (*B′*, *B″*), flagellar roots (*R 1–R 4*), and flagellar swelling (*FS*) in zoospores of the eustigmatophycean species *Vischeria stellata*. Root R 4 possibly runs around the eyespot (not shown). View from the ventral to the dorsal side of the zoospore. *F* Emergent flagellum, *M* mastigonemes, *Rh* rhizoplast

Figs. 12–20. *Vischeria stellata*. Selected longitudinal and transverse sections of the flagellar apparatus of the zoospores, showing the different microtubular roots. View from the ventral side of the zoospore, except for Figs. 17–20 where specific orientation is given

Fig. 12. Roots R 1, R 3, and R 4 shown in relation to the basal bodies. × 42,000

Fig. 13. Roots R 1 and R 4. *N* Nucleus, *SV* spiral vesicle. × 70,000

Fig. 14. Flagellar swelling (*FS*) in oblique section. Note the three MTs of root R 1, passing around it. × 48,000

Figs. 15 and **16.** Two consecutive sections of a series, showing root R 2 (Fig. 15), dense material connecting the basal bodies and root R 3 with basal body B′ (arrows, Fig. 16) and root R 4. *E* Eyespot. Fig. 15, × 54,000; Fig. 16, × 60,000

Figs. 17 and **18.** Sections showing the five MTs of root R 3 in a semicircular arc near the basal bodies (Fig. 17) and lower in the cell when the basal bodies are no longer seen and the tip of the nucleus (*N*) appears. Dorsal side of the zoospore to the top of the page. × 95,000

Figs. 19 and **20.** Two consecutive sections, showing root R 4. View from the left side of the zoospore. × 37,000

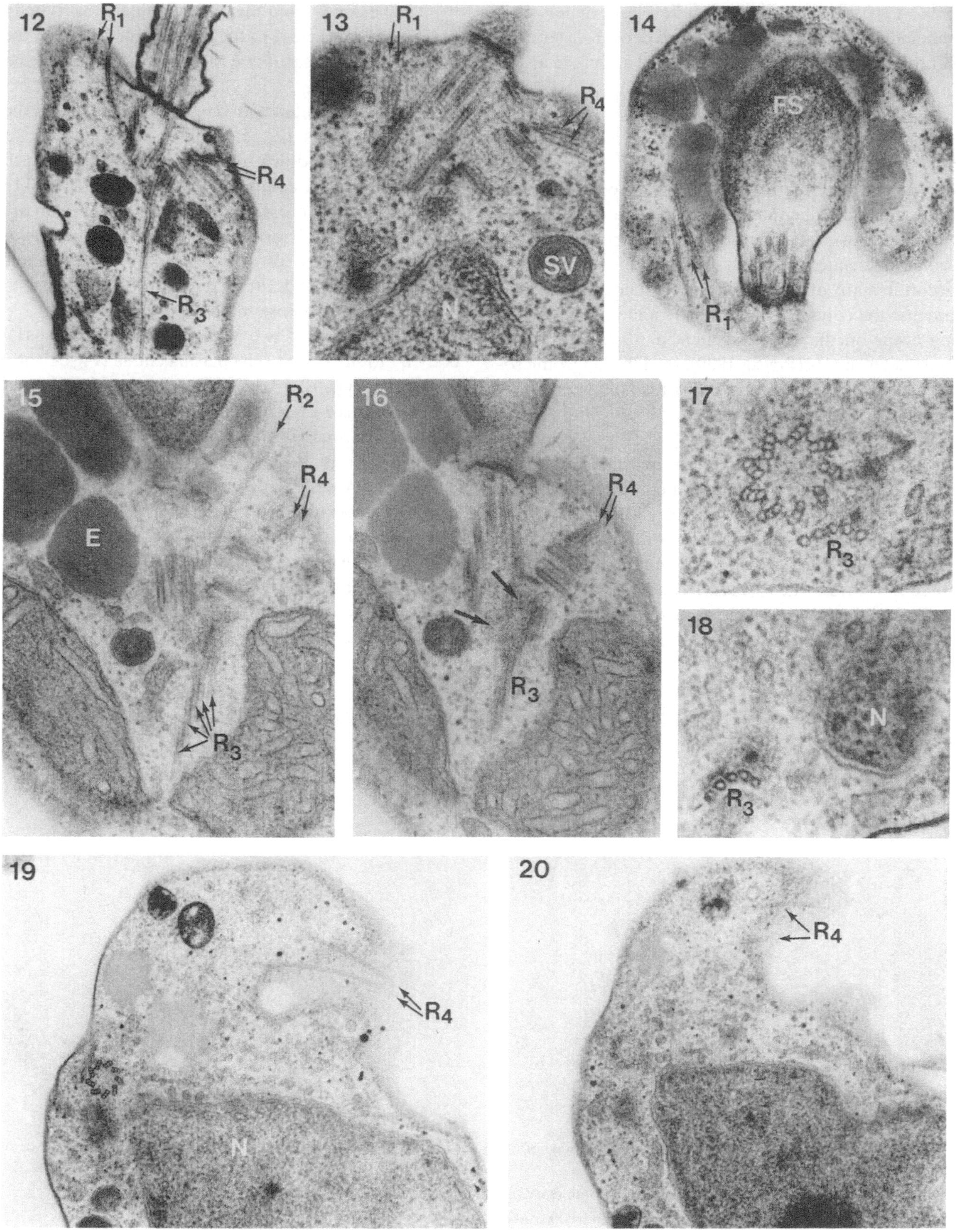
12
R_1
R_4
R_3
13
R_1
R_4
SV
N
14
FS
R_1
15
R_2
R_4
E
R_3
16
R_4
R_3
17
R_3
18
N
R_3
19
R_4
N
20
R_4

bres. A second basal body (B″) lies more or less perpendicular to the first and is connected to it by a striated band (Figs. 7–9). There is a cartwheel structure at the proximal end of basal body B′ (Fig. 10).

Flagellar roots

A system of four microtubular roots and one rhizoplast is present and is diagrammatically illustrated in Fig. 11. Because cytoskeletal microtubules were not clearly observed associated with the R 1 root of *Vischeria*, as occurs in many other heterokonts (Andersen 1989), and because the tripartite tubular hairs and flagellar swelling occur on the same flagellum in *Vischeria*, unlike most heterokonts (Preisig 1989), the labeling of microtubular roots R 1–R 4 for *Vischeria* is not as certain as for many other heterokonts.

Root R 1 originates at basal body B′ (Figs. 12 and 13) and runs anteriorly, close to the plasmalemma and around the flagellar swelling for a considerable extent, almost completing a loop around it (Fig. 14). Transverse sections of the zoospores show this root to consist of two MTs (Figs. 4–9) but in longitudinal sections three MTs are evident (Figs. 13 and 14); the leftmost in Figs. 13 and 14 is widely separated from the other two MTs and is probably part of root R 2.

Root R 2 extends from the opposite side of basal body B′ (Fig. 15), runs close to the plasmalemma around the swelling and consists of two MTs at the origin, reducing to one (Figs. 4–7).

Root R 3 is associated with dense material between the basal bodies (Figs. 12, 15, and 16) and runs close to the plasmalemma down to the posterior end of the cell (not shown). It is composed of a band of five MTs which appear in cross-section as a semicircular arc (Figs. 17 and 18). It originates between the two basal bodies (Fig. 17; see also Figs. 9 and 10) and, in some sections (Fig. 12 and 16), it seems to be connected to basal body B′ by dense material.

Root R 4 consists of two MTs and extends from basal body B″ to run close to the plasmalemma (Figs. 19 and 20; see also Figs. 12, 13, 15, and 16) and possibly around the eyespot (see Figs. 6 and 8).

The only fibrous root seen in these zoospores is the rhizoplast (Figs. 21–24). It originates near the basal bodies and passes deeply into the cell, forming a characteristic association with the nucleus (Figs. 21 and 22). Near the nucleus, the rhizoplast splits into several branches spreading over the nuclear surface (Figs. 23 and 24). Near the basal bodies it appears also branched and divided into two parts that connect with the basal bodies (Fig. 22).

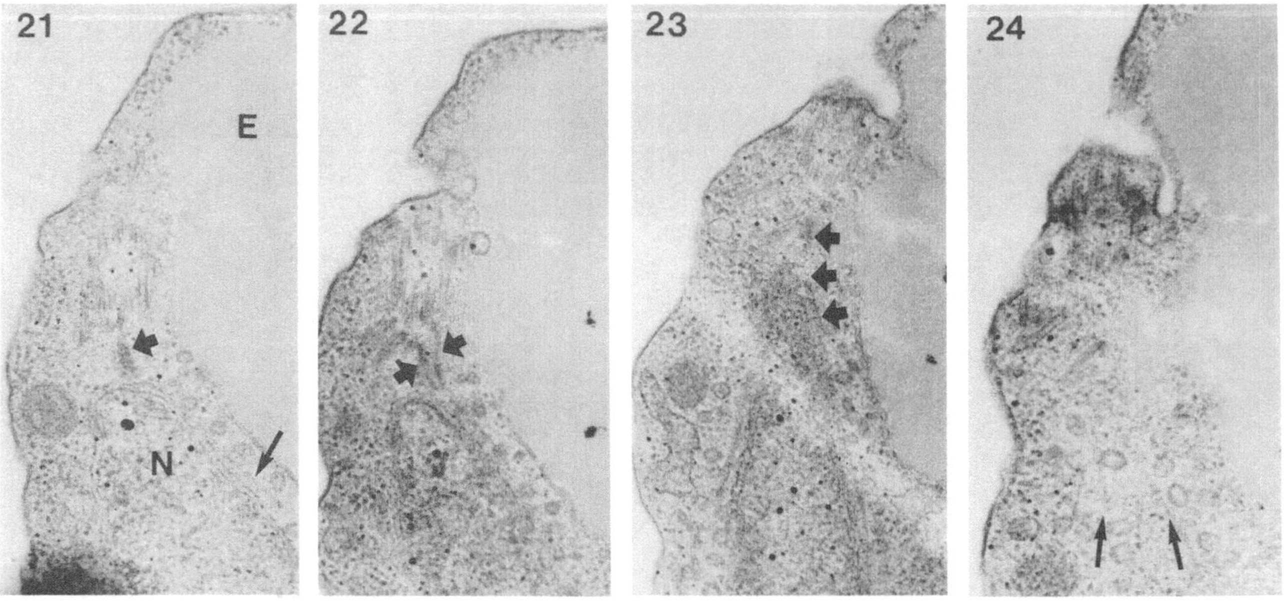

Figs. 21–24. *Vischeria stellata.* Selected longitudinal sections of a series, showing the rhizoplast in the zoospores. View from the left side of the zoospore

Fig. 21. Cross-banded structure (large arrow) and a fibre (small arrows) near the nuclear envelope. *E* Eyespot, *N* nucleus. × 50,000

Fig. 22. Two branches that connect with the basal bodies (arrows). × 50,000

Fig. 23. The rhizoplast divides into several parts (arrow) near the nucleus. × 47,500

Fig. 24. Fibres (arrows) immediately above the plane of the nuclear surface. × 60,000

Discussion

The flagellar roots

Four basic types of microtubular roots are now recognized in heterokont algae and fungi, referred to as roots R 1, R 2, R 3, and R 4 (for reviews, see Andersen 1987, 1989; Preisig 1989). These roots are also present in the zoospores of *Vischeria stellata* and other eustigmatophycean species observed (Santos 1990) in a system that, so far, seems typical for the class and similar to that displayed by the oomycete fungi (compare Fig. 11 with Fig. 1 d of Andersen 1987).

Root R 1 has the same origin and consists of two or three MTs as it does in the heterokont fungi (Barr and Désaulniers 1989) and most of the other heterokont algae (Andersen 1987, Preisig 1989) but the evidence that it also nucleates cytoskeletal MTs is confined to an occasional "whiskery" appearance in *Vischeria* (not shown). A difference is that this root R 1, together with root R 2, seems to loop around the anterior flagellum, as do roots R 3 and R 4 around the posterior flagellum in the Chrysophyceae. The roots perhaps have a similar function of maintaining a close association between the flagellar swelling and the eyespot in both groups and, if this is so, it may be a feature common to other organisms with related types of photoreceptors (e.g., species of Tribophyceae and Paeophyceae). Root R 1 in the Synurophyceae loops around both flagella but the photoreceptor consists only of paired flagellar swellings that are not associated with the cell membrane or chloroplast envelope, and an eyespot is absent (Andersen 1985, 1987). The loop may therefore help to maintain the parallel position of the flagella and swellings, possibly important in the photoreception of these organisms.

Root R 3 is associated with dense material between the two basal bodies and runs directly to the posterior end of the cell whereas in many heterokonts it extends laterally to the ventral surface. Transverse sections of the root show that the five MTs are arranged in a semicircular manner as in *Thraustochytrium aureum* Goldstein (Barr and Allan 1985) but it was not possible to determine with certainty if it originates from basal body B″, basal body B′, or both; in some sections (Fig. 16) it seems to be connected to basal body B′ by two bands of dense material. A function in maintaining the flask-shaped form typical of the eustigmatophycean zoospores seems likely since some of the MTs are seen running to the posterior end of the cell.

Finally, root R 4 also originates from basal body B″ but seems to curve around the bulging eyespot when entire transverse sections are examined (shown partially in Figs. 6 and 8); if this is so, it presumably maintains the characteristic shape of the eyespot and also of the anterior end of the cell, together with roots R 1 and R 2.

The rhizoplast seems basically similar to that described for other heterokont algae (Chrysophyceae, Synurophyceae, and Raphidophyceae; for reviews, see Moestrup 1982, Preisig 1989).

Division level classification

The class Eustigmatophyceae has been generally accepted by phycologists but its proposed elevation to a division as the Eustigmatophyta has been controversial. Several authors place the class in this separate division (e.g., Leedale, 1974, 1980; Loeblich and Loeblich 1978; Hibberd 1981) whereas some include it in the same division as other heterokont algal classes and heterokont fungi (e.g., Cavalier-Smith 1986).

Flagellar differences have long been used for taxonomic and phylogenetic purposes and the ultrastructure of the motile cell, in particular the flagellar apparatus, has recently become the most useful indication of relationships within the protists in general and the algae in particular (Hibberd 1979, Moestrup 1982, Melkonian 1984, O'Kelly and Floyd 1984, Preisig 1989). Hibberd considers the most consistent classification to be that of van den Hoek (1978), where the eustigmatophytes are raised to the level of division; conversely, Preisig (1989) maintains the Eustigmatophyceae within the Heterokonta sensu Cavalier-Smith (1986). The main ultrastructural features shared by the Eustigmatophyceae and several of the classes included in this division are as follows.

1. Heterokont flagellation, with tripartite tubular hairs (mastigonemes) on the anteriorly directed flagellum.
2. Transition region, with a transitional helix in both flagella of the eustigmatophycean species having biflagellate zoospores [e.g., *Pseudocharaciopsis minuta* (Braun) Hibberd (Santos 1990); *Pseudocharaciopsis ovalis* (Chodat) Hibberd (Hibberd 1969)].
3. Basal bodies more or less perpendicular to one another.
4. Similar patterns of flagellar roots.
5. Photoreceptor system involving a flagellar swelling.
6. Chloroplast surrounded by endoplasmic reticulum, sometimes connected with the nuclear envelope.
7. Chloroplast lamellae consisting of three thylakoids.

The main differences of cell ultrastructure that separate the Eustigmatophyceae from those same groups are the following.

1. Unique type of photoreceptor system, consisting of a flagellar swelling located at the base of the anteriorly directed, mastigoneme-bearing long flagellum (even if the species is biflagellate) in association with an eyespot that is very large, extraplastidial and not membrane-bounded. In other heterokont algae where a photoreceptor is present, it is of the chrysophycean or synurophycean types (Andersen 1987); the first includes a swelling on the smooth short flagellum associated with an intraplastidial eyespot and the second consists of paired flagellar swellings.
2. Absence of a girdle lamella in the Eustigmatophyceae, though some rapidophycean (see Heywood 1989), tribophycean (Massalski and Leedale 1969, Hibberd and Leedale 1971 b), synurophycean and chrysophycean species (see Andersen 1986) also lack the girdle lamella.

In our opinion, the new information presented on the system of flagellar roots displayed by *Vischeria stellata* and other eustigmatophycean species (Santos 1990) supports its relationship to the heterokont algal classes Raphidophyceae, Tribophyceae, Chrysophyceae, Synurophyceae, Phaeophyceae, Bacillariophyceae, and the fungal classes Oomycetes and Hyphochytriomycetes. The Eustigmatophyceae should therefore be placed with these classes in a common division, the Heterokontophyta (van den Hoek 1978, Leedale 1980).

Acknowledgements

This research was supported by the Portuguese National Institute for Scientific Investigation (INIC). We are also grateful to Dr. R. A. Andersen for some very informative discussions.

References

Andersen RA (1985) The flagellar apparatus of the golden alga *Synura uvella*: four absolute orientations. Protoplasma 128: 94–106

– (1986) Some new observations on *Saccochrysis piriformis* Korsh. emend. Andersen (Chrysophyceae). In: Kristiansen J, Andersen RA (eds) Chrysophytes: aspects and problems. Cambridge University Press, Cambridge, pp 107–118

– (1987) Synurophyceae classis nov., a new class of algae. Amer J Bot 74: 337–353

– (1989) Absolute orientation of the flagellar apparatus of *Hibberdia magna* comb. nov. (Chrysophyceae). Nord J Bot 8: 653–669

Barr DJS, Allan PME (1985) A comparison of the flagellar apparatus in *Phytophthora, Saprolegnia, Thraustochytrium* and *Rhizidiomyces*. Can J Bot 63: 138–154

– Désaulniers NL (1989) The flagellar apparatus of the Oomycetes and Hyphochytriomycetes. In: Green JC, Leadbeater BSC, Diver WL (eds) The chromophyte algae: problems and perspectives. Clarendon Press, Oxford, pp 343–355

Cavalier-Smith T (1986) The kingdom Chromista: origin and systematics. Prog Phycol Res 4: 309–347

Heywood P (1989) Some affinities of the Raphidophyceae with other chromophyte algae. In: Green JC, Leadbeater BSC, Diver WL (eds) The chromophyte algae: problems and perspectives. Clarendon Press, Oxford, pp 279–293

Hibberd DJ (1969) Cytological studies on the coccoid Xanthophyceae. PhD thesis, University of Leeds, Leeds, U.K.

– (1979) The structure and phylogenetic significance of the flagellar transition region in the chlorophyll *c*-containing algae. BioSystems 11: 243–261

– (1980) Eustigmatophytes. In: Cox ER (ed) Phytoflagellates. Developments in marine biology, vol. 2. Elsevier North-Holland, New York, pp 319–334

– (1981) Notes on taxonomy and nomenclature of the algal classes Eustigmatophyceae and Tribophyceae (synonym Xanthophyceae). Bot J Linn Soc 82: 93–119

– Leedale GF (1970) Eustigmatophyceae – a new algal class with unique organization of the motile cell. Nature 225: 758–760

– – (1971 a) A new algal class – the Eustigmatophyceae. Taxon 20: 523–525

– – (1971 b) Cytology and ultrastructure of the Xanthophyceae. II. The zoospore and vegetative cell of coccoid forms, with special reference to *Ophiocytium majus* Naegeli. Br Phycol J 6: 1–23

– – (1972) Observations on the cytology and ultrastructure of the new algal class Eustigmatophyceae. Ann Bot 36: 49–71

Lee KW, Bold HC (1973) *Pseudocharaciopsis texensis* gen. et sp. nov., a new member of the Eustigmatophyceae. Br Phycol J 8: 31–37

Leedale GF (1974) How many are the kingdoms of organisms? Taxon 23: 261–270

– (1980) Algal classes: delimitation, interrelationships and phylogeny. In: Desikachary TV, Raja Rao VN (eds) Taxonomy of algae: 1974 Symposium. University of Madras, Madras, pp 1–19

Loeblich AR III, Loeblich LA (1978) Division Eustigmatophyta. In: Laskin AI, Lechevalier HA (eds) CRC Handbock of microbiology. Fungi, algae, protozoa, and viruses, vol. 2. CRC Press, West Palm Beach, FL, pp 481–487

Massalski A, Leedale GF (1969) Cytology and ultrastructure of the Xanthophyceae. I. Comparative morphology of the zoospores of *Bumilleria sicula* Borzi and *Tribonema vulgare* Pascher. Br Phycol J 4: 159–180

Mattox KR, Stewart KD (1984) Classification of the green algae; a concept based on comparative cytology. In: Irvine DEG, John DM (eds) Systematics of the green algae. Academic Press, London, pp 29–72

Melkonian M (1984) Flagellar apparatus ultrastructure in relation to green algal classification. In: Irvine DEG, John DM (eds) Systematics of the green algae. Academic Press, London, pp 73–120

Moestrup Ø (1982) Flagellar structure in algae: a review, with new observations particularly on the Chrysophyceae, Phaeophyceae (Fucophyceae), Euglenophyceae, and *Reckertia*. Phycologia 21: 427–528

O'Kelly CJ, Floyd GL (1984) Flagellar apparatus absolute orientations and the phylogeny of the green algae. BioSystems 16: 227–251

Preisig HR (1989) The flagellar base ultrastructure and phylogeny of chromophytes. In: Green JC, Leadbeater BSC, Diver WL (eds) The chromophyte algae: problems and perspectives. Clarendon Press, Oxford, pp 167–187

Santos LMA (1990) Cytology and ultrastructure of Eustigmatophyceae. PhD Thesis, University of Leeds, U.K.

Schlösser UG (1982) List of strains. Ber Deutsch Bot Ges B 95: 181–276

van den Hoek C (1978) Algen. Einführung in die Phykologie. Thieme, Stuttgart

Verleger: Springer-Verlag KG, Sachsenplatz 4-6, A-1201 Wien. – Herausgeber: Prof. D. J. Morré, Arthur G. Hansen Life Sciences Research Building, Purdue University, West Lafayette, IN 47907, U.S.A. – Redaktion: Sachsenplatz 4-6, A-1201 Wien. – Hersteller: Adolf Holzhausens Nachfolger, Kandlgasse 19-21, A-1070 Wien. – Verlagsort: Wien. – Herstellungsort: Wien.

GPSR Compliance

The European Union's (EU) General Product Safety Regulation (GPSR) is a set of rules that requires consumer products to be safe and our obligations to ensure this.

If you have any concerns about our products, you can contact us on ProductSafety@springernature.com

In case Publisher is established outside the EU, the EU authorized representative is:

Springer Nature Customer Service Center GmbH
Europaplatz 3
69115 Heidelberg, Germany

Batch number: 10731490

Printed by Printforce, the Netherlands